ELECTRICAL AND ELECTRONIC PRINCIPLES III

FIRST EDITION

ANTHONY NICOLAIDES

B.Sc. (Eng.), C. Eng. M.I.E.E.
SENIOR LECTURER

P.A.S.S. PUBLICATIONS

PRIVATE ACADEMIC & SCIENTIFIC STUDIES LTD

First Published in Great Britain 1996 by
Private Academic & Scientific Studies Limited

Electrical and Electronic Principles III

ISBN 1 872684 18 1
FIRST EDITION 1996

Published also in 1996
Electrical and Electronic Principles II
Second Edition 1 872684 34 3

PREFACE

This book covers the topic of Electrical and Electronic Principles III.

This books is designed to assist the student wishing to master the subject of Electrical and Electronic Principles III. This book is easy to follow with minimum help. It can be easily adopted by a student who wishes to study it in the comforts of his home at his pace without having to attend classes formally; it is ideal for the working person who wishes to enhance his knowledge and qualification. Electrical and Electronic Principles II book, is divided into two parts. In Part I, the theory is comprehensively dealt with, together with many worked examples and exercises. A step by step approach is adopted in all the worked examples. Part II of the book, a special and unique feature acts as a problem solver for all the exercises set at the end of each chapter in Part I.

This book is extremely useful for the first year degree students studying for Electrical and Electronic Engineering. This book develops the basic concepts and skills that are essential for the Electrical and Electronic Principles II.

I am grateful to Mr. Myat Thaw Kaung, an excellent ex-student of mine, who typeset the manuscript superbly with great care on a desktop publishing system.

I am also grateful to Mr. Norman Turvey for his contribution in the D.C. Machines.

I am also grateful to Mr. Alex Yau for checking thoroughly this book.

I am also grateful to Mr. Constantin and Mr.Andrew Chan for drawing the diagrams in this book.

This book covers the advanced GNVQ syllabus.

First Edition
A. Nicolaides

ELECTRICAL AND ELECTRONIC PRINCIPLES III

PART I

CONTENTS

PART II

ELECTRICAL AND ELECTRONIC PRINCIPLES III

PART I

1. PARALLEL CIRCUIT CONTAINING *C* AND *R*

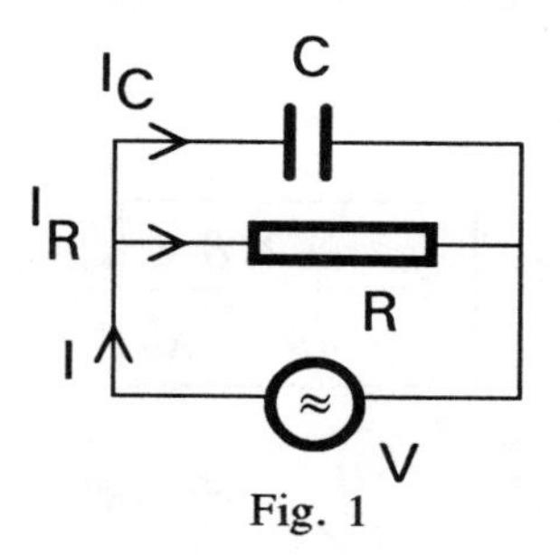

Fig. 1

An a.c. supply is connected across *C* and *R* in parallel as shown, the current drawn from the supply can be found from the phasor diagram of Fig. 2

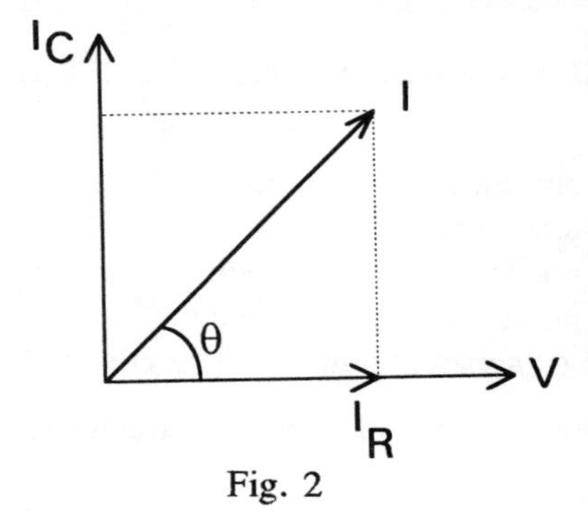

Fig. 2

The current triangle is shown in Fig. 3, if *Z* is the impedance of the circuit, $I = \frac{V}{Z}$, $I_C = \frac{V}{X_C}$, $I_R = \frac{V}{R}$.

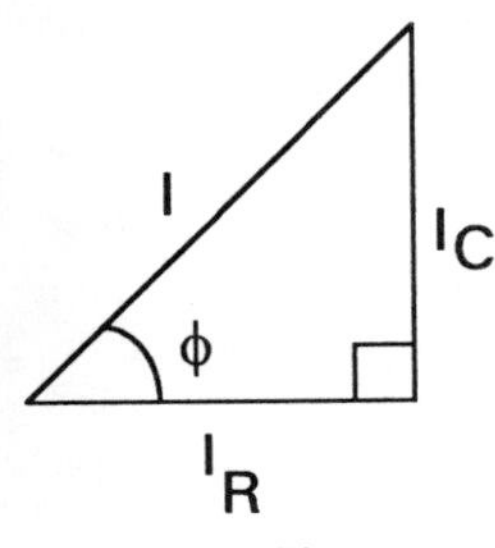

Fig. 3

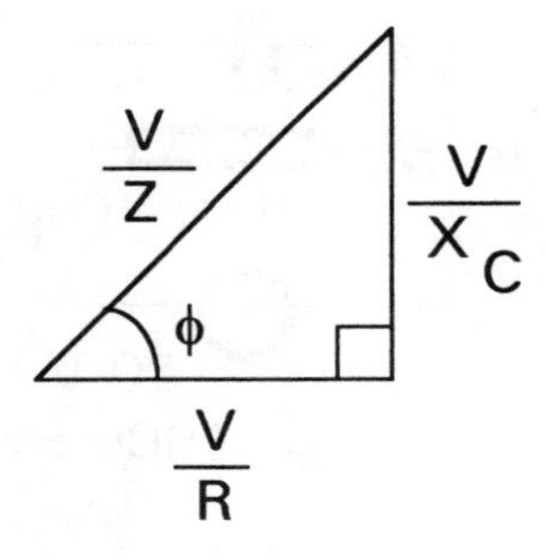

Fig. 4

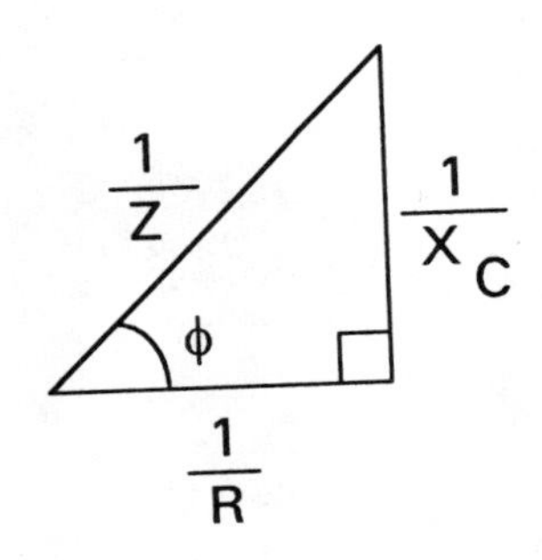

Fig. 5

From Fig. 3 $I^2 = I_C^2 + I_R^2$

From Fig. 4 $\left(\frac{V}{Z}\right)^2 = \left(\frac{V}{X_C}\right)^2 + \left(\frac{V}{R}\right)^2$

From Fig. 5 $$\frac{1}{Z^2} = \frac{1}{X_C^2} + \frac{1}{R^2}$$

The impedance, Z of the circuit is given

$$Z = \frac{X_C R}{\sqrt{R^2 + X_C^2}}$$

The power factor, $\cos \phi = \frac{I_R}{I} = \frac{1/R}{1/Z} = \frac{Z}{R}$.

WORKED EXAMPLE 1

A resistance of 100 Ω is connected in parallel with a 1 μF capacitor to a 20 V, 1000 Hz supply.

Calculate: (a) the current in each branch
(b) the total current
(c) the impedance of the circuit
(d) the phase angle of the circuit
(e) the (i) apparent power (ii) reactive volt ampere (iii) true power.

(**Hint:** Use the power, current and impedance triangles)

SOLUTION 1

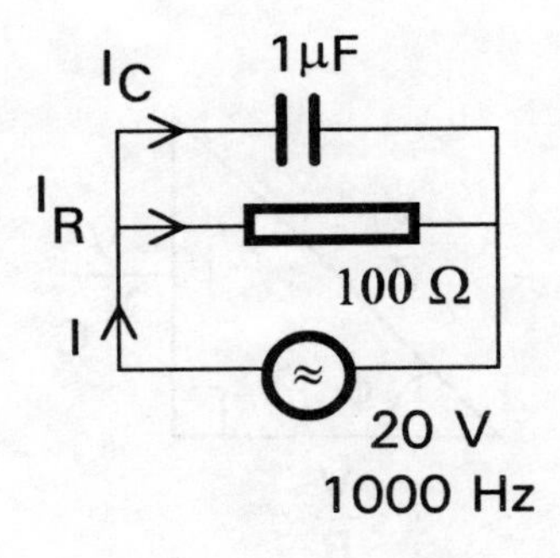

Fig. 6

(a) $I_C = \frac{V}{X_C} = \frac{20}{159.2} = \boxed{126 \text{ mA}}$

where $X_C = \frac{1}{2\pi fC} = \frac{1}{2\pi 10^3 \times 1 \times 10^{-6}} = 159.2\ \Omega$

$I_R = \frac{V}{R} = \frac{20}{100} = 0.2 \text{ A or } \boxed{200 \text{ mA}}$

(b)

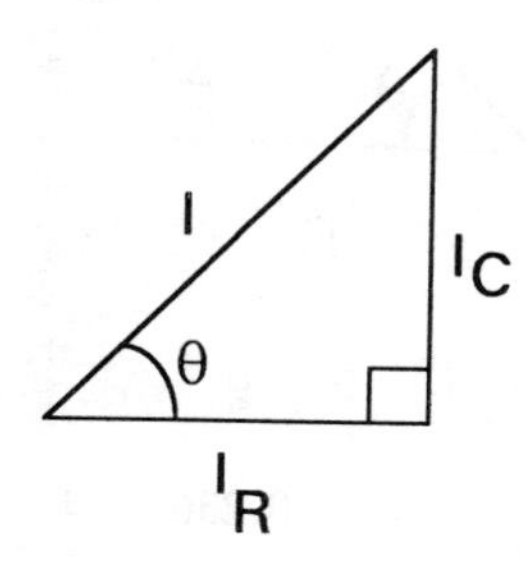

Fig. 7

$I = \sqrt{I_C^2 + I_R^2} = 236 \text{ mA}$

(c)

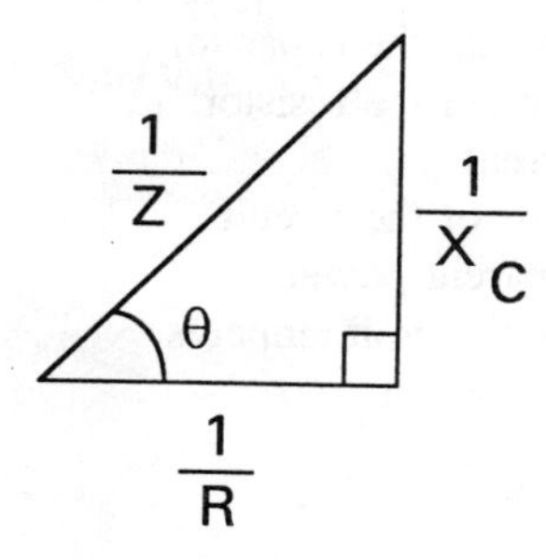

Fig. 8

$$\frac{1}{Z} = \sqrt{\frac{1}{X_C^2} + \frac{1}{R^2}} = \sqrt{\frac{1}{159.2^2} + \frac{1}{100^2}} = 0.0118$$

$$\boxed{Z = 84.7\ \Omega}$$

(d) $\tan\theta = \frac{I_C}{I_R} = \frac{126 \times 10^{-3}}{200 \times 10^{-3}} = 0.63$

$$\boxed{\theta = 32.3^{\circ}}$$

(e)

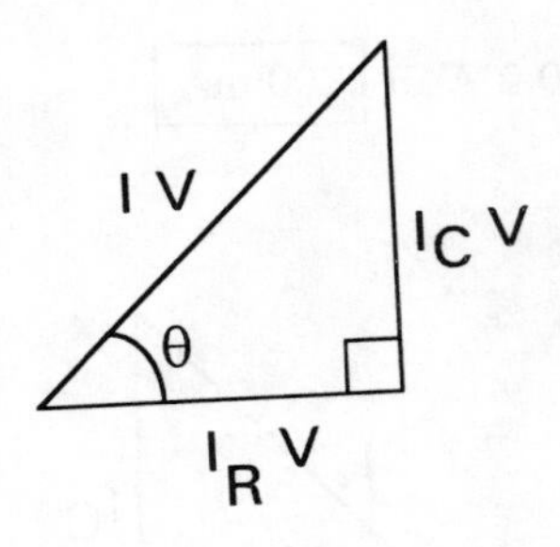

Fig. 9

(i) Power (VA) $= IV = 20 \times 0.236 = 4.72$ VA

(ii) Power (VAr) $= I_C V = 0.126 \times 20 = 2.52$ VAr

(iii) Power (true) $= I_R V = 0.2 \times 20 = 4$ W.

WORKED EXAMPLE 2

A capacitor has a reactance of 200 Ω and is placed in parallel with a 200 Ω resistor. The combination is connected across a 240 V supply of infinite impedance.

Calculate: (a) the current through the capacitor
(b) the current through the resistor
(c) the supply current
(d) the power factor of the circuit
(e) the (i) apparent power
(ii) reactive volt amperes
(iii) true power.

SOLUTION 2

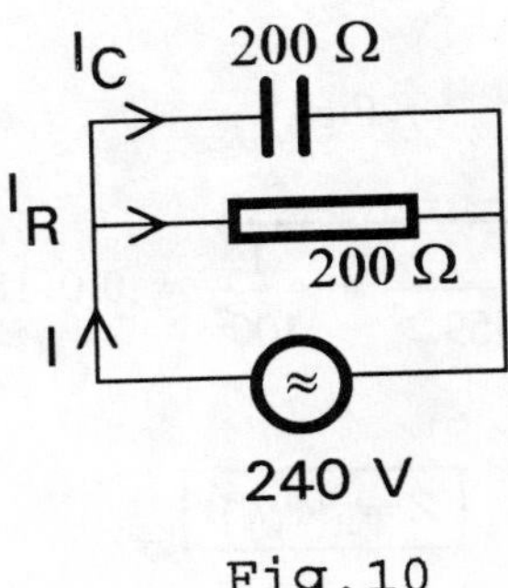

Fig.10

(a) $I_C = \frac{V}{X_C} = \frac{240}{200} = 1.2\text{ A}$

(b) $I_R = \frac{V}{R} = \frac{240}{200} = 1.2\text{ A}$

(c) $I = \sqrt{I_C^2 + I_R^2} = \sqrt{1.2^2 + 1.2^2} = 1.7\text{ A}$

(d) $\cos \phi = \frac{I_R}{I} = \frac{1.2}{1.7} = 0.707$

(e) (i) $IV = 1.7 \times 240 = 408$ volt amperes

(ii) $IV \sin \theta = 408 \times \frac{I_C}{I} = 408 \times \frac{1.2}{1.7} = 288\text{ VAr}$

(iii) $IV \cos \theta = 288\text{ W}$.

PARALLEL CIRCUIT CONTAINING *C* AND *L*

The circuit diagram of C and L and the phasor diagram are shown in Fig. 11. The capacitor and inductor are assumed to be pure components. I_C leads V by 90° and I_L lags V by 90° as shown in Fig. 12, Fig. 13 and Fig. 14.

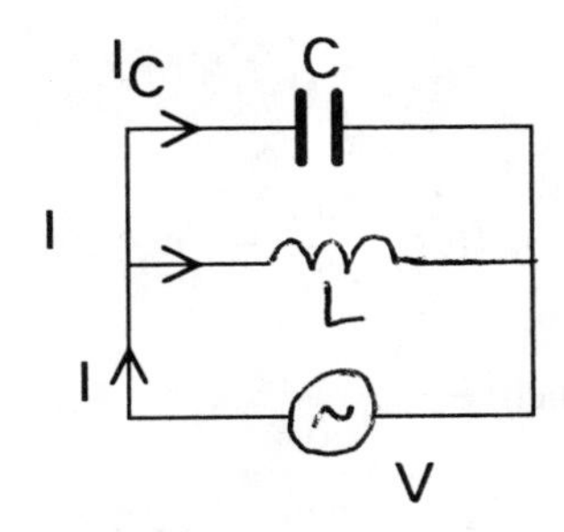

Fig. 11

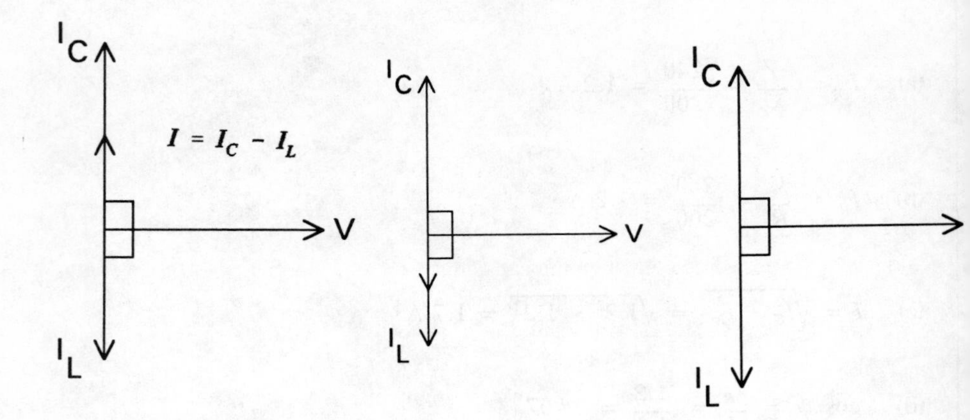

Fig. 12 Fig. 13 Fig. 14

There are three cases to consider (i) $I_C > I_L$ capacitive (Fig. 12)

(ii) $I_C < I_L$ inductive (Fig. 13)

(iii) $I_C = I_L$ resonance. (Fig. 14)

WORKED EXAMPLE 3

An LC parallel circuit has the following circuit components:-

(a) $X_C = 10\ \Omega,\ X_L = 5\ \Omega$

(b) $X_C = 2\ \Omega,\ X_L = 10\ \Omega$

(c) $X_C = X_L = 100\ \Omega$.

If the r.m.s. supply voltage is 240 V, calculate the currents in each case.

SOLUTION 3

(a) $I_C = \dfrac{V}{X_C} = \dfrac{240}{10} = 24\text{ A},\ I_L = \dfrac{240}{5} = 48\text{ A}$

$I = I_L - I_C = 24\text{ A}$ inductive

(b) $I_C = \dfrac{V}{X_C} = \dfrac{240}{2} = 120\text{ A},\ I_L = \dfrac{240}{10} = 24\text{ A}$

$I = I_C - I_L = 120 - 24 = 96\text{ A}$ capacitive

(c) $I_C = I_L = \dfrac{240}{100} = 2.4\text{ A},\ I = 0.$

WORKED EXAMPLE 4

A pure capacitor C is connected in parallel with a pure inductor of value 15.9 mH, the combination is supplied with a voltage of 10 V r.m.s. The supply current is 1 A r.m.s. and the current through L is 2 A r.m.s.

Calculate: (i) the current through the capacitor
(ii) the frequency of the supply
(iii) the capacitance value
(iv) the resonant frequency

SOLUTION 4

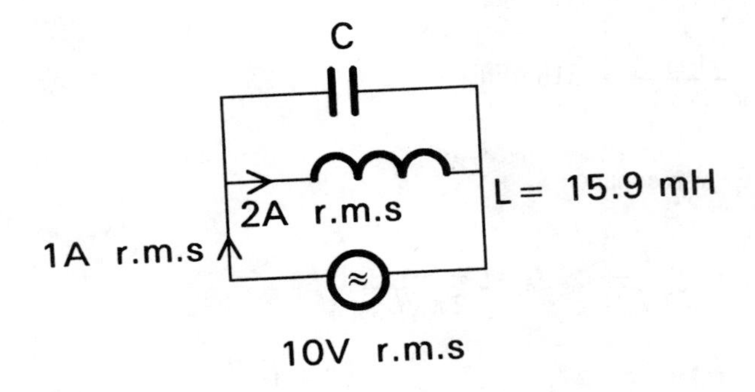

Fig. 15

(i)

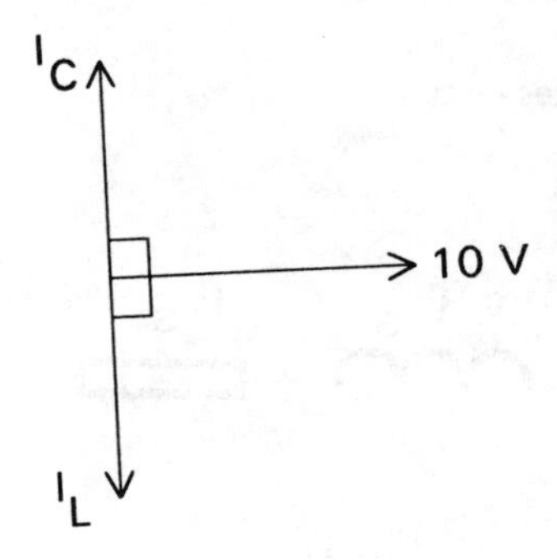

Fig. 16

$$I_L - I_C = I = 1 \text{ A}$$

$$I_L - I = I_C = 2 - 1 = 1 \text{ A}$$

(ii) $$X_L = \frac{V}{I_L} = \frac{10}{2} = 5\,\Omega, \; X_L = 2\pi f L = 5$$

$$f = \frac{5}{2\pi L} = \frac{5 \times 1000}{2\pi\, 15.9} = 50 \text{ Hz}$$

(iii) $$X_C = \frac{10}{1} = 10\,\Omega = \frac{1}{2\pi f C}$$

$$C = \frac{1}{2\pi\, 50 \times 10} = 318\,\mu\text{F}$$

(iv) $X_L = X_C$ at resonance

$$2\pi f_o L = \frac{1}{2\pi f_o C}, \quad f_o = \frac{1}{2\pi\sqrt{LC}}$$

$$= \frac{1}{2\pi\sqrt{15.9 \times 10^{-3} \times 318 \times 10^{-6}}}$$

$$= 70.78 \approx 71 \text{ Hz}$$

PARALLEL CIRCUITS AND RESONANCE

A coil is represented by the series circuit.

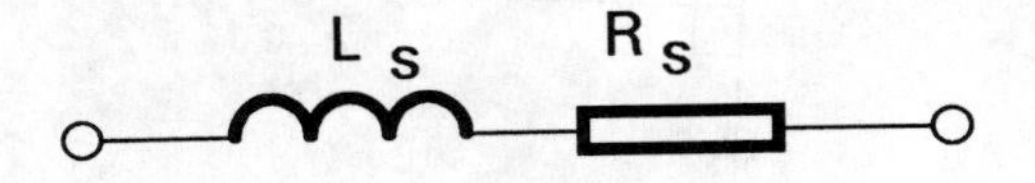

Fig. 17

where L_s represents the pure coil and R_s is the resistance of the winding. A coil can also be represented by the parallel circuit

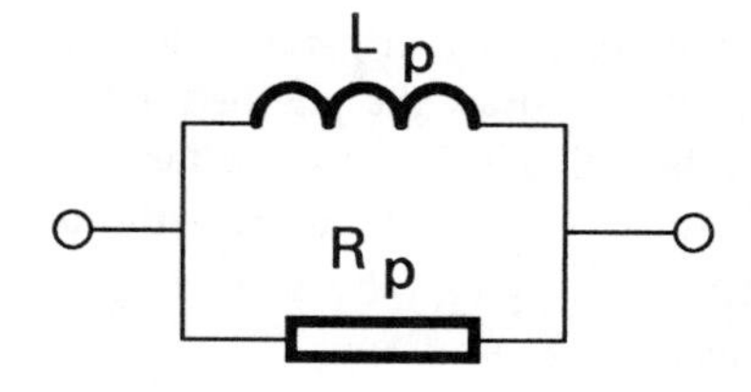

Fig. 18

where L_p represents the pure coil again and R_p the equivalent parallel resistance which is different in value from that of R_s. A coil is now connected in parallel with a pure capacitor as shown in Fig. 19.

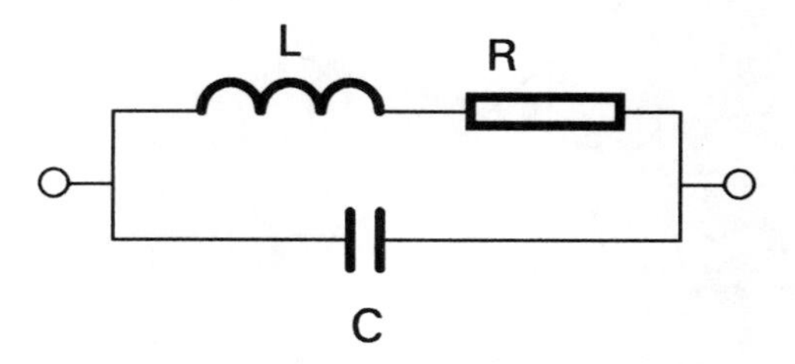

Fig. 19

Therefore a coil and a capacitor when they are connected in parallel it is normal to assume the equivalent circuit is as shown in Fig. 19.

ANALYSIS OF PARALLEL CIRCUIT

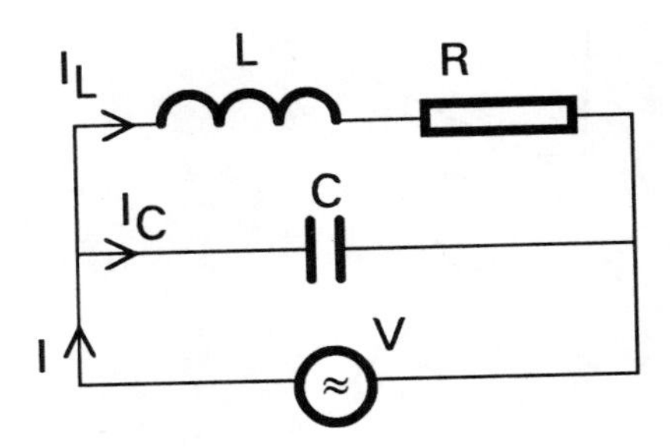

Fig. 20

A coil and a capacitor are connected in parallel across an a.c. supply of V volts r.m.s. as shown in Fig. 20. To draw the phasor for Fig. 20, V is common across the capacitor and across the coil and is therefore taken as a reference and is drawn horizontally. The p.d. across C is V, the current through C is 90° leading V, the current through the coil, I_L, is lagging V by an angle ϕ, this angle would have been 90° if there were no resistance R.

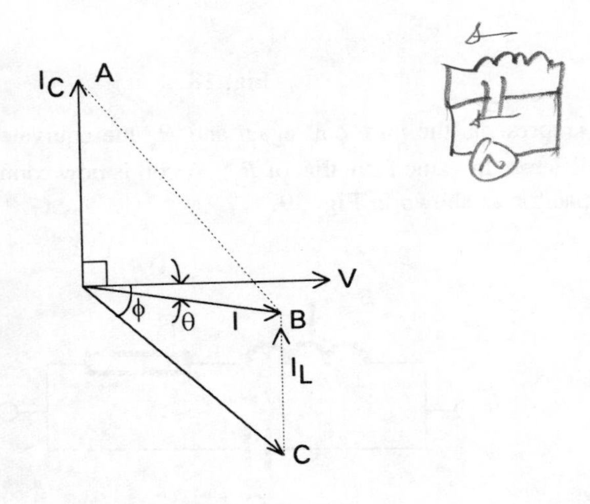

Fig. 21

The total current I is the resultant or the diagonal of the parallelogram so formed, AB is drawn parallel to OC and BC is drawn parallel to OA.
I lags V by an angle θ and the circuit is lagging.
Let us redraw the phasor where I leads V by angle θ and the circuit is leading.

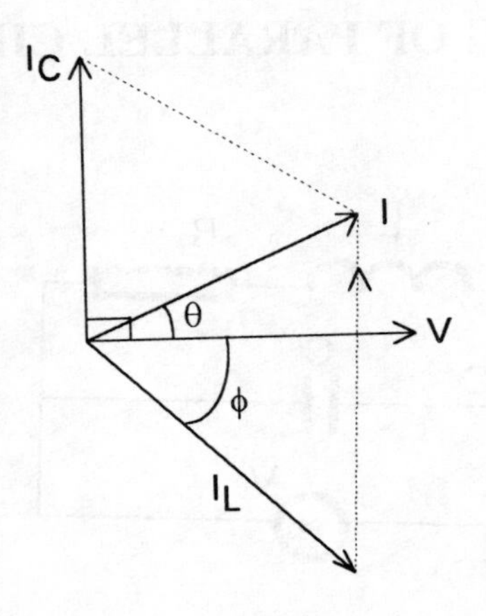

Fig. 22

Parallel Resonance, or tuned circuit. The general definition of resonance is when the total current, I is in phase with the supply voltage, V. The circuit is neither inductive nor capacitive. The parallel resonant or tuned circuit is resistive at resonance. The current is minimum and the resistance is maximum, $\theta = 0$.

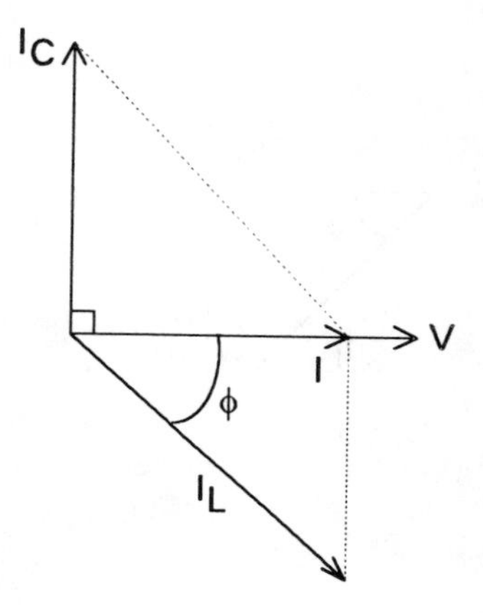

Fig. 23 Phasor at Resonance

If the resistance R were zero I_L will be lagging V by 90° and $I_C = I_L$, the resultant current $I = 0$.

This situation is shown in Fig. 24

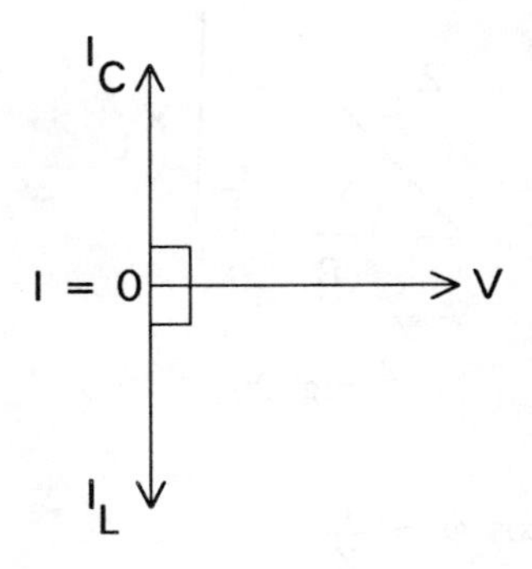

Fig. 24

RESONANT FREQUENCY AND DYNAMIC RESISTANCE OF PARALLEL TUNED CIRCUIT.

The current triangle is as shown in Fig. 25.

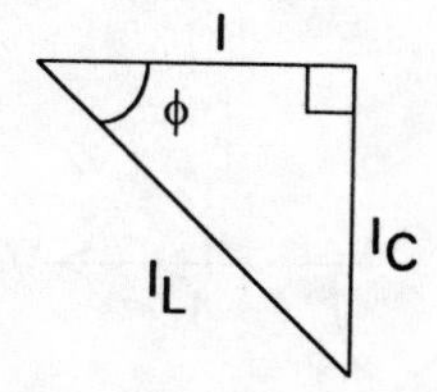

Fig. 25

$I = I_L \cos \phi$

$I_C = I_L \sin \phi$

The impedance triangle for the coil is shown in Fig. 26.

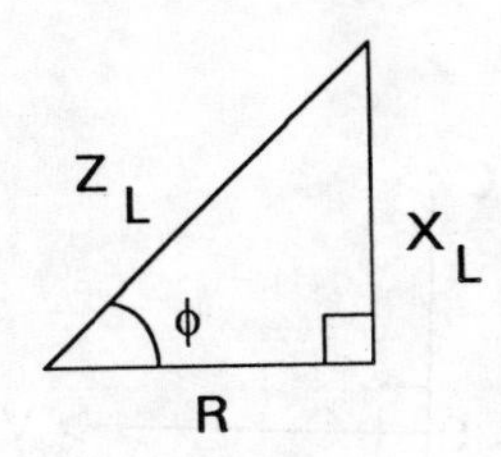

Fig. 26

$$\sin \phi = \frac{X_L}{Z_L} \qquad \cos \phi = \frac{R}{Z_L}$$

where ϕ is the same angle as that of Fig. 26.

From the circuit diagram

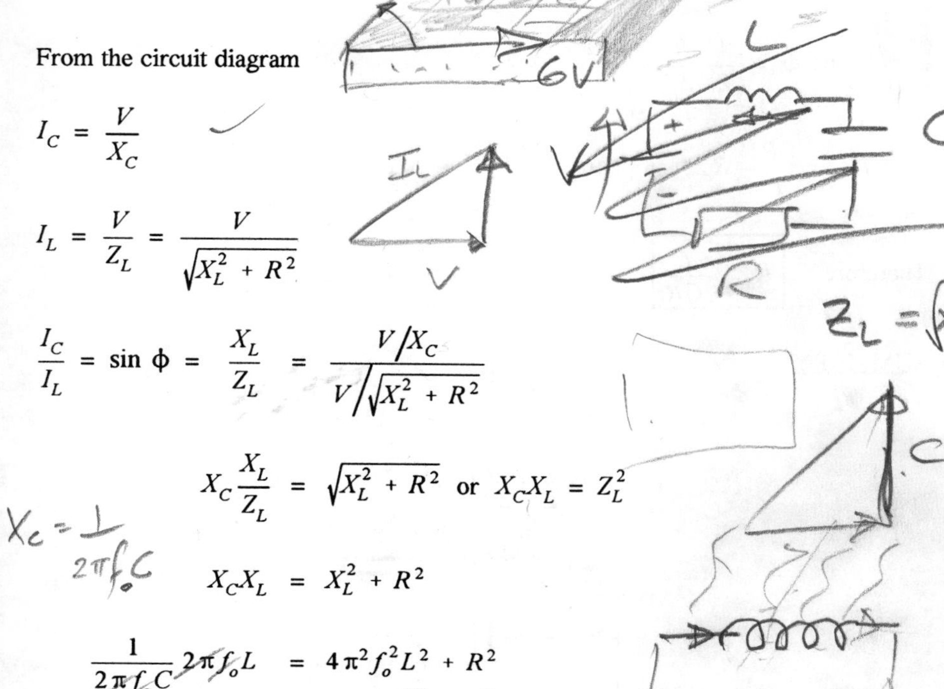

$$I_C = \frac{V}{X_C}$$

$$I_L = \frac{V}{Z_L} = \frac{V}{\sqrt{X_L^2 + R^2}}$$

$$\frac{I_C}{I_L} = \sin\phi = \frac{X_L}{Z_L} = \frac{V/X_C}{V/\sqrt{X_L^2 + R^2}}$$

$$X_C \frac{X_L}{Z_L} = \sqrt{X_L^2 + R^2} \text{ or } X_C X_L = Z_L^2$$

$$X_C X_L = X_L^2 + R^2$$

$$\frac{1}{2\pi f_o C} 2\pi f_o L = 4\pi^2 f_o^2 L^2 + R^2$$

$$\frac{L}{C} = 4\pi^2 f_o^2 L^2 + R^2$$

$$4\pi^2 f_o^2 L^2 = \frac{L}{C} - R^2$$

$$f_o^2 = \frac{L}{4\pi^2 L^2 C} - \frac{R^2}{4\pi^2 L^2} = \frac{1}{4\pi^2}\left(\frac{1}{LC} - \frac{R^2}{L^2}\right)$$

$$\boxed{f_o = \frac{1}{2\pi}\sqrt{\frac{1}{LC} - \frac{R^2}{L^2}}}$$

$$I^2 + I_C^2 = I_L^2$$

$I = \frac{V}{R_d}$ where R_d is the dynamic resistance of the circuit

(the resistance of the circuit at resonance)

$$\frac{V^2}{R_d^2} + \frac{V^2}{X_C^2} = \frac{V^2}{Z_L^2}$$

$$Z_L^2 = X_L X_C = \frac{2\pi f L}{2\pi f C}$$

$$I = I_L \cos\phi = \frac{V}{Z_L}\frac{R}{Z_L} = \frac{RV}{L/C}$$

$$V = IR_d = \frac{RV}{L/C}R_d.$$

Therefore $\boxed{R_d = \frac{L}{CR}}$

<u>SUMMARY</u>

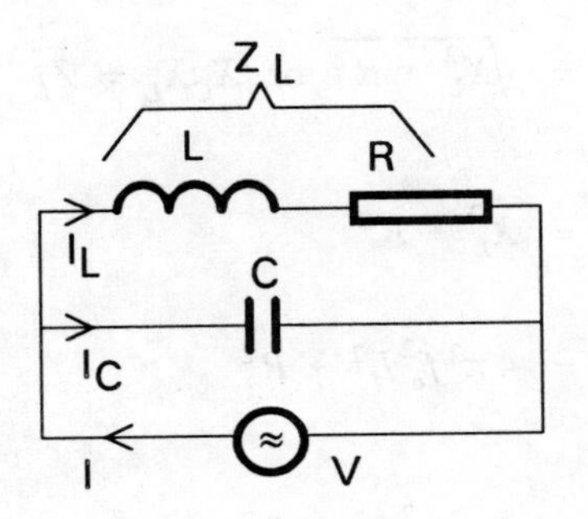

Fig. 27

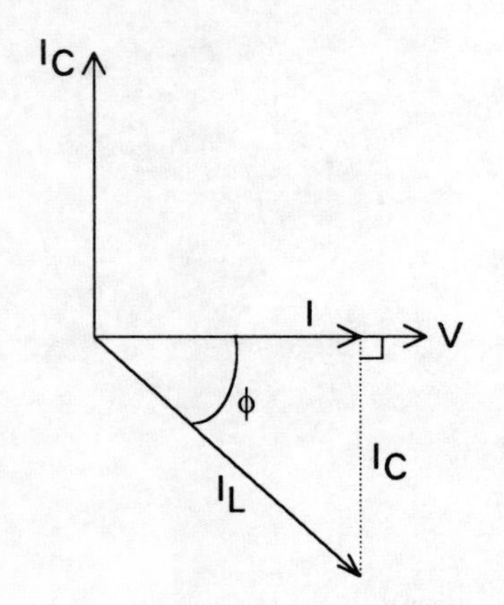

Fig. 28

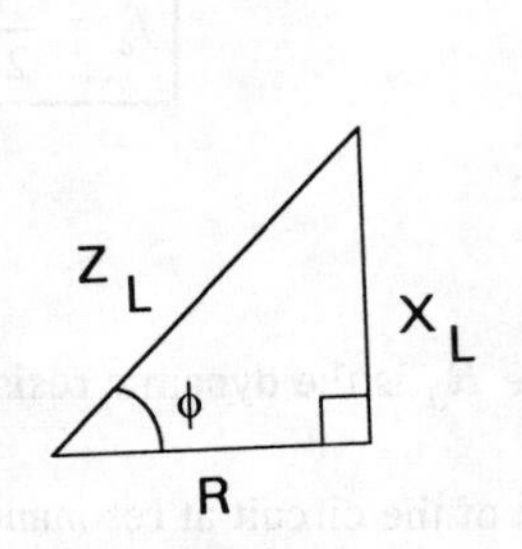

Fig. 29

$$f_o = \frac{1}{2\pi}\sqrt{\frac{1}{LC} - \frac{R^2}{C^2}} \qquad R_d = \frac{L}{CR}$$

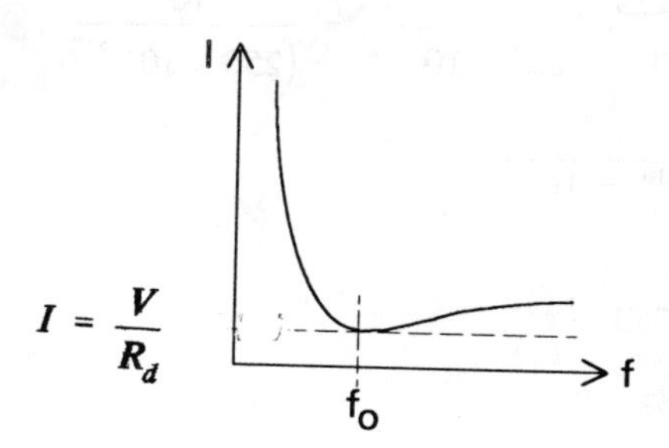

Fig. 30

The impedance at resonance is resistive and it is maximum, $\boxed{R_d = \frac{L}{CR}}$ since V is fixed and the current is minimum.

WORKED EXAMPLE 5

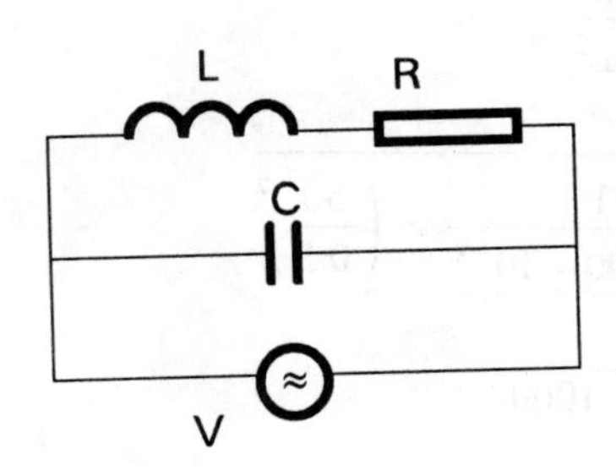

Fig. 31

(a) If $L = 250$ mH, $R = 10\ \Omega$, $C = 250$ pF. Calculate the resonant frequency.

(b) If $L = 0.5$ H, $R = 50\Omega$ and $C = 100\ \mu$F. Recalculate the resonant frequency.

SOLUTION 5

(a) $$f_o = \frac{1}{2\pi}\sqrt{\frac{1}{LC} - \frac{R^2}{L^2}}$$

$$= \frac{1}{2\pi}\sqrt{\frac{1}{250 \times 10^{-3} \times 250 \times 10^{-12}} - \frac{10^2}{(250 \times 10^{-3})^2}}$$

$$= \frac{1}{2\pi}\sqrt{1.6 \times 10^{10} - 1600}$$

$$\approx \frac{1}{2\pi}\sqrt{1.6 \times 10^{10}}$$

$$\approx \frac{1}{2\pi} 1.265 \times 10^5$$

$$\approx 20.1 \text{ KHz.}$$

In this example $\frac{R^2}{L^2}$ is much less than $\frac{1}{LC}$ and may be neglected.

An approximate expression would be $\boxed{f_o \approx \frac{1}{2\pi\sqrt{LC}}}$

(b) $$f_o = \frac{1}{2\pi}\sqrt{\frac{1}{LC} - \frac{R^2}{L^2}}$$

$$= \frac{1}{2\pi}\sqrt{\frac{1}{0.5 \times 100 \times 10^{-6}} - \left(\frac{50}{0.5}\right)^2}$$

$$= \frac{1}{2\pi}\sqrt{20000 - 10000}$$

$$= \frac{1}{2\pi} 100$$

$$= 15.9 \text{ Hz.}$$

In this case $\frac{R^2}{L^2}$ is substantial and must be taken into account.

WORKED EXAMPLE 6

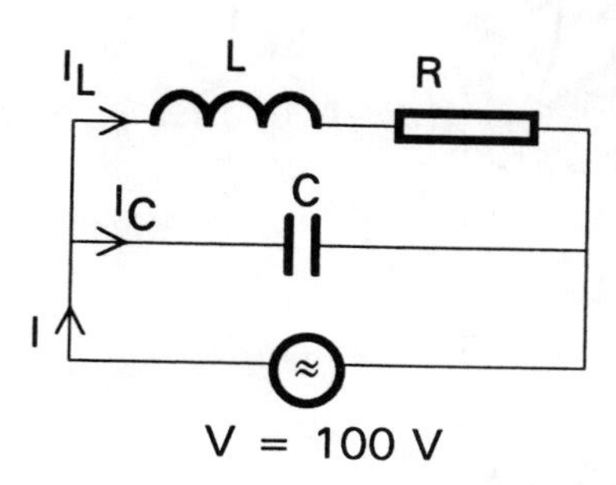

ω_o = 20000 rad/s resonant angular frequency

Fig. 32

The impedance of the circuit shown at resonance is 1000 Ω and the magnification factor Q = 10 (quality factor)

Calculate (i) the total current drawn from the supply
(ii) the currents in the inductive and capacitive branches
(iii) the values of R, L and C.

SOLUTION 6

R_d = dynamic resistance = impedance at resonance = $\dfrac{L}{CR}$

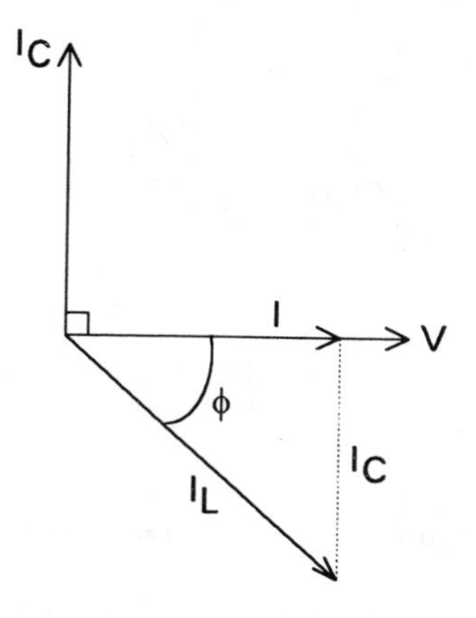

Fig. 33 The phasor diagram

$$Q = \tan\phi = \frac{I_C}{I}$$

$$I_C = QI$$

$$Q = \frac{I_C}{I}$$

$$V = IR_d$$

(i) $I = \dfrac{100}{1000} = 0.1 \text{ A}$

(ii) $I_C = QI = 10\ (0.1) = 1 \text{ A}$

$$I_L^2 = I^2 + I_C^2$$

$$I_L = \sqrt{0.1^2 + 1^2} = \sqrt{1.01} = 1.005 \text{ A}$$

(iii) $I_C = \dfrac{100}{X_C} = \dfrac{100}{\dfrac{1}{2\pi f_o C}} = 100\omega_o C$

$$C = \frac{I_C}{100\omega_o} = \frac{1}{100 \times 20000} = 0.5 \times 10^{-6} \text{ F}$$

$$Z_L = \sqrt{X_L^2 + R^2} = \frac{100}{1.005} = 99.5\ \Omega$$

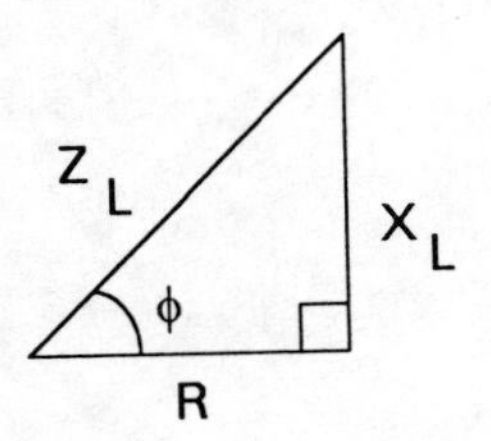

Fig. 34

$$X_L^2 + R^2 = 9901$$

$R_d = \dfrac{L}{CR} = 1000$ and $\sin\phi = \dfrac{X_L}{Z_L} = 0.995$ or $X_L = 0.995 \times 1000$

$$X_L = 995\ \Omega$$

where $\tan \phi = 10, \phi = 84.289407$

$\omega_o L = 995$ or $L = 49.8$ mH and $C = \dfrac{L/R}{1000} = \dfrac{49.8 \times 10^{-3}}{1000R} = 0.5 \times 10^{-6}$

$$\frac{49.8}{R} = 0.5$$

$R = 99.6\ \Omega$

$I = 0.1$ A, $I_C = 1$ A, $I_L = 1.005$

$C = 0.5\ \mu$F, $L = 49.8$ mH and $R = 99.6\ \Omega$

WORKED EXAMPLE 7

An inductor of inductance 0.75 H takes a current of 2 mA from a 10 V, 1000 Hz supply. Calculate the capacitance of the capacitor required to be connected in parallel with the inductor to produce resonance at 1000 Hz, and determine the total current taken from the supply for this voltage.

SOLUTION 7

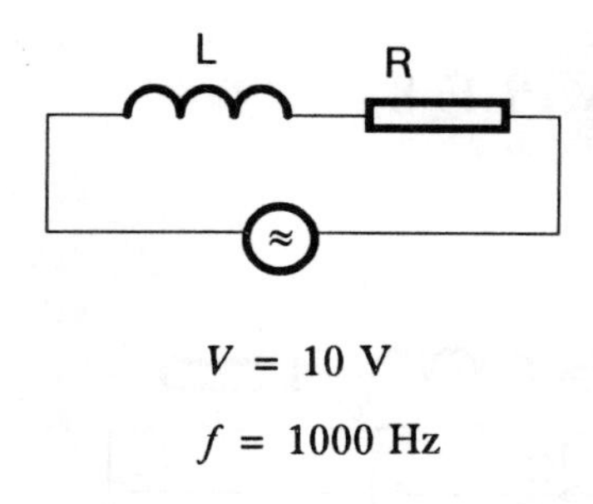

$V = 10$ V

$f = 1000$ Hz

Fig. 35

Let $Z = \sqrt{R^2 + X_L^2} = \dfrac{10}{2 \times 10^{-3}}$

$Z = 5000\ \Omega$

$R^2 + 4\pi^2\, 10^6\, 0.75^2 = 5000^2$

$R^2 = 5000^2 - 22206609.9$

$R = 1671\ \Omega$

$$f_o = \frac{1}{2\pi}\sqrt{\frac{1}{LC} - \frac{R^2}{L^2}}$$

$$(2\pi\,1000)^2 = \frac{1}{LC} - \frac{R^2}{L^2} = \frac{1}{0.75 \times C} - \frac{1671^2}{0.75^2}$$

$$4\pi^2\,1000000 + \frac{1671^2}{0.75^2} = \frac{1}{0.75C}$$

$$39478417.6 + 4963984 = \frac{1}{0.75C}$$

$$C = \frac{1}{44442401.6 \times 0.75}$$

$$C = 3.0 \times 10^{-8} = 30 \text{ nF}$$

$$R_d = \frac{L}{CR} = \frac{0.75}{30 \times 10^{-9}\,1671} = 14961\ \Omega$$

$$I = \frac{V}{R_d} = \frac{10}{L/CR} = \frac{10\,CR}{L} = \frac{10 \times 30 \times 10^{-9}\,1671}{0.75} = 0.668 \text{ mA}$$

WORKED EXAMPLE 8

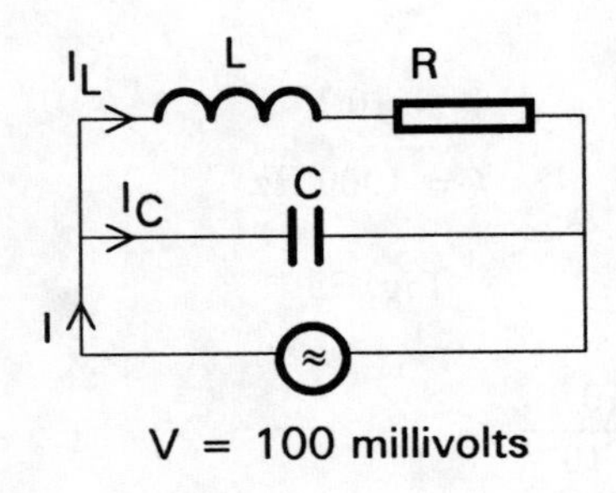

Fig. 36

$L = 50\ \mu\text{H}$, $R = 10\ \Omega$, $C = 100 \times 10^{-12}$ F

Calculate (i) the resonant frequency
(ii) the dynamic resistance
(iii) the power dissipated in the circuit
(iv) the currents: I, I_C, I_L.
(v) the Q-factor

SOLUTION 8

(i) $$f_o = \frac{1}{2\pi}\sqrt{\frac{1}{LC} - \frac{R^2}{L^2}}$$

$$= \frac{1}{2\pi}\sqrt{\frac{1}{50 \times 10^{-6} \times 100 \times 10^{-12}} - \left(\frac{10}{50 \times 10^{-6}}\right)^2}$$

$$= \frac{1}{2\pi}\sqrt{2 \times 10^{14} - 0.04 \times 10^{12}} = \frac{10^7}{2\pi}\sqrt{2 - 0.0004}$$

$$= \frac{10^7}{2\pi} 1.41435 = 2.25 \text{ MHz}$$

(ii)

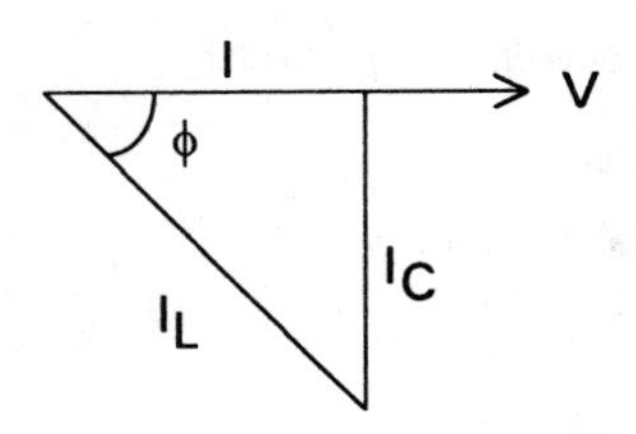

Fig. 37

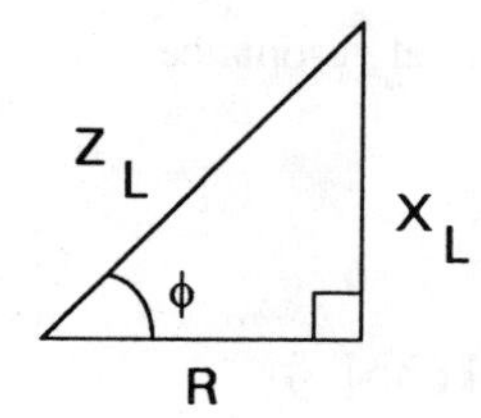

Fig. 38

$$R_d = \frac{L}{CR} = \frac{50 \times 10^{-6}}{100 \times 10^{-12} 10} = 50000 \ \Omega$$

(iii) $$P = \frac{V^2}{R_d} = \frac{100^2 \times 10^{-6}}{5 \times 10^4} = 0.2 \ \mu\text{W}$$

(iv) $I = \frac{V}{R_d} = \frac{100 \times 10^{-3}}{5 \times 10^4} = 2\ \mu A$

$$I_C = \frac{V}{X_C} = \frac{100 \times 10^{-3}}{707} = 1.41 \times 10^{-4}\ A$$

$$X_C = \frac{1}{2\pi f_o C} = \frac{1}{2\pi \times 2.25 \times 10^6 \times 100 \times 10^{-12}} = 707\ \Omega$$

$$Q = \frac{I_C}{I} = \frac{1.41 \times 10^{-4}}{2 \times 10^{-6}} = 70.7.$$

WORKED EXAMPLE 9

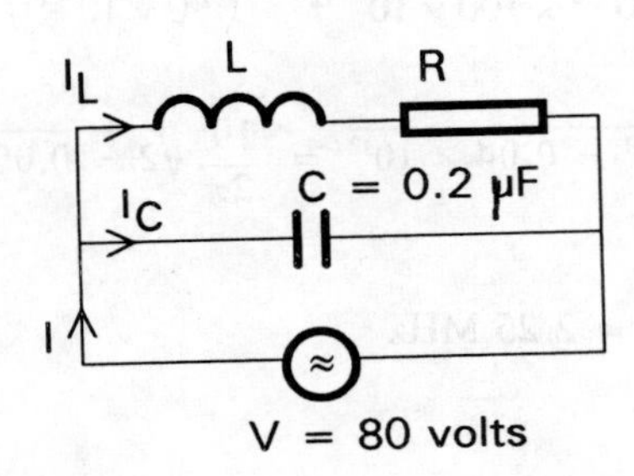

Fig. 39

The circuit is at resonance (i) Determine the currents I, I_C and I_L.

(ii) Calculate R, L and f_o

Data: $R_d = 400\ \Omega$, $Q = 1.732$.

SOLUTION 9

(i) $V = IR_d = 80 = I400 \Rightarrow I = 0.2\ A$

$$I_C = IQ = 0.2 \times 1.732 = 0.3464\ A$$

From Fig. 33 $I_L^2 = I^2 + I_C^2 = 0.2^2 + 0.3464^2 = 0.16$

$$I_L = 0.4\ A$$

(ii) $Z_L = \dfrac{V}{I_L} = \dfrac{80}{0.4} = 200\ \Omega$. From Fig. 34 we have

$$\tan\phi = 1.732 \Rightarrow \phi = 60°,\ \cos 60° = \frac{R}{Z_L} \Rightarrow R = Z_L 0.5 = 100\ \Omega$$

$$\tan\phi = \frac{X_L}{R} = 1.732 \Rightarrow X_L = 100 \times 1.732 = 173.2\ \Omega$$

$$R_d = 400 = \frac{L}{CR} = \frac{L}{0.2 \times 10^{-6} \times 100} \Rightarrow L = 8\ \text{mH}.$$

$$X_L = 2\pi f_o L = 173.2 \Rightarrow f_o = \frac{173.2}{2\pi 8 \times 10^{-3}} = 3446\ \text{Hz}.$$

WORKED EXAMPLE 10

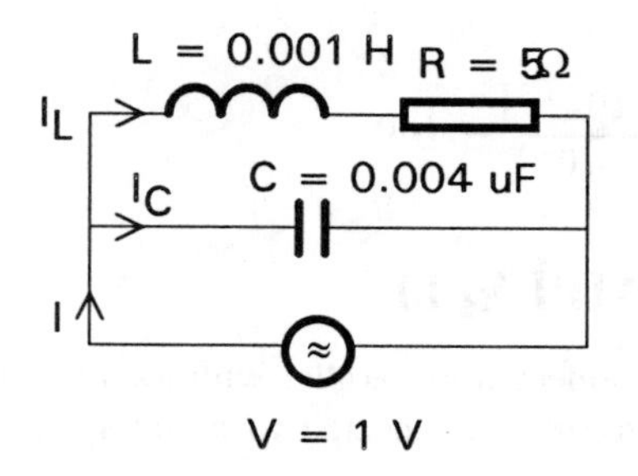

Fig. 40

Calculate (i) the frequency of the supply that will cause minimum current to flow
(ii) the current through C
(iii) the Q-factor.

Sketch a graph of the impedance against frequency.

SOLUTION 10

(i) $$f_o = \frac{1}{2\pi}\sqrt{\frac{1}{LC} - \frac{R^2}{L^2}}$$

$$= \frac{1}{2\pi}\sqrt{\frac{1}{0.001 \times 0.004 \times 10^{-6}} - \frac{(5)^2}{(0.001)^2}}$$

$$= \frac{1}{2\pi}\sqrt{\frac{10^{12}}{4} - 25 \times 10^6}$$

$$= \frac{1}{2\pi}\sqrt{25 \times 10^{10} - 25 \times 10^6}$$

$$\approx \frac{500000}{2\pi} \approx 79.6 \text{ KHz}$$

(ii) $$I_C = \frac{1}{X_C} = \frac{1}{1/2\pi f_o C} = 2\pi f_o C$$

$$= 2\pi\, 79.6 \times 10^3 \times 0.004 \times 10^{-6} = 2 \text{ mA}$$

(iii) $$I = \frac{1}{R_d} = \frac{1}{50 \times 10^3} = 0.02 \text{ mA}$$

where $$R_d = \frac{L}{CR} = \frac{0.001}{0.004 \times 10^{-6} \times 5} = \frac{10^6}{20} = 50000\ \Omega$$

$$Q = \frac{I_C}{I} = \frac{2 \times 10^{-3}}{0.02 \times 10^{-3}} = 100.$$

WORKED EXAMPLE 11

A 1000 pF capacitor is connected in parallel with a coil of inductance 2.5 mH and resistance 75 Ω. The circuit is connected to a supply of 240 V at the resonant frequency.

Calculate
(i) the resonant frequency
(ii) the dynamic impedance
(iii) the magnification factor
(iv) the supply current
(v) the capacitor current.

SOLUTION 11

(i) $$f_o = \frac{1}{2\pi}\sqrt{\frac{1}{LC} - \frac{R^2}{L^2}}$$

$$= \frac{1}{2\pi}\sqrt{\frac{1}{2.5 \times 10^{-3} \times 1000 \times 10^{-12}} - \frac{75^2}{(2.5 \times 10^{-3})^2}} = 100 \text{ KHz}$$

(ii) $R_d = \frac{L}{RC} = \frac{2.5 \times 10^{-3}}{75 \times 1000 \times 10^{-12}} = 33.3 \text{ K}\Omega$

(iii) $Q = \frac{X_L}{R} = \frac{2\pi f_o L}{R} = \frac{2\pi\, 100 \times 10^3 \times 2.5 \times 10^{-3}}{75} = 20.9$

(iv) $I = \frac{V}{R_d} = \frac{240}{33.3 \times 10^3} = 7.21 \text{ mA}$

(v) $I_c = IQ = 7.21 \times 10^{-3} \times 20.9 = 151 \text{ mA}.$

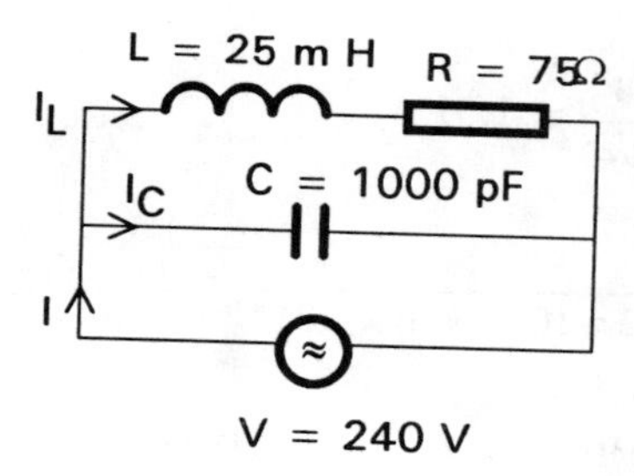

Fig. 41

WORKED EXAMPLE 12

A coil of inductance 250 mH and resistance 100 Ω is connected with a 1 μF capacitor. The parallel combination is supplied at 240 V r.m.s. at a frequency of 1000 Hz. Calculate: (i) I_C

(ii) I_L

(iii) I

(iv) the phase angle between I and V, ϕ

(v) the power consumed.

SOLUTION 12

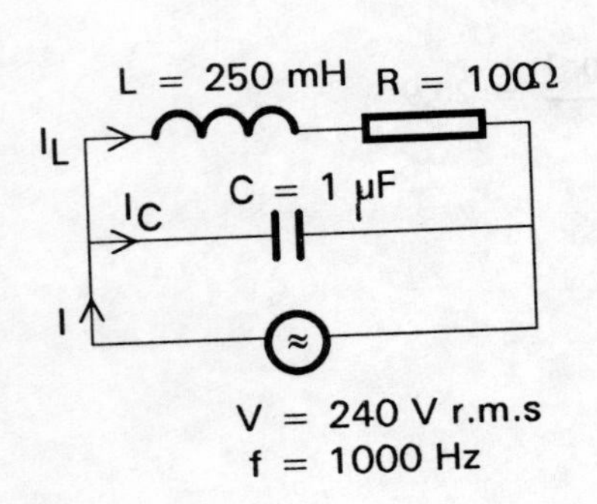

Fig. 42

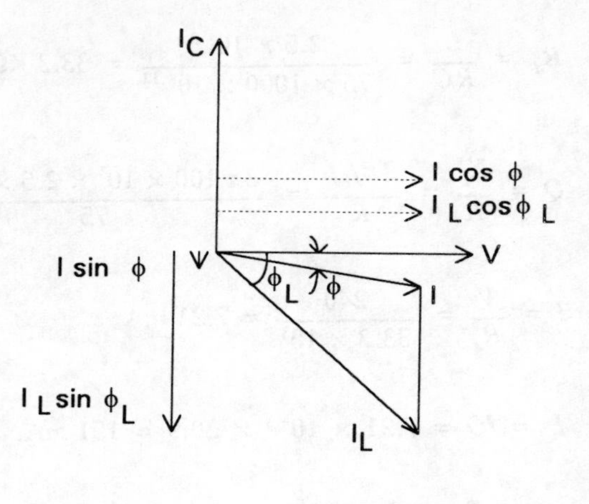

Fig. 43

(i) $I_C = \frac{240}{X_C} = \frac{240}{159.2} = 1.51$ A

$$X_C = \frac{1}{2\pi fC} = \frac{1}{2\pi\, 1000 \times 1 \times 10^{-6}} = 159.2\ \Omega$$

(ii) $X_L = 2\pi fL = 2\pi 1000 \times 0.25 = 1571\ \Omega$

$R = 100\ \Omega$

$$\phi_L = \tan^{-1}\frac{1571}{100} = \tan^{-1} 15.71 = 86.4°$$

$$Z_L = \sqrt{R^2 + X_L^2} = \sqrt{100^2 + 1571^2} = 1574.2\ \Omega$$

$$I_L = \frac{V}{1574.2} = \frac{240}{1574.2} = 0.15 \text{ A}$$

(iii) $I_L \sin\phi_L - I_C = I\sin\phi$ ... (1)

$I\cos\phi = I_L\cos\phi_L$ ... (2)

Substituting the values of $I_L = 0.15$ A, $I_C = 1.51$ A, $\phi_L = 86.4°$ in (1) and (2) we have

$$0.15 \sin 86.4° - 1.51 = I\sin\phi = -1.36$$

$$I\cos\phi = 0.15\cos 86.4° = 9.42 \times 10^{-3}$$

$$\tan \phi = -\frac{1.36}{9.42 \times 10^{-3}} \Rightarrow \phi = -89.7^\circ \text{ hence } -I \sin 89.7^\circ = -1.36$$

$$\boxed{I = 1.36 \text{ mA}}$$

(iv) The power consumed = $IV \cos \phi = 1.36 \times 10^{-3} \times 240 \times \cos 89.7^\circ = 1.71$ mW.

PARALLEL CIRCUIT CONTAINING C IN PARALLEL WITH LR

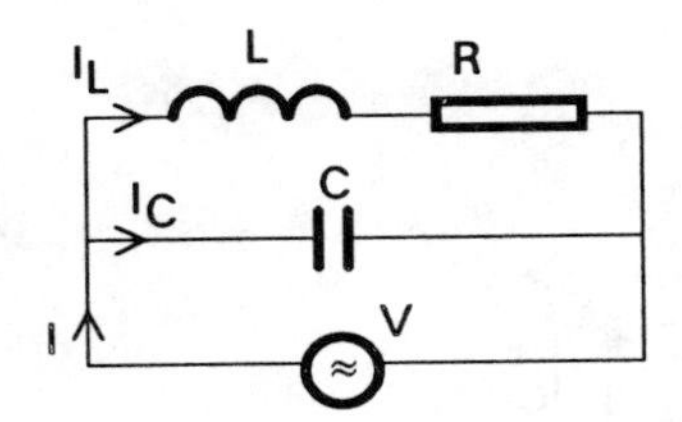

Fig. 44

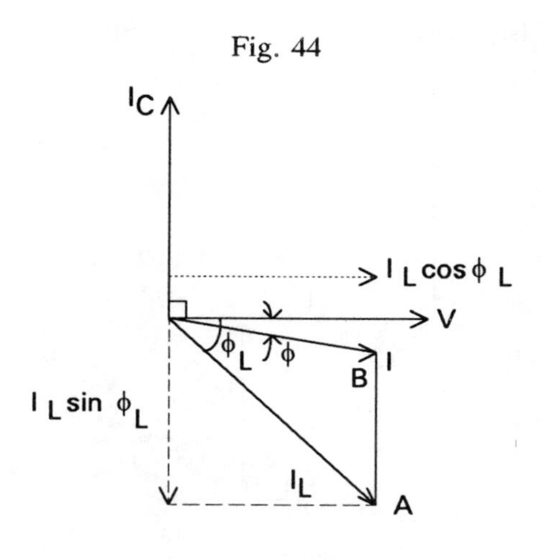

Fig. 45

The current through C, I_C, leads the supply voltage, V, by 90°, the current through LR, I_L, lags the supply voltage, V, by an angle ϕ_L. Drawing a line parallel and equal to I_C from the end of OA, then $OA = I_L$, $AB = I_C$ then $OB = I$.

Resolving $OA = I_L$ into a vertical and horizontal components, $I_L \sin \phi_L$, and $I_L \cos \phi_L$. Resolving $OB = I$ into a vertical and horizontal components, $I \sin \phi$ and $I \cos \phi$. It can be seen that $I \cos \phi = I_L \cos \phi_L$.

There are three cases to consider.

(i) $I_L \sin \phi > I_C$ (I lags V by ϕ)

(ii) $I_L \sin \phi < I_C$ (I leads V by ϕ)

(iii) $I_L \sin \phi = I_C$ (I is in phase with V)

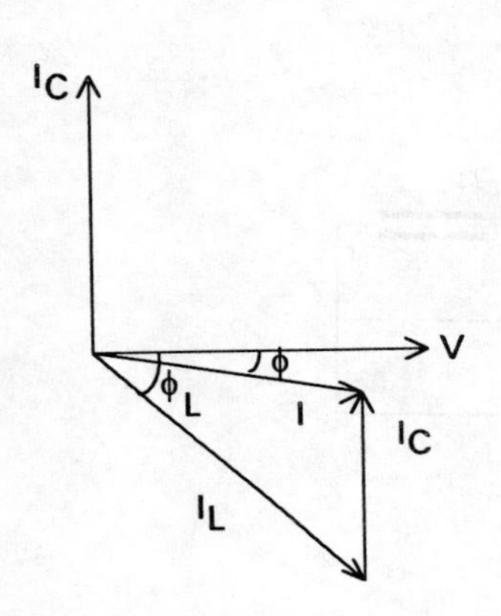

Fig. 46 I lags V by ϕ

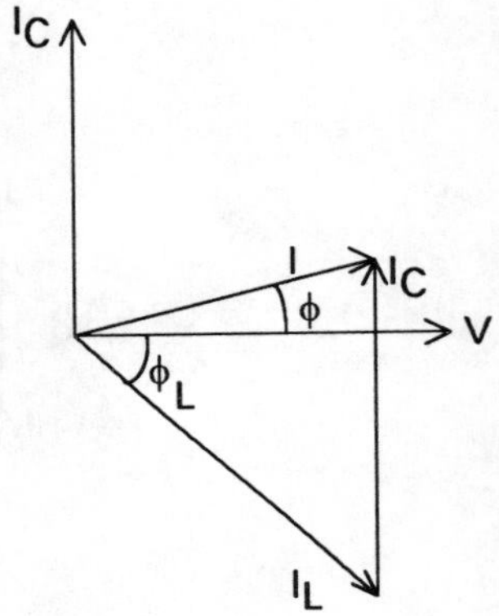

Fig. 47 I leads V by ϕ

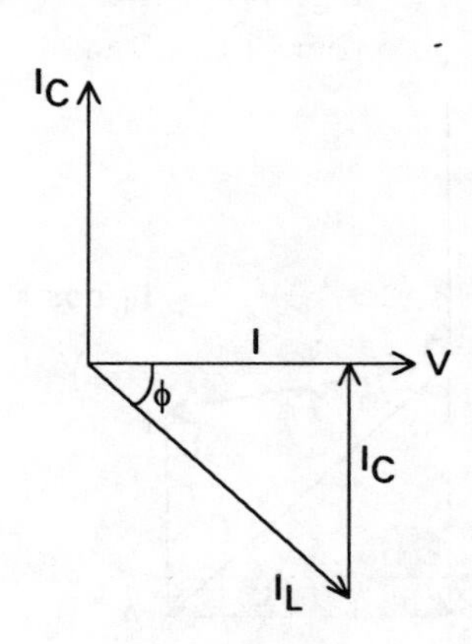

Fig. 48 I is in phase with V

The circuits are inductive, capacitive, resistive.

WORKED EXAMPLE 13

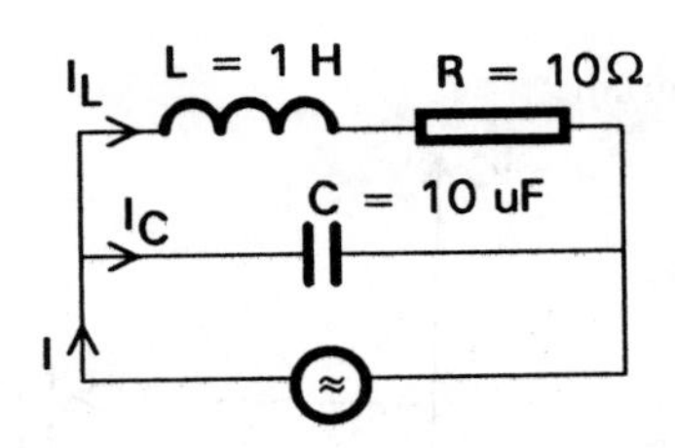

$V = 24$ V r.m.s.

$f = 50$ Hz

Fig. 49

Calculate: (i) the current through C, I_C,

(ii) the current through LR, I_L

(iii) the phase angle of I_L with V

(iv) the phase angle of I with V, is it lagging or leading?

(v) the total current, I

SOLUTION 13

(i) $$I_C = \frac{V}{X_C} = \frac{24}{318} = \boxed{75.4 \text{ mA}}$$

where $$X_C = \frac{1}{2\pi f C} = \frac{1}{2\pi 50 \times 10 \times 10^{-6}} = 318\ \Omega$$

(ii) $$I_L = \frac{24}{\sqrt{R^2 + X_L^2}} = \frac{24}{\sqrt{10^2 + 314.2^2}} = \frac{24}{314.3} = 76.4 \text{ mA}$$

where $X_L = 2\pi f L = 2\pi 50 \times 1 = 314.2\ \Omega$

(iii) $$\tan \phi_L = \frac{314.2}{10} = 31.42$$

$$\phi_L = \tan^{-1} 31.42 = \boxed{88.2^\circ}$$

(iv) From the phasor diagram

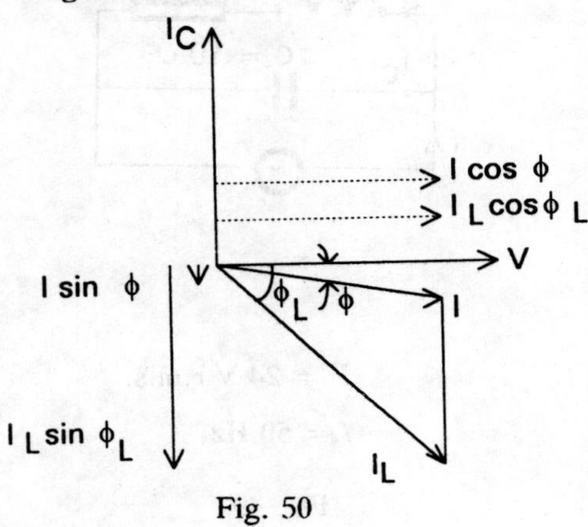

Fig. 50

I_L is resolved in the components

$I_L \sin \phi_L$ vertical $I_L \cos \phi_L$ horizontal

I is resolved in the components

$I \sin \phi$ vertical $I \cos \phi$ horizontal

$I \cos \phi = I_L \cos \phi_L$ ... (1)

$I_L \sin \phi_L - I_C = I \sin \phi$ ... (2)

$$\boxed{\tan \phi = \frac{I_L \sin \phi_L - I_C}{I_L \cos \phi_L}}$$

$$\tan \phi = \frac{76.4 \times 10^{-3} \sin 88.2^\circ - 75.4 \times 10^{-3}}{76.4 \times 10^{-3} \cos 88.2^\circ} = 0.4,$$

$$\boxed{\phi = 21.8^\circ.}$$

(v) $I \cos \phi = I_L \cos \phi_L, \quad I = \dfrac{I_L \cos \phi_L}{\cos \phi}$

$$I = \boxed{2.58 \text{ mA}}$$

Q-FACTOR

Q stands for Quality

Q-factor is a measure of the quality of an electrical component.

Definition

$$Q = 2\pi \times \frac{\text{Maximum energy stored in one cycle}}{\text{Energy dissipated during one cycle}} \quad \dots (1)$$

Q-factor of a capacitor.

The equivalent circuit of a capacitor may be represented a pure capacitance, C, in series with a non reactive resistor, R_1, as shown in Fig. 51

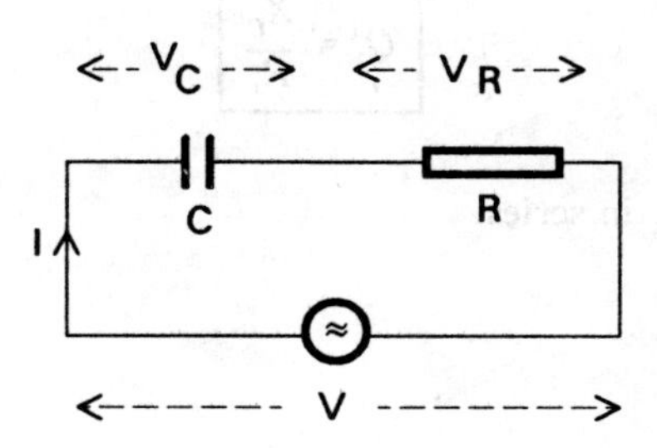

Fig. 51

Let V be the r.m.s. value of the voltage of an a.c. power supply and I is the r.m.s. current drawn.

$$V_C = I \times \frac{1}{\omega C}, \quad V_{c\max} = \sqrt{2}\, I \times \frac{1}{\omega C}.$$

The maximum energy stored in one cycle in the capacitor $= \frac{1}{2} C \left(\frac{\sqrt{2}\, I}{\omega C} \right)^2$

$$= \frac{1}{2} C V_C^2$$

$$= \frac{I^2}{\omega^2 C} \text{ joules.}$$

The energy dissipated during one cycle $= \frac{I^2 R}{f}$ joules.

Applying formula (1)

$$Q = 2\pi \cdot \frac{I^2}{\omega^2 C} \cdot \frac{f}{I^2 R_1} = \frac{1}{\omega C R_1}$$

$$\boxed{Q = \frac{1}{2\pi f C R_1}}$$

Quality factor of the capacitor $= \frac{\text{reactance}}{\text{resistance}}$ where reactance $= \frac{1}{2\pi f C} = X_C$

$$\boxed{Q = \frac{X_C}{R_1}}$$

Phasor of the C and R_1 in series.

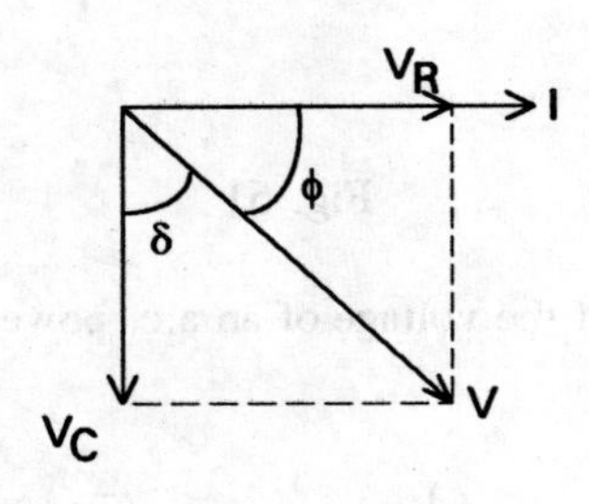

Fig. 52

The phase angle between I and V ideally is 90°, in practice this is slightly less than 90° by an angle δ, called the loss angle of the dielectric.

The impedance of the circuit is $\sqrt{R^2 + X_C^2}$

$$\cos \phi = \frac{V_R}{V} = \frac{IR_1}{I\sqrt{R_1^2 + X_C^2}} = \frac{R_1}{\sqrt{R_1^2 + X_C^2}}$$

R is small compared to X_C and $R_1^2 \ll \frac{1}{\omega^2 C^2}$

$$\text{p.f.} = \cos \phi \approx R_1 \omega C$$

$\cos \phi = \sin \delta \approx \delta$ when δ is expressed in radians

$$\boxed{\delta \approx R_1 \omega C}$$

The capacitor may be represented also by a pure capacitor and a parallel non-reactive resistor as shown in Fig. 53

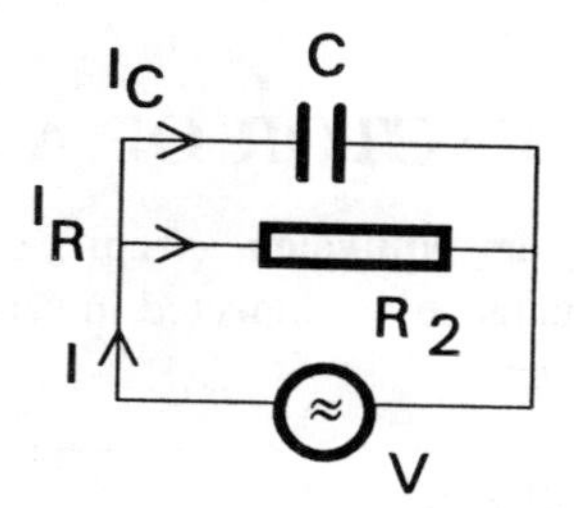

Fig. 53

The phasor diagram is shown in Fig. 54

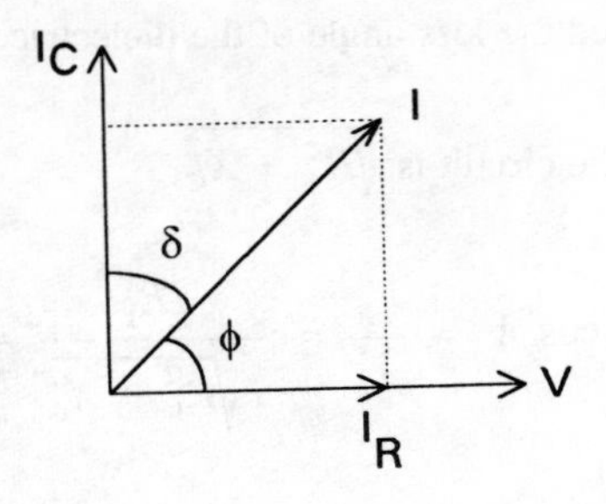

Fig. 54

$$\cos\phi = \frac{I_R}{I} = \frac{V/R_2}{V\sqrt{\frac{1}{R_2^2} + \omega^2 C^2}}$$

R_2 is very large and $\left(\frac{1}{R_2}\right)^2$ is very small $\sin\delta = \cos\phi \approx \frac{1}{\omega C R_2}$

$$\boxed{\delta \approx \frac{1}{\omega C R_2}}$$

The losses in dielectrics usually increase with increase in frequency then the loss angle also increases with increase in frequency and the loss angle does not approximately equal to the power factor.

THE *Q*-FACTOR OF A COIL

The coil can be replaced by an equivalent circuit, consisting of a non-reactive resistance R and an ideal inductance L connected in series. The values of R and L will generally vary with frequency.

The losses in an inductor (coil) are:-

(i) Copper losses (the resistance of the windings)
(ii) hysteresis losses (the hysteresis loop)
(iii) the eddy current losses (changing of magnetic flux of the coil).

All the above losses can be lumped together to an effective resistance of the coil, which is represented by a series resistor

Equivalent circuit of a coil.

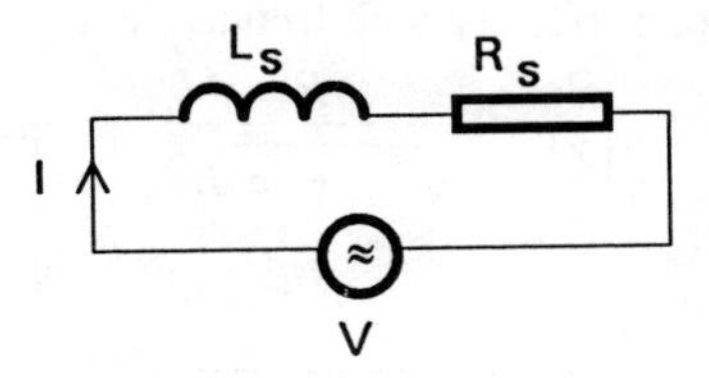

Fig. 55

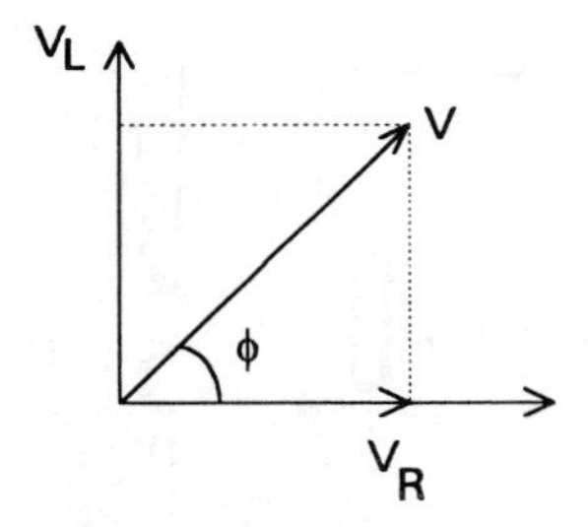

Fig. 56 Phasor diagram

The effective resistance of a coil will vary with frequency. Its value at higher frequencies will be much greater than the resistance measured with a direct current.

$$Q = 2\pi \frac{\text{Maximum energy stored in one cycle}}{\text{Energy dissipated during one cycle}}$$

$$= 2\pi \cdot \frac{\frac{1}{2} L I_m^2}{\frac{I^2 R}{f}} = \frac{2\pi \frac{1}{2} L \left(\sqrt{2}\, I\right)^2}{\frac{I^2 R}{f}} = \frac{2\pi f L}{R} = \frac{X_L}{R}$$

$$\boxed{Q = \frac{\text{reactance}}{\text{resistance}}}$$

The reactance and resistance of a coil vary with frequency, thus Q varies with frequency

$$\cos\phi = \frac{V_R}{V} = \frac{R}{|Z|} = \frac{R}{\sqrt{R^2 + \omega^2 L^2}} = \frac{1}{\sqrt{1 + \frac{\omega^2 L^2}{R^2}}}$$

$$\boxed{\cos\phi = \frac{1}{\sqrt{1 + Q^2}}}$$

If $Q \geq 10$, $\cos\phi \approx \frac{1}{Q}$ = p.f.

The equivalent circuit of a coil can be represented by a pure inductance in parallel by a non-reactive resistor as shown in Fig. 57.

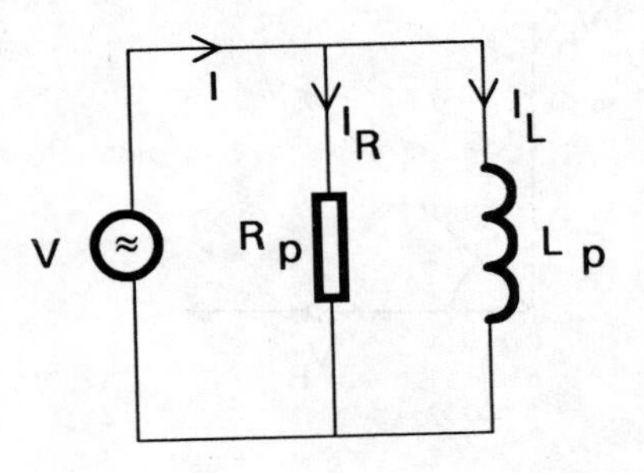

Fig. 57

The phasor is as shown

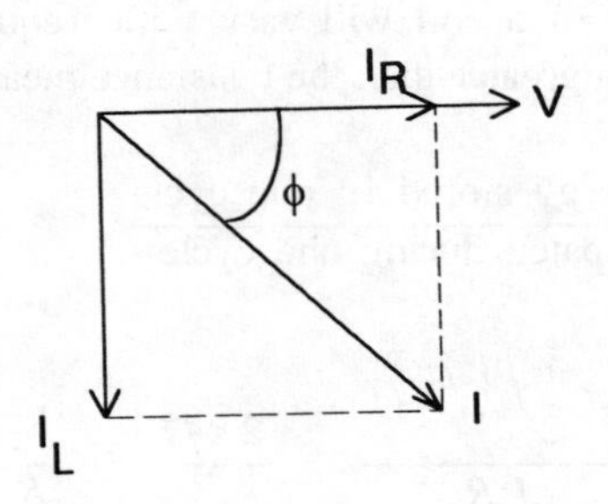

Fig. 58

$$I_R = \frac{V}{R_p}, \quad I_L = \frac{V}{X_L} = \frac{V}{\omega L},$$

$$I = \sqrt{I_R^2 + I_L^2} = \sqrt{\frac{V^2}{R_P^2} + \frac{V^2}{\omega^2 L^2}} = V\sqrt{\frac{1}{R_p^2} + \frac{1}{\omega^2 L^2}} = \frac{V}{Z},$$

$$\tan\phi = \frac{I_L}{I_R} = \frac{V/\omega L}{V/R} = \frac{R}{\omega L}$$

$$\boxed{\tan\phi = \frac{R}{\omega L}}$$

$$P = VI_R = \frac{V^2}{R_p}, \ \cos\phi = \frac{VI_R}{VI} = \frac{VI\cos\phi}{VI} = \text{power factor.}$$

EQUIVALENT CIRCUITS OF AN INDUCTOR

The inductor can either be represented by a pure inductance L_S and a series resistor R_S or a pure inductance L_p and a parallel resistor R_p.

The circuits are also equivalent.

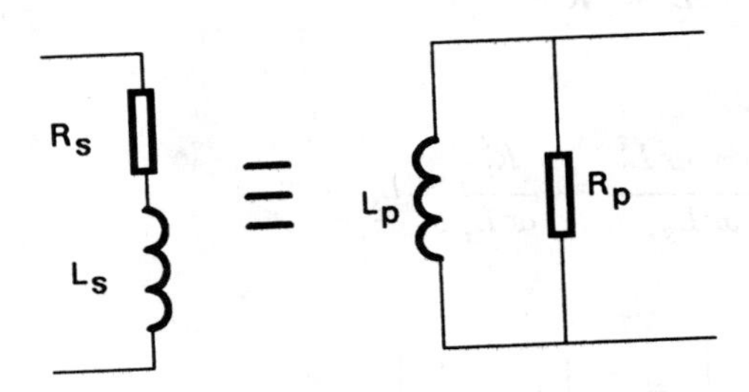

Fig. 59

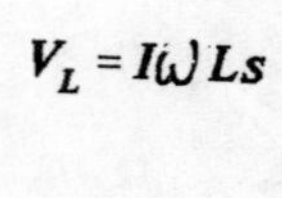

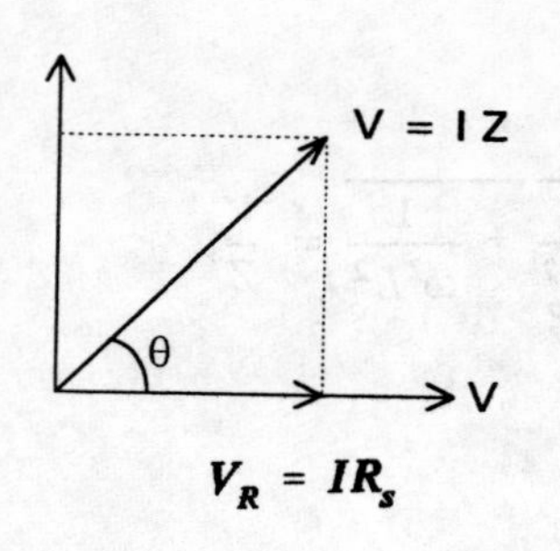

Fig. 60

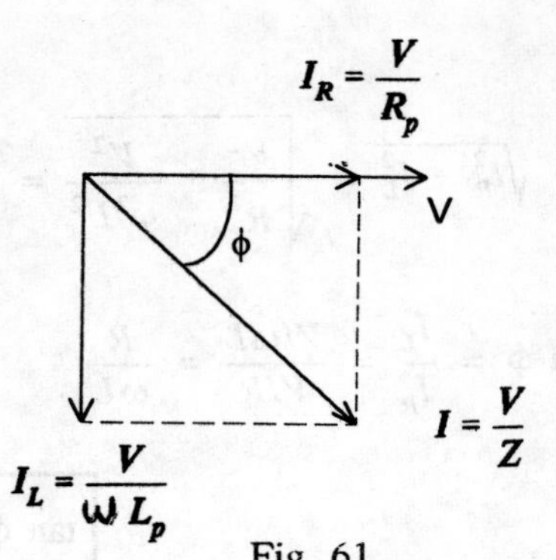

Fig. 61

$$\sin\theta = \frac{I\omega L_s}{IZ} = \frac{\omega L_s}{Z} \quad \ldots (1) \qquad \sin\theta = \frac{V/\omega L_p}{V/Z} = \frac{Z}{\omega L_p} \quad \ldots (2)$$

$$\cos\theta = \frac{IR_s}{IZ} = \frac{R_s}{Z} \quad \ldots (3) \qquad \cos\theta = \frac{V/R_p}{V/Z} = \frac{Z}{R_p} \quad \ldots (4)$$

Equating (1) and (2), (3) and (4)

$$\frac{\omega L_s}{Z} = \frac{Z}{\omega L_p} \qquad Z^2 = \omega^2 L_s L_p$$

$$\frac{R_s}{Z} = \frac{Z}{R_p} \qquad Z^2 = R_p R_s$$

$$L_p = \frac{Z^2}{\omega^2 LS} = \frac{R_s^2 + \omega^2 L_S^2}{\omega^2 L_S} = \frac{R_S^2}{\omega^2 L_s} + L_s$$

$$L_p = L_s\left(1 + \frac{R_s^2}{\omega^2 L_s^2}\right) = L_s\left(1 + \frac{1}{Q^2}\right)$$

$$\boxed{L_p = L_s\left(1 + \frac{1}{Q^2}\right)} \quad \ldots (1)$$

$$R_p = \frac{Z^2}{R_s} = \frac{R_s^2 + \omega^2 L_S^2}{R_s^2} = R_s + \frac{\omega^2 L_S^2}{R_S^2} = R_S\left(1 + \frac{\omega^2 L_S^2}{R_s^2}\right)$$

$$\boxed{R_p = R_s\left(1 + Q^2\right)} \quad \ldots (2)$$

In practice, a good inductor has a quality factor of the order of 10

$$\boxed{Q \sim 10}$$

From (1) $L_p \approx L_s$ since $\frac{1}{Q^2} \ll 1$

From (2) $R_p \approx R_s$ since $1 \ll Q^2$.

EXERCISES 1

1. For the circuit shown

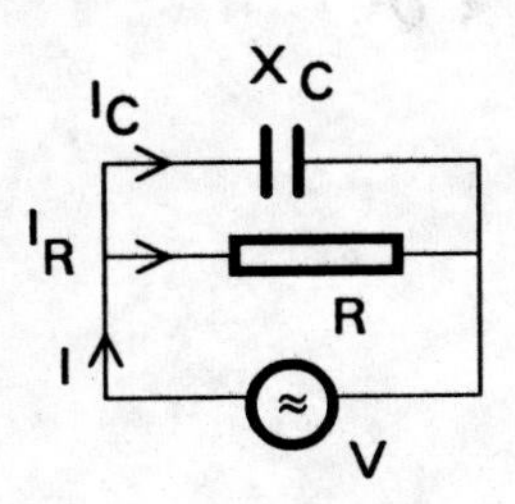

Fig. 62

Draw the phasor diagram and show that the magnitude of the impedance of the circuit is given by the formula

$$Z = \frac{RX_C}{\sqrt{R^2 + X_C^2}}$$

2. For the circuit shown

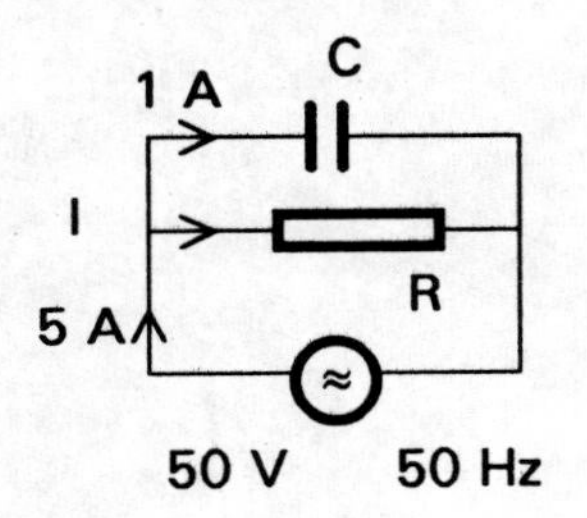

Fig. 63

Calculate:- (a) the value of the capacitance
(b) the current through the resistance
(c) the value of the resistance
(d) the active power and hence the power factor.

3. A pure capacitor is connected in parallel with a pure inductor. It is required to obtain a resonant frequency of 1 MHz, determine the values for the capacitance and inductance.

4. For the circuit in Fig. 64 give the phasor diagram and from it derive the condition for parallel resonance.

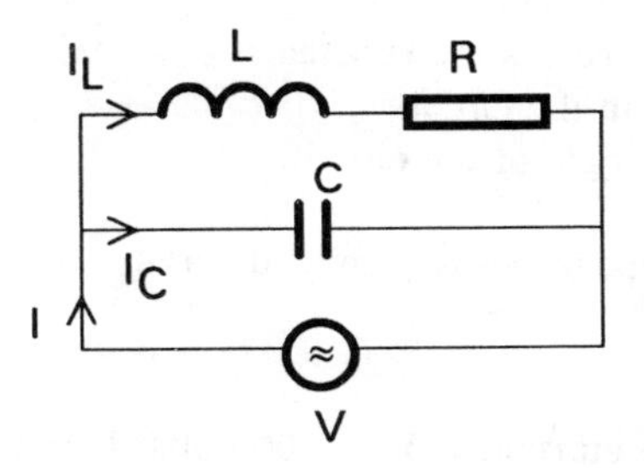

Fig. 64

Also give the formulae for calculating
(i) the frequency at resonance
(ii) the impedance at resonance.

Draw the impedance-frequency curve and show how you can obtain Q from that graph.

5. For the figures shown, calculate the currents and phase angles of the circuits and draw the relevant phasor diagram.

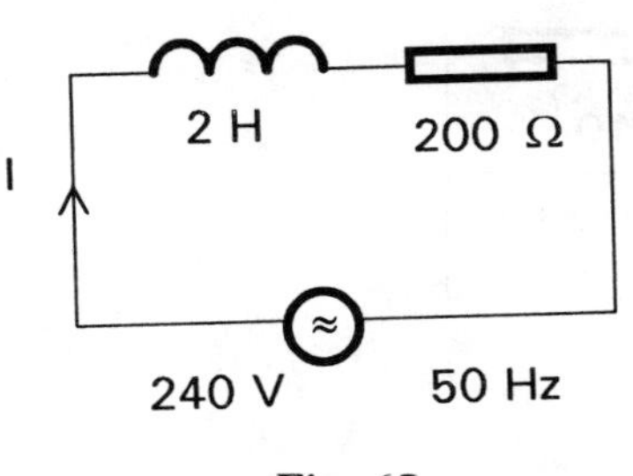

Fig. 65

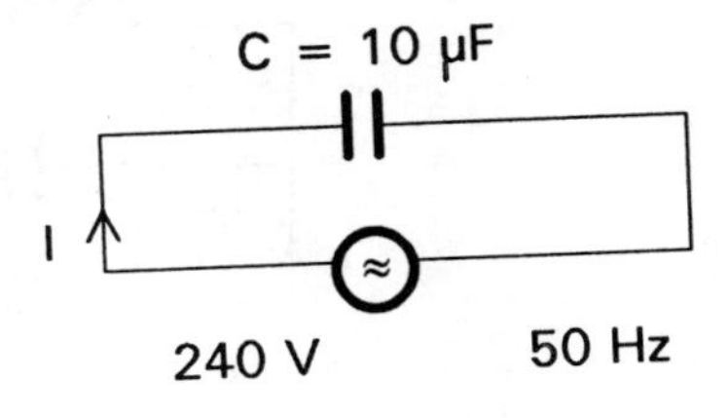

Fig. 66

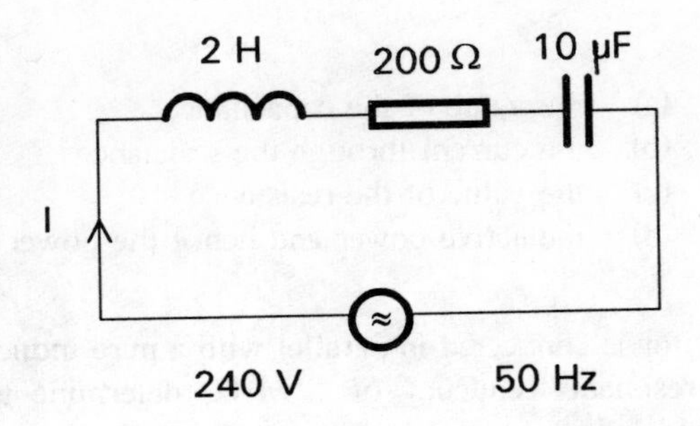

Fig. 67

For the *LCR* circuit the frequency is varied until $X_L = X_C$. Calculate the frequency at which this occurs. Also, at this frequency determine.

(i) the supply current
(ii) the voltage across the resistance
(iii) the power in the circuit
(iv) the phase angle of the circuit.

Also draw the impedance-frequency diagram and the phasor diagram (at resonance).

6. Find the maximum current, if $V = 100$ volts, $L = 0.2$ H, $C = 10\ \mu$F, $R = 10\ \Omega$. Find the p.d. across C if $\tan\theta = 5$.

7. For the circuit shown, I and V are in phase. Draw the phasor diagram and

show that $I_L = \sqrt{I_C^2 + I^2}$, $Q = \tan\theta = \dfrac{I_C}{I}$ and $f_o = \dfrac{1}{2\pi}\sqrt{\dfrac{1}{LC} - \dfrac{R^2}{L^2}}$.

Calculate f_o and the current amplification factor.

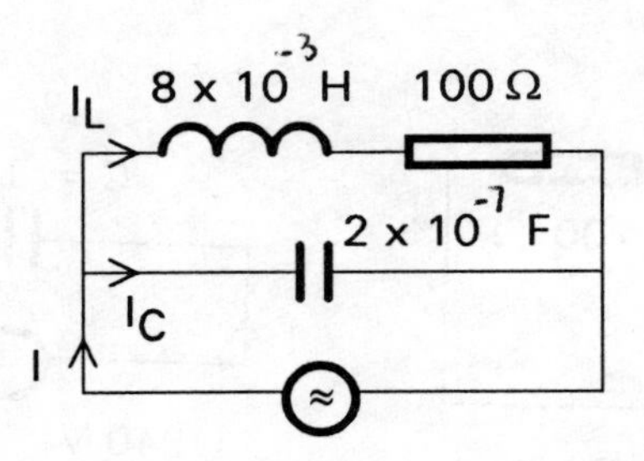

Fig. 68

8. Draw phasor diagrams to indicate the current and voltage when a sinusoidal voltage 10 V r.m.s. and frequency f hertz is applied across

   (a) a non reactive resistance, *R*, of 100 Ω
   (b) a pure inductance, *L*, of 0.5 henry
   (c) a pure capacitance, *C*, of 1 microfarad.

   Hence or otherwise construct the phasor diagram for a circuit in which the three components are connected in parallel.

   What is the frequency at resonance for the circuit?

   Draw to scale a phasor diagram of the currents and voltage at this frequency.

   Determine the impedance and the *Q*-factor of the circuit at resonance.

9. A 10 μF capacitor is connected in parallel with a 0.05 H inductor. The *Q*-factors of the two components are 500 and 10 at the resonant frequency. You may represent the losses of these components by a shunt resistance.

   Determine:- (a) the resonant frequency
   (b) the impedance of the circuit at resonance
   (c) the *Q*-factor of the total circuit.

10. An inductor of 1 mH is connected in parallel with a 1000 pF loss free capacitor. The dynamic resistance of the circuit is 200 KΩ.

    Calculate:- (a) the frequency
    (b) the current through the capacitor
    (c) the *Q*-factor of the coil.

    The voltage applied across the circuit is 20 V r.m.s.

11. A coil of inductance 100 mH and resistance of 25 Ω is connected in series with a capacitor of 1 μF of infinite *Q*-factor.

    Calculate:- (i) the resonant frequency
    (ii) the impedance at resonance
    (iii) the *Q*-factor of the circuit at this frequency.

12. **Determine the total reactance of the circuit and the total current.**

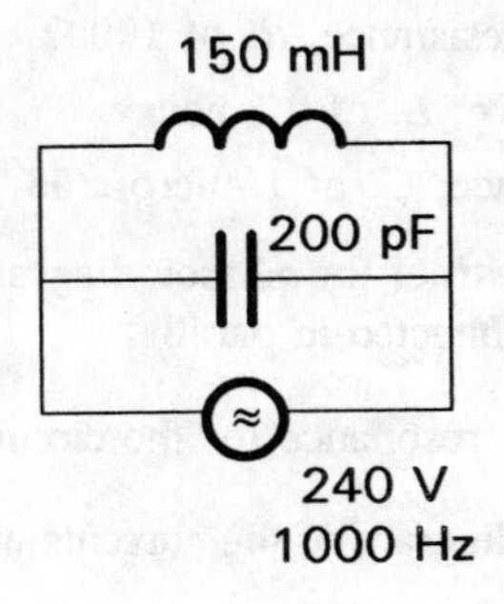

Fig. 69

13. **Determine the impedance of the circuit.**

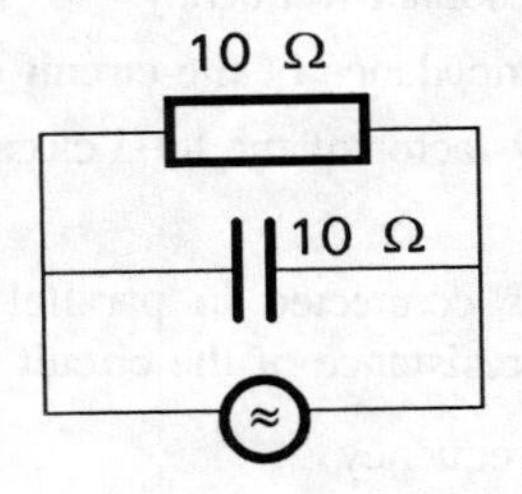

Fig. 70

14. **Determine the reactance of the circuit.**

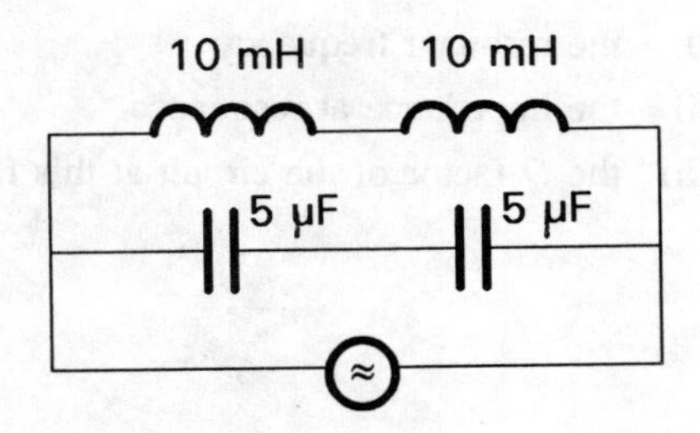

Fig. 71

15. Determine the dynamic resistance of the circuit.

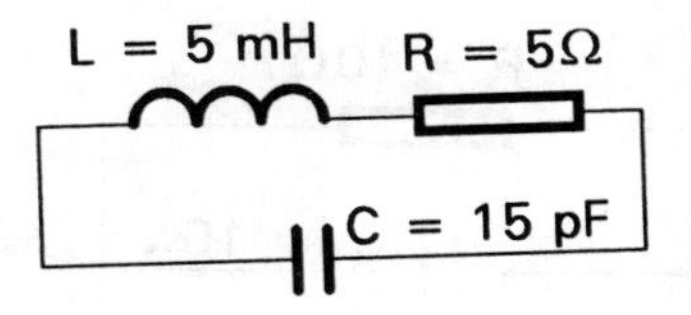

Fig. 72

16. Determine the current drawn by the a.c. supply when the circuit is at resonance.

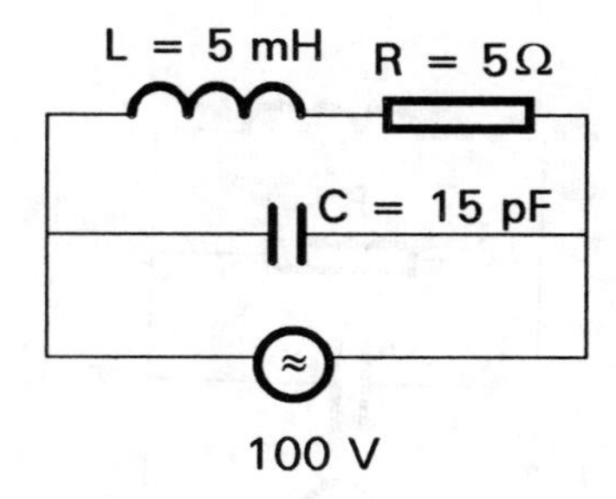

Fig. 73

17. Determine the magnitude of the supply current in the circuit.

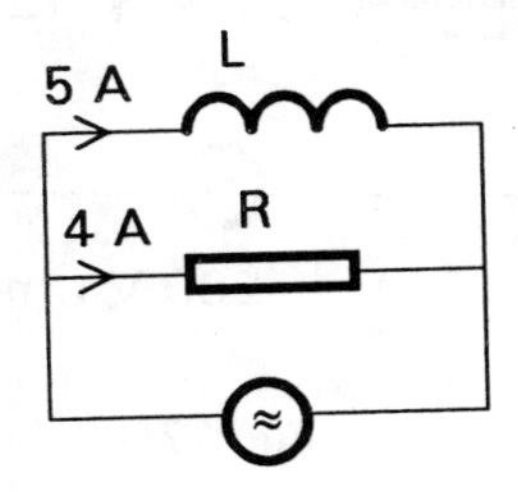

Fig. 74

18. Determine the magnitude of the impedance of the circuit.

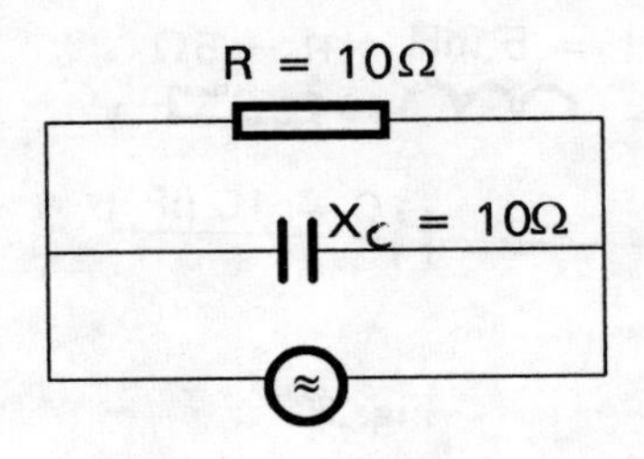

Fig. 75

19. Determine the magnitude of the supply current.

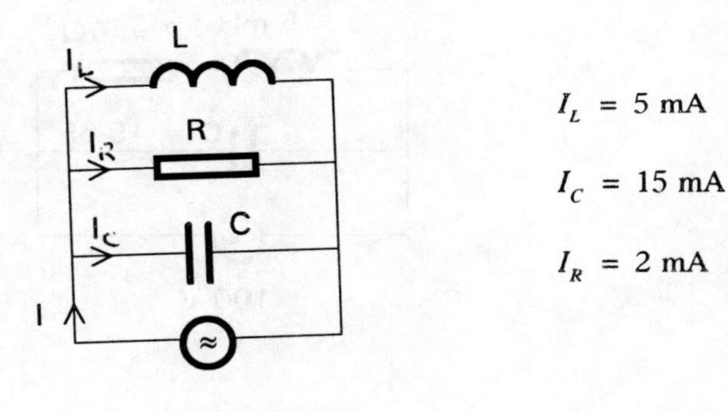

I_L = 5 mA

I_C = 15 mA

I_R = 2 mA

Fig. 76

20. Calculate the total current when the frequency is 1000 Hz.

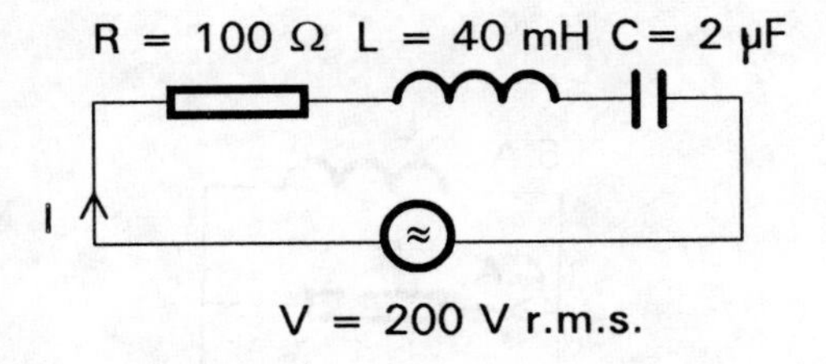

V = 200 V r.m.s.

Fig. 77

21. (i) What is the phase angle between the current I_o and the voltage V (taking current as the reference)

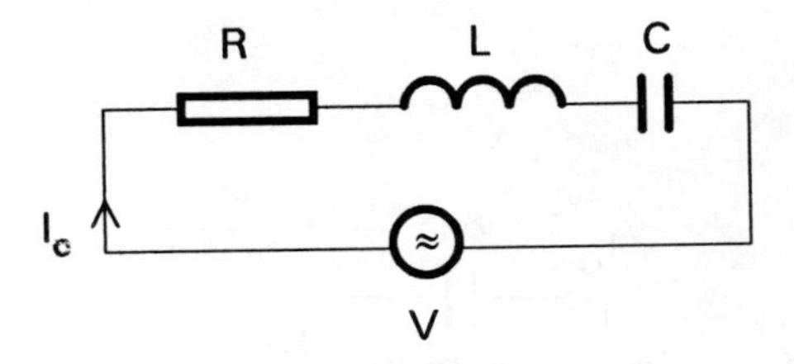

Fig. 78

(ii) If V = 40 mV and the Q-factor is 120, find the voltage across the capacitor.

22.

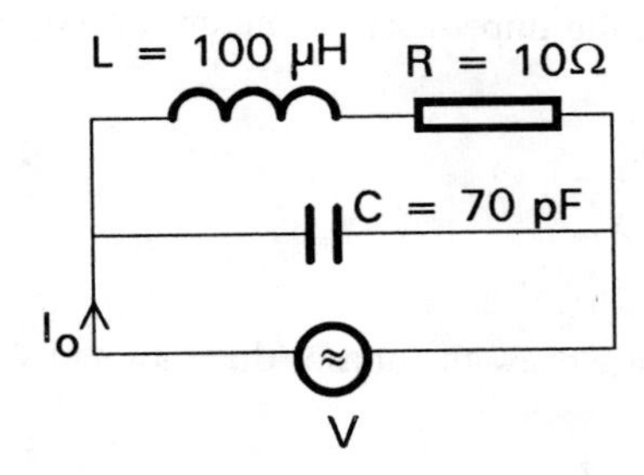

R = 10 Ω

L = 100 μH. At resonant frequency

C = 70 pF

Fig. 79

(i) Determine the resonant frequency of the circuit

$$\left(f_o = \frac{1}{2\pi} \sqrt{\frac{1}{LC} - \frac{R^2}{L^2}} \right).$$

(ii) What is $\dfrac{L}{RC}$?

(iii) If Q = 180, I = 25 μA at resonance, calculate the value of I_C.

(iv) If V = 100 volts, then find the supply current of the circuit.

23. For the circuit shown

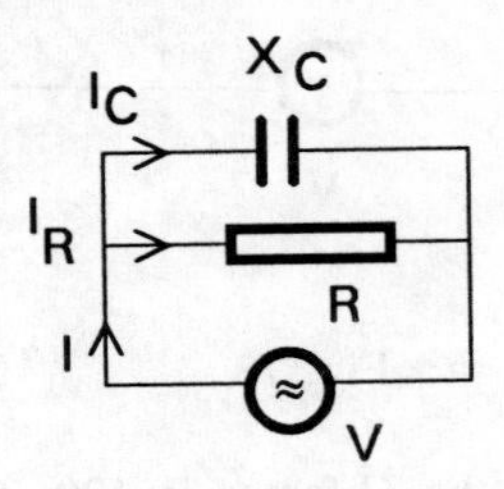

Fig. 80

Draw the phasor diagram and show that the impedance of the circuit is given by the formula

$$Z = \frac{RX_c}{\sqrt{R^2 + X_C^2}}$$

24. For the circuit shown, I and V are in phase draw the phasor diagram and show that $I_L = \sqrt{I_C^2 + I^2}$ and $Q = \tan\theta = \frac{I_C}{I}$.

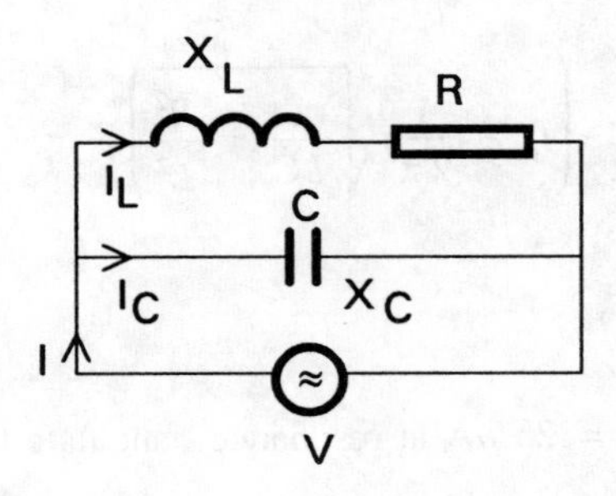

Fig. 81

$$\text{Find } f_o = \frac{1}{2\pi}\sqrt{\frac{1}{LC} - \frac{R^2}{L^2}}.$$

If $R = 100\ \Omega$, $C = 2 \times 10^{-7}$ F, $L = 8 \times 10^{-3}$ H, what is the current amplification factor?

25. Find the maximum current, if $V = 100$ volts, $L = 0.2$ H, $C = 10\ \mu$F, $R = 10\ \Omega$ and V_C if $\tan\theta = 5$.

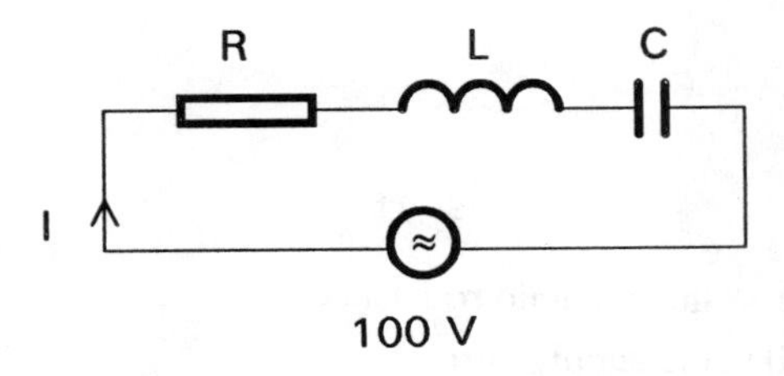

Fig. 82

26.

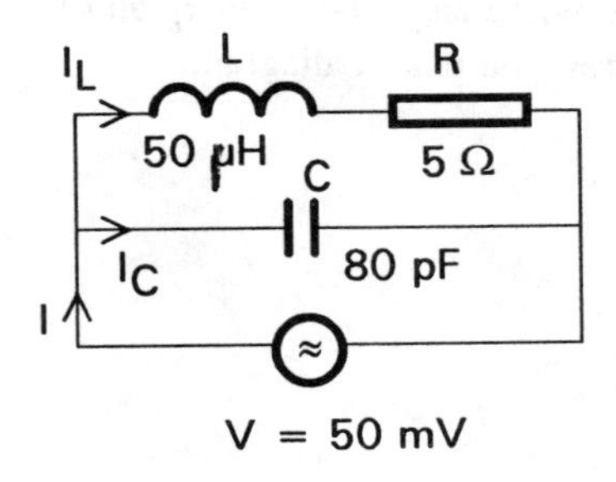

Fig. 83

The circuit is at resonance.

Determine:-
(a) the resonant frequency
(b) the dynamic resistance
(c) the supply current
(d) the current amplification factor
(e) the power absorbed.

27.

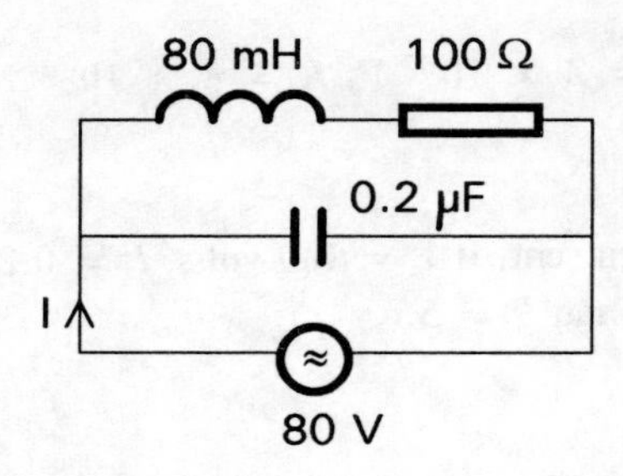

Fig. 84

Determine:-
(i) the dynamic resistance
(ii) the supply current
(iii) the resonant frequency
(iv) the capacitive reactance
(v) the current through the capacitor
(vi) the current through the inductive circuit
(vii) the *Q*-factor of the circuit
(viii) the phase angle between I_L and I (draw the phasor diagram).

28.

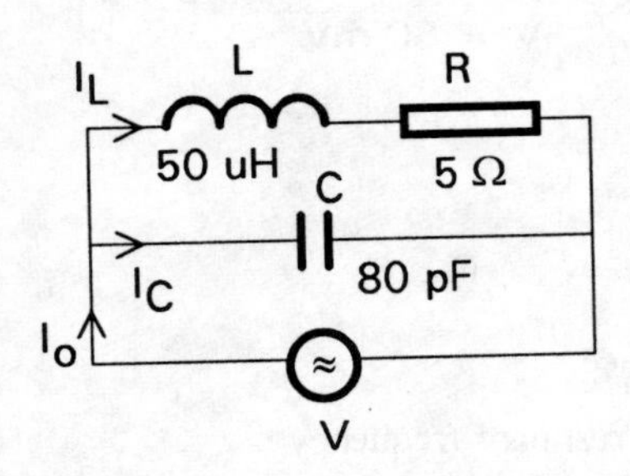

Fig. 85

(i) The frequency for parallel resonance is given by

$$f_o = \frac{1}{2\pi}\sqrt{\frac{1}{LC} - \left(\frac{R}{L}\right)^2}.$$

Which of the resonant frequency of the above circuit will be:-

19.6 Hz 1.95 MHz 2.52 MHz 15.8 MHz

(ii) Which symbol represents the ratio I_C / I_o?

V X_C Z Q

(iii) The dynamic impedance of the series-parallel combination is given by

$$Z_o = \frac{L}{CR}$$

If $V = 50$ mV the current will be:-

0.4 μA 4.0 μA 10 μA 0.4 A

(iv) If $V = 50$ mV the power absorbed by the combination will be:-

0.8 pW 20 pW 20 nW 500 μW.

29.

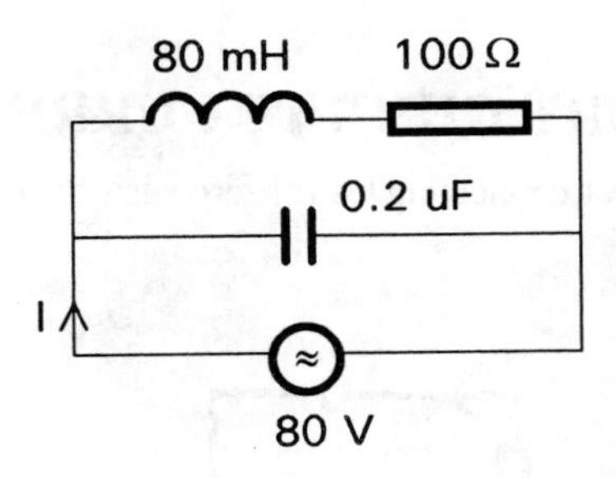

Fig. 86

Determine:- (i) the dynamic resistance
(ii) the resonant frequency
(iii) the supply current
(iv) the capacitor current
(v) the inductor current.

2. NETWORK THEOREMS

Thévenin-Helmholtz's and Norton's Theorems

OPEN CIRCUIT VOLTAGE

A cell is represented by an e.m.f., E, and an internal resistance, r, as shown in Fig. 87.

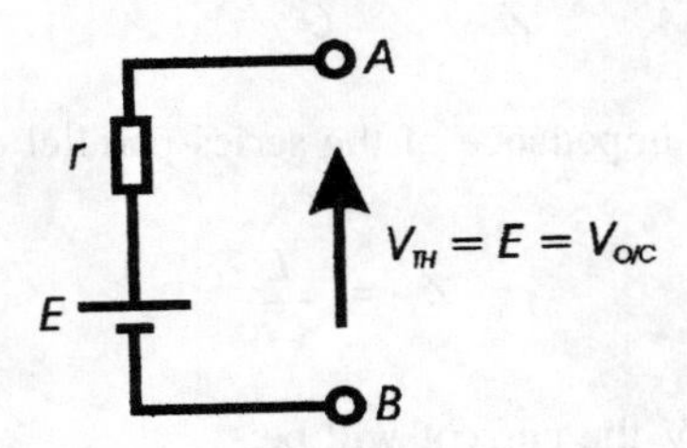

Fig. 87 Open circuit voltage.

The voltage across the terminals A and B is equal to E since the current is zero and there is no p.d. across r. Therefore the e.m.f., E, is called the open circuit voltage or the Thévenin's voltage, V_{TH}

$$V_{O/C} = V_{TH} = E \quad ... (1)$$

SHORT CIRCUIT CURRENT

If A and B are connected by a connecting lead of zero resistance as shown in Fig. 88

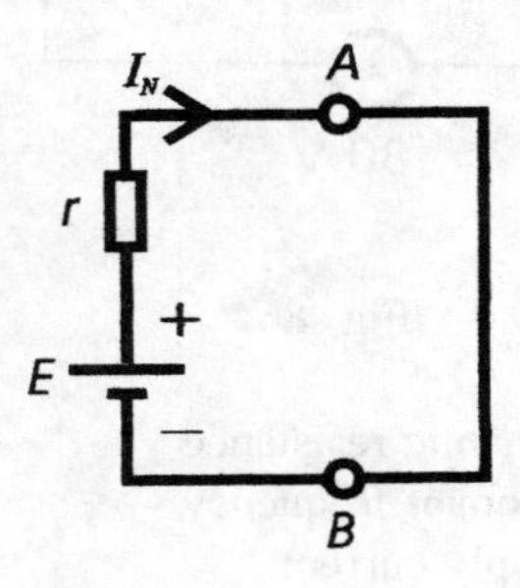

Fig. 88 Short Circuit Current.

Applying Ohm's law, $I_N = \frac{E}{r} = I_{S/C}$ ... (2)

where I_N denotes the Norton's current, or the short circuit current. From equations (1) and (2) we have ohm's law again in terms of open circuit voltage and short circuit current.

$$r = \frac{V_{O/C}}{I_{S/C}} = \frac{V_{TH}}{I_N}$$

$$= \text{internal resistance}$$

$$= R_{TH} = R_N$$

$$\boxed{R_{TH} = R_N = \frac{V_{TH}}{I_N}}$$

where R_{TH} denotes the Thévenin's equivalent resistance of the network or R_N denotes the Norton's equivalent resistance of the network.

Therefore, $R_N = R_{TH}$ is equal to the internal resistance of the network or the output resistance of the network.

$$\text{Output resistance} = \frac{\text{Open Circuit Voltage}}{\text{Short Circuit Current}}.$$

If a resistive load, R, is now connected between the terminals A and B; we have a network on load as shown in Fig. 89

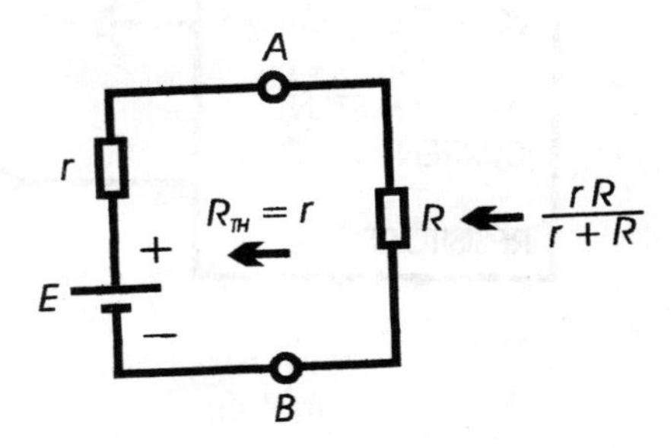

Fig. 89

The internal resistance or output resistance looking into the terminal A and B to the left is R_{TH} and the output resistance looking into the resistive load R to the left is $R + r$.

These are simple concepts which will be extremely useful in more advanced treatment of the subject.

Thévenin-Helmholtz's Theorem

This theorem is applicable to a.c. and d.c. supplies.

In this book the treatment will be restricted to d.c. supplies only.

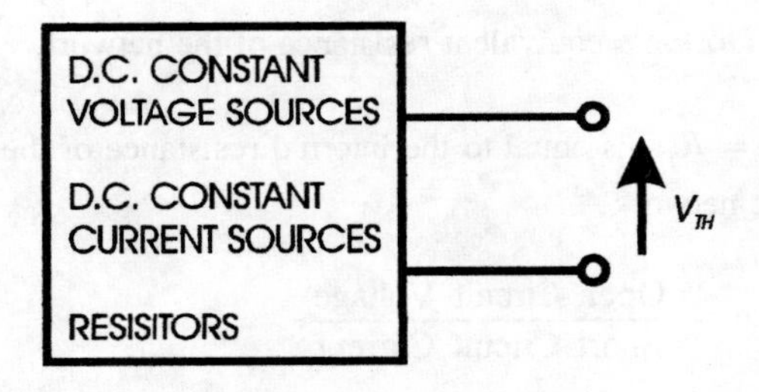

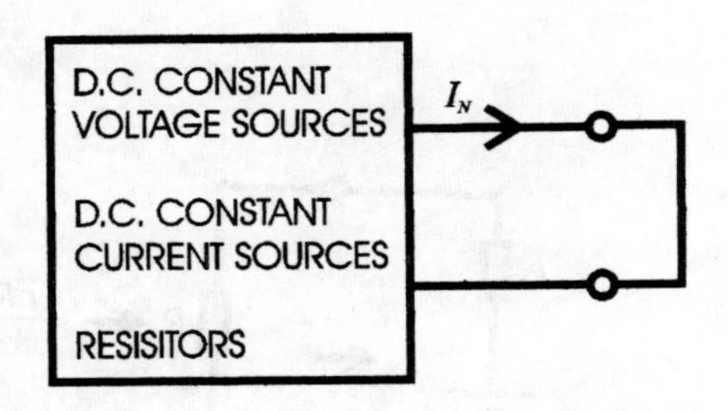

Fig. 90

$$R_{TH} = \frac{V_{TH}}{I_N}$$

Let us now illustrate the theorem with a few practical examples.

WORKED EXAMPLE 14

Determine the open circuit voltage, V_{TH}, across AB and the equivalent resistance viewed at AB terminals when the e.m.f.s are replaced by their internal resistances.

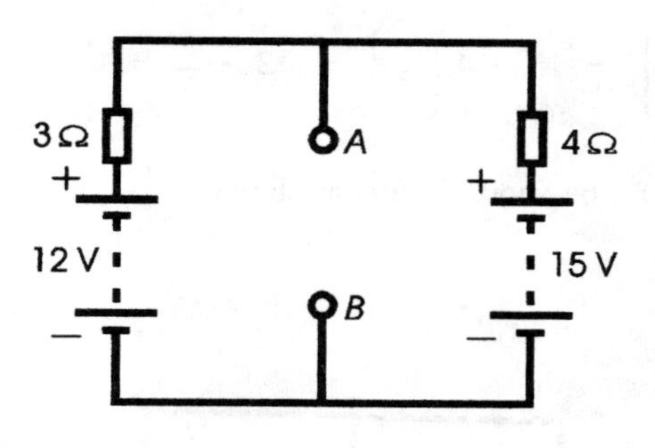

Fig. 91

If a load resistance, $\frac{81}{7}$, is connected between A and B, determine the load current.

SOLUTION 14

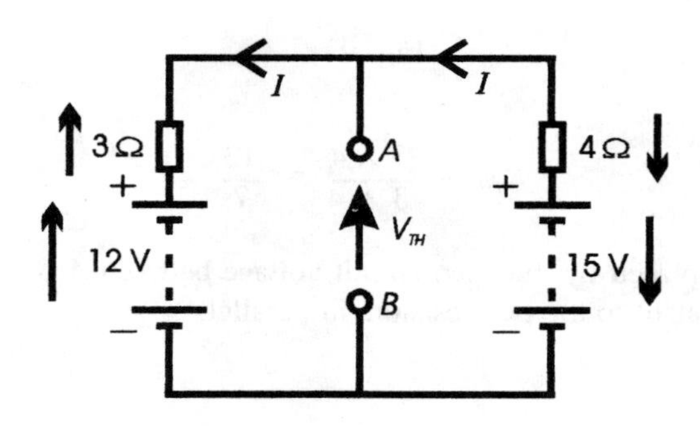

Fig. 92

Using Kirchhoff's, voltage law for one loop, we have

$$15 - 4I - 3I - 12 = 0$$

$$3 = 7I$$

$$I = \frac{3}{7}\text{ A}$$

$$V_{TH} = 12 + 3I = 15 - 4I$$

$$V_{TH} = 12 + 3\left(\frac{3}{7}\right) = 15 - 4\left(\frac{3}{7}\right) = 12 + \frac{9}{7} = \frac{93}{7}.$$

We replace the e.m.f.s by short circuits as shown.

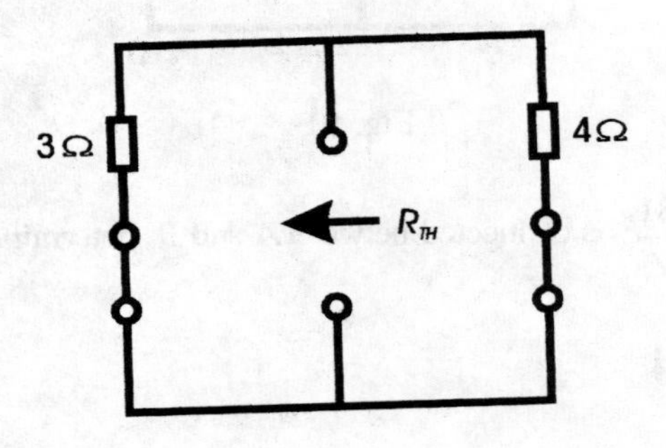

Fig. 93

$$R_{TH} = \frac{3 \times 4}{3 + 4} = \frac{12}{7}$$

The network is replaced by the open circuit voltage between *A* & *B* in series with a resistance equivalent to the two resistors in parallel.

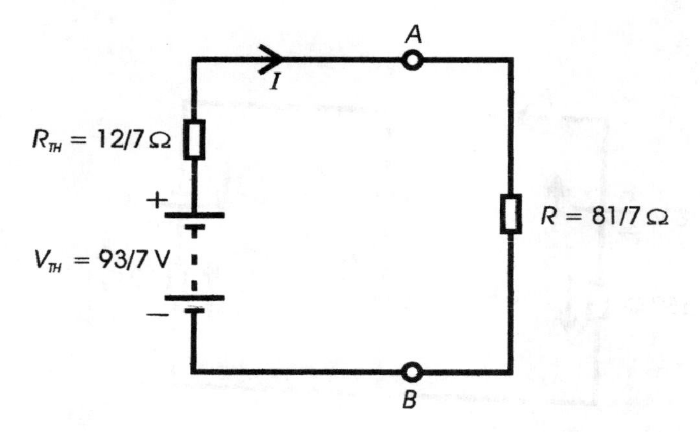

Fig. 94

$$I = \frac{V_{TH}}{R_{TH} + R} = \frac{93/7}{12/7 + 81/7} = 1 \text{ A}$$

WORKED EXAMPLE 15

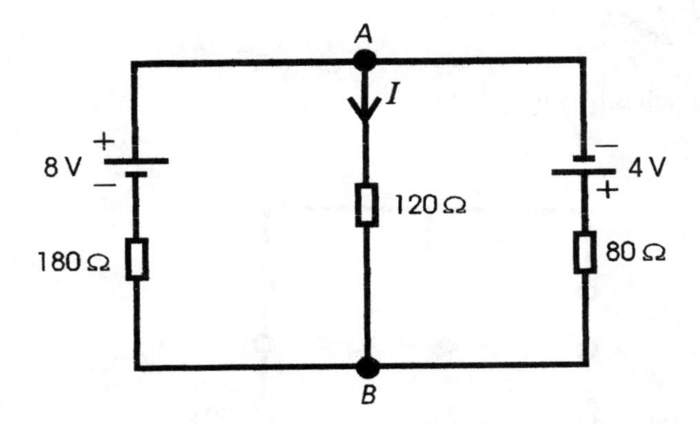

Fig. 95

Use Thévenin's theorem to determine the load current I.

SOLUTION 15

Remove the load 120 Ω. Determine the V_{TH}.

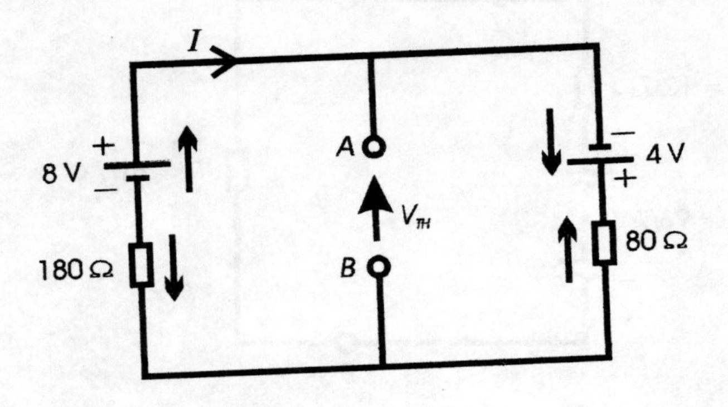

Fig. 96

$V_{TH} = 8 - 180I = -4 + 80I$

$I = \dfrac{12}{260} = 46 \text{ mA}$

$V_{TH} = 8 - 180 \times \dfrac{12}{260} = -0.308$ volts.

Replace the d.c. supplies by short circuits.

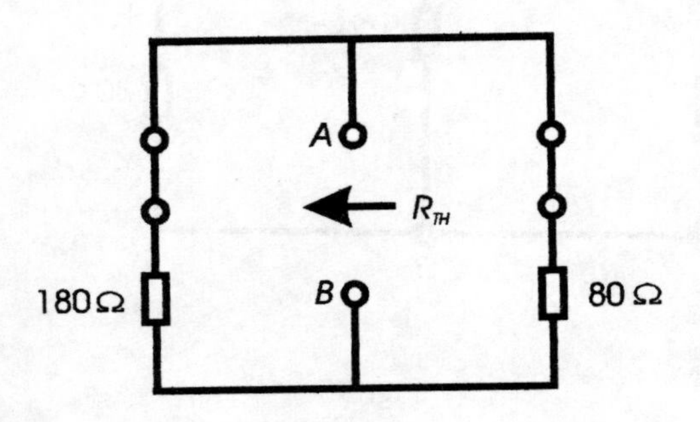

Fig. 97

$R_{TH} = \dfrac{180 \times 80}{180 + 80} = 55.4\ \Omega$

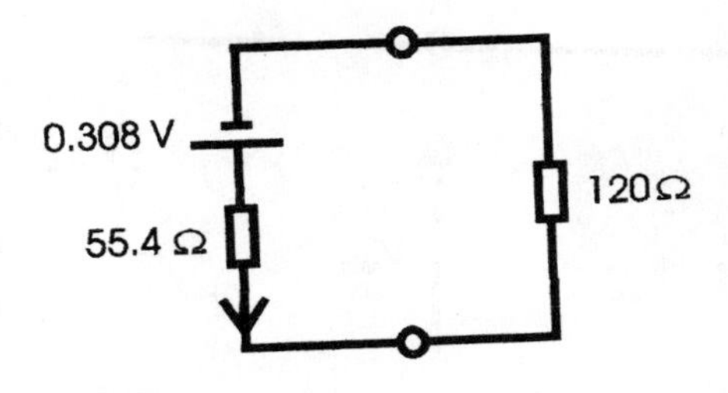

Fig. 98

$$I = \frac{0.308}{120 + 55.4} = 1.76 \text{ mA}$$

WORKED EXAMPLE 16

Use the Thévenin's theorem to determine the current through the 3 Ω resistor.

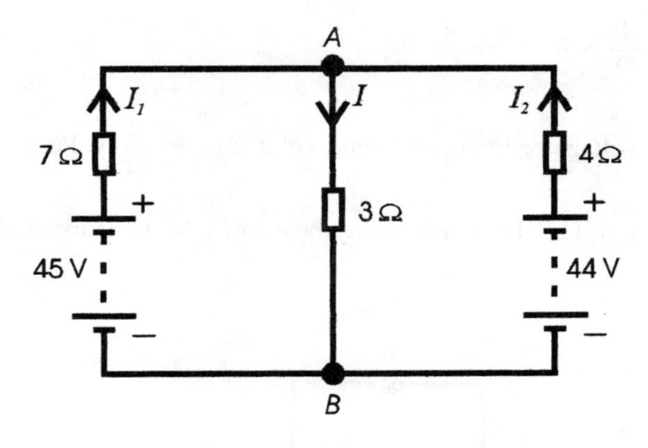

Fig. 99

SOLUTION 16

Remove the 3 Ω resistor and determine the open circuit voltage between A and B.

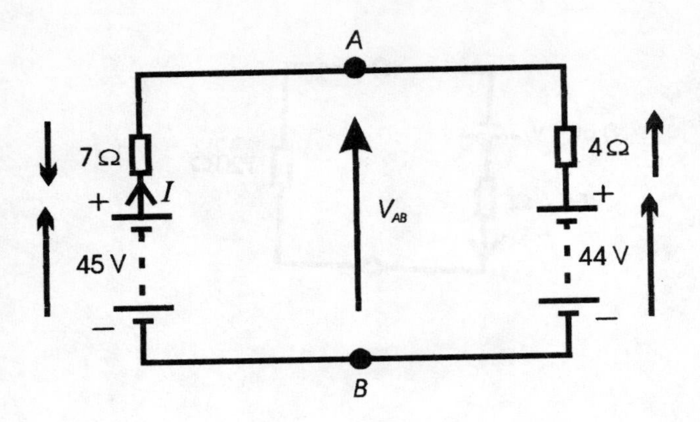

Fig. 100

$$I = \frac{45 - 44}{11} = \frac{1}{11} \text{ A}$$

$$V_{AB} = 45 - \frac{1}{11} \times 7 = 44\frac{4}{11} \text{ volts}$$

$$= 44 + \frac{1}{11} \times 4 = 44\frac{4}{11} \text{ volts.}$$

The Thévenin's voltage or the open circuit voltage is $44\frac{4}{11}$ volts.

Suppressing the e.m.f.s. that is replacing the e.m.f.s by their internal resistances, we have

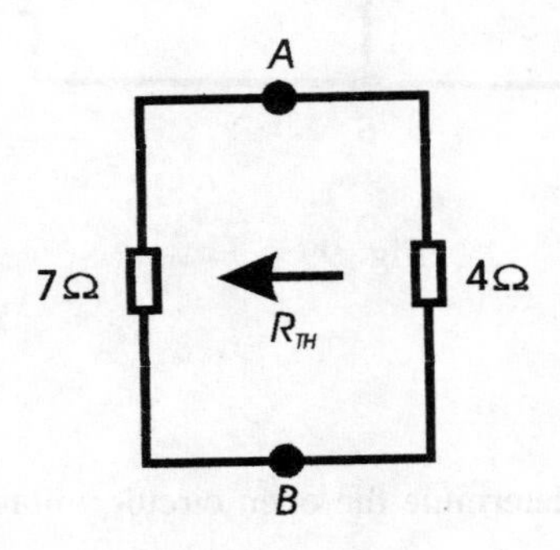

Fig. 101

$$R_{TH} = \frac{7 \times 4}{7 + 4} = \frac{28}{11} = 2\frac{6}{11}\ \Omega.$$

The circuit, now is replaced as follows:-

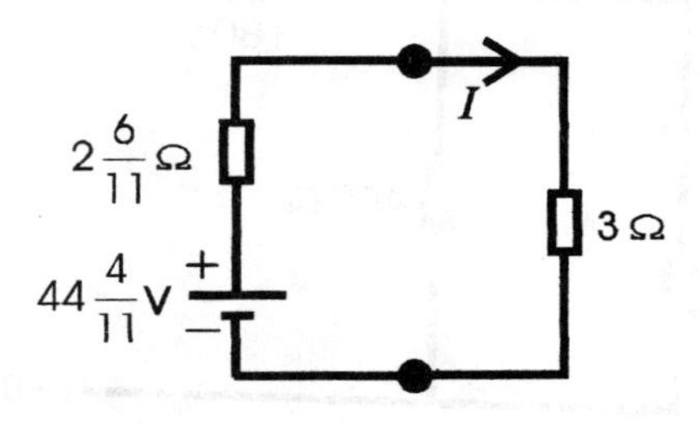

Fig. 102

$$I = \frac{44\frac{4}{11}}{2\frac{6}{11} + 3} = \frac{\frac{488}{11}}{\frac{61}{11}} = 8\ \text{A}.$$

The current through the 3 Ω resistor is 8 A.

WORKED EXAMPLE 17

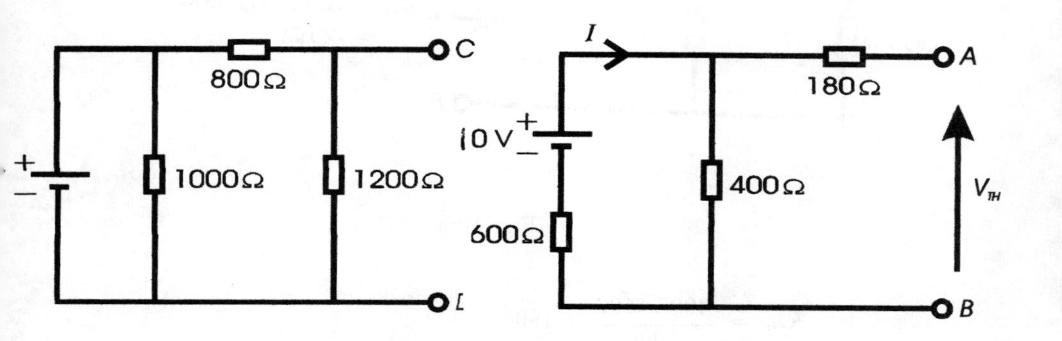

Fig. 103

Determine the Thévenin's equivalent circuits for terminals AB and CD of the two circuits shown in Fig. 103.

SOLUTION 17

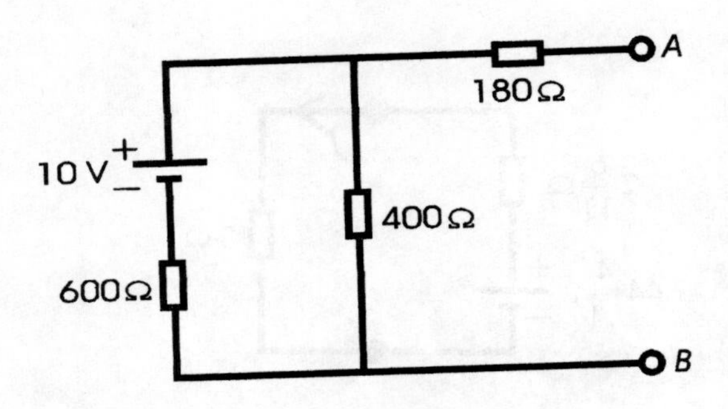

Fig. 104

$$I = \frac{10}{600 + 400} = \frac{10}{1000} = 0.01 \text{ A}$$

$$V_{TH} = 0.01 \times 400 = 4 \text{ V}$$

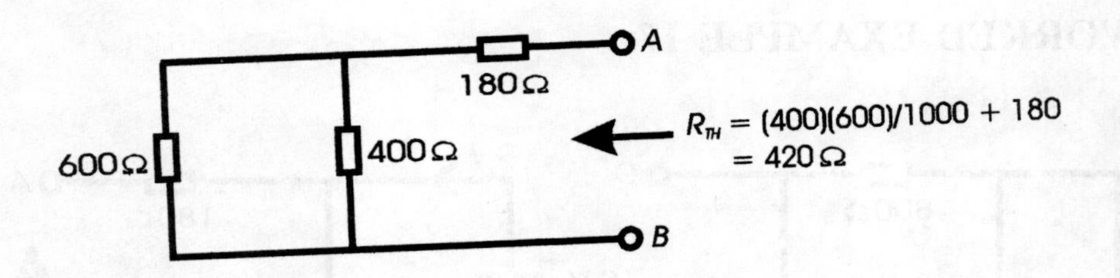

Fig. 105

$$R_{TH} = \frac{400 \times 600}{1000} + 180$$

$$= 420 \ \Omega.$$

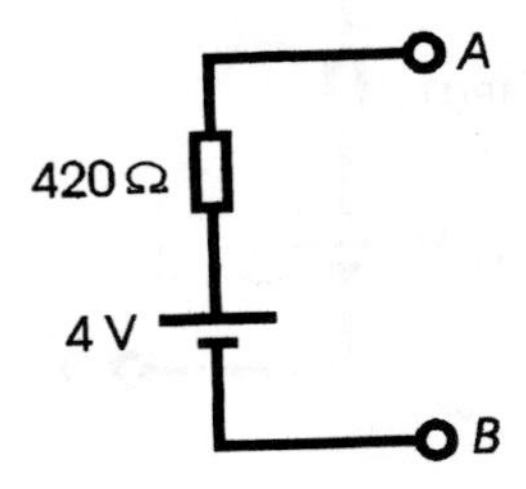

Fig. 106

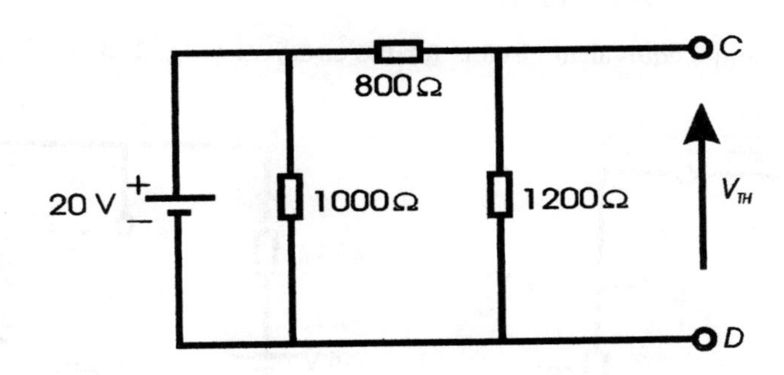

Fig. 107

$$V_{TH} = \frac{1200}{1200 + 800} \times 20 = 12 \text{ V}$$

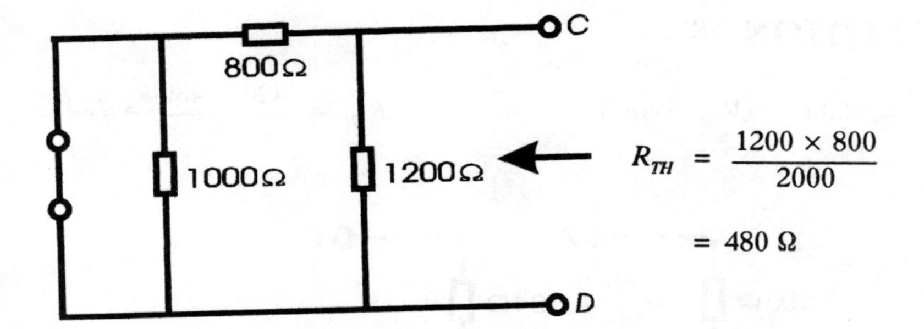

$$R_{TH} = \frac{1200 \times 800}{2000}$$

$$= 480 \ \Omega$$

Fig. 108

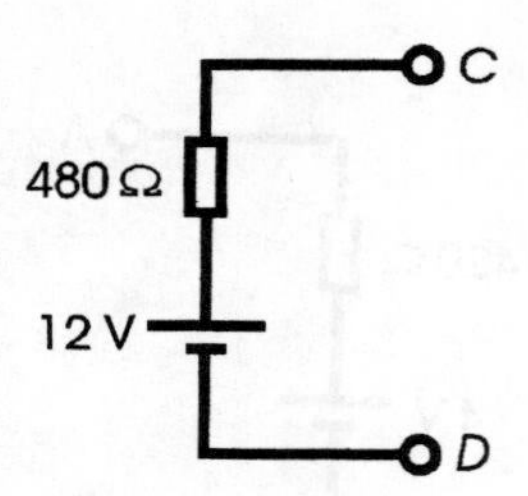

Fig. 109

WORKED EXAMPLE 18

Obtain the Thévenin's equivalent circuits for the circuits in Fig. 110.

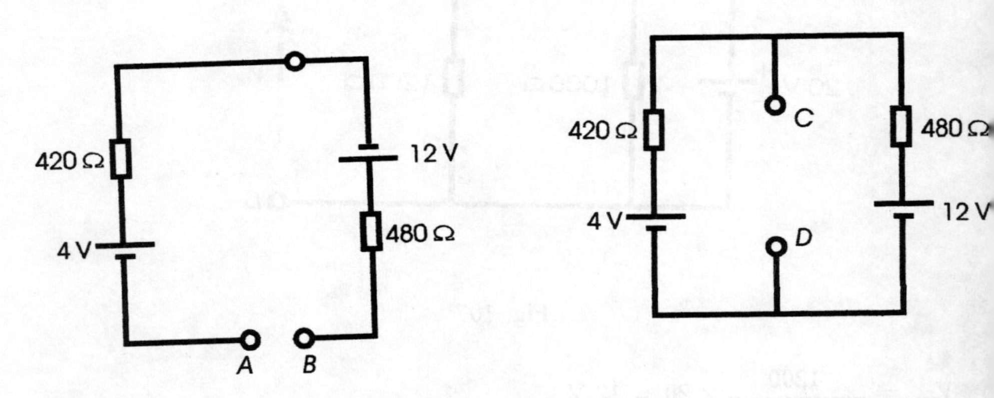

Fig. 110

SOLUTION 18

$R_{TH} = 900\ \Omega \qquad V_{TH} = 16\ \text{V}$

$$R_{TH} = \frac{420 \times 480}{900} = 224\ \Omega$$

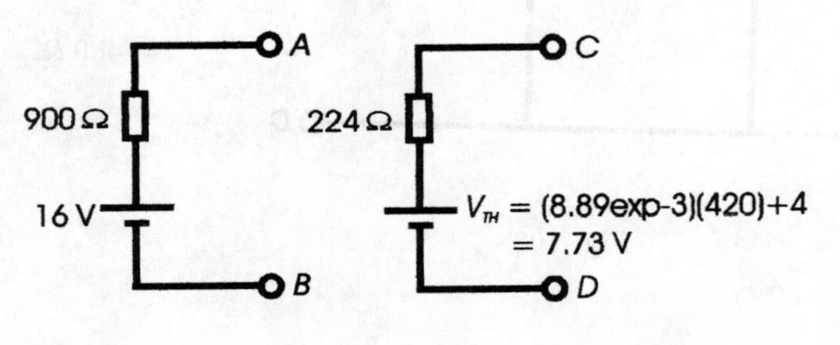

WORKED EXAMPLE 19

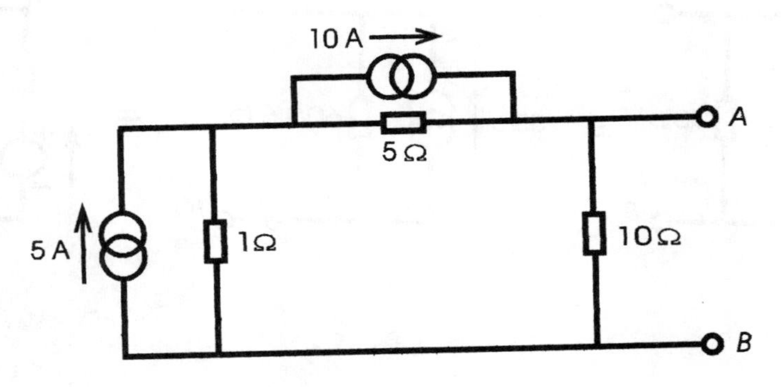

Fig. 111

Determine the Thévenin's equivalent circuit.

SOLUTION 19

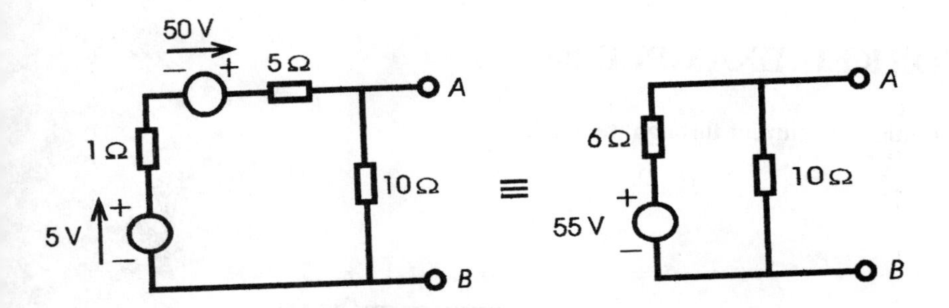

Fig. 112

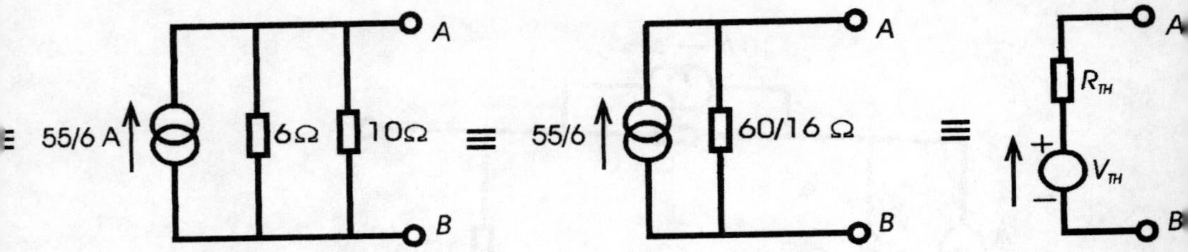

≡ 55/6 A 6Ω 10Ω A B ≡ 55/6 60/16 Ω A B ≡ R_{TH} V_{TH} + − A B

Fig. 113

$$V_{TH} = \frac{55}{6} \times \frac{60}{16} = \frac{550}{16} \text{ volts} = 34.375 \text{ volts}$$

$$R_{TH} = \frac{60}{16} = \frac{30}{8} = \frac{15}{4} = 3.75\ \Omega.$$

WORKED EXAMPLE 20

Determine the current through the load.

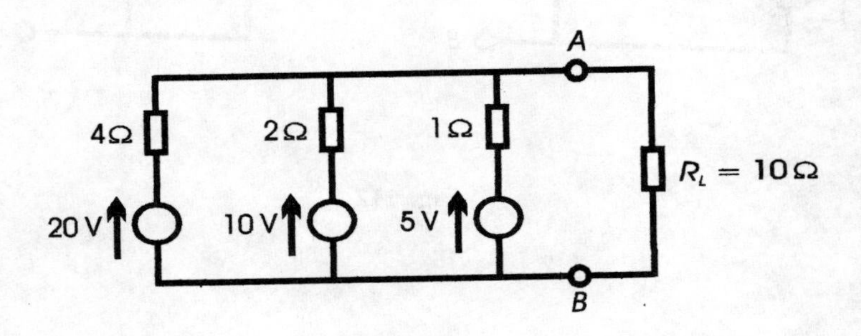

Fig. 114

SOLUTION 20

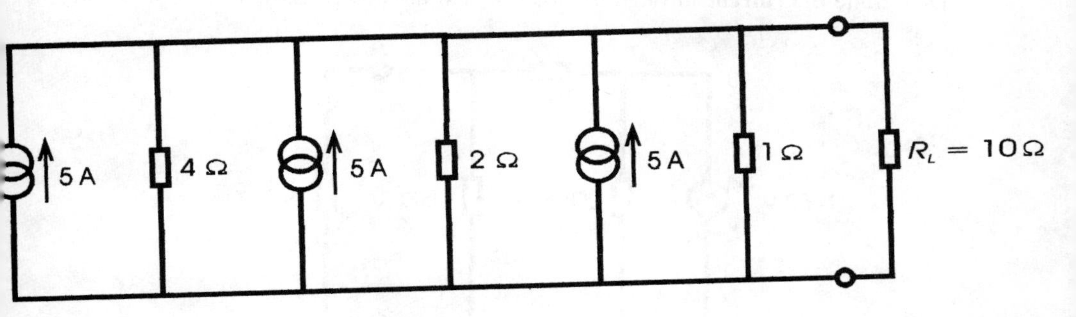

Fig. 115

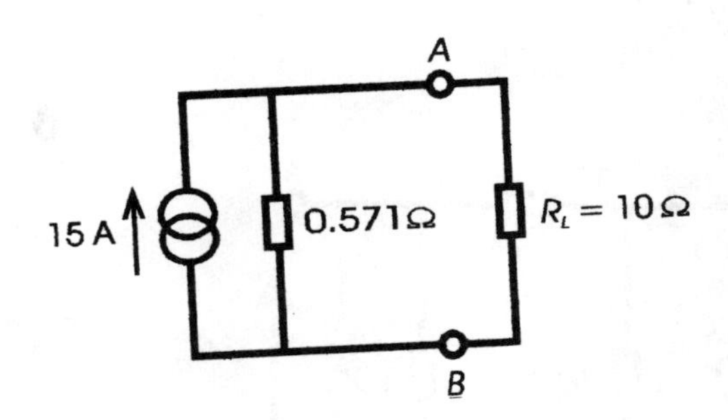

Fig. 116

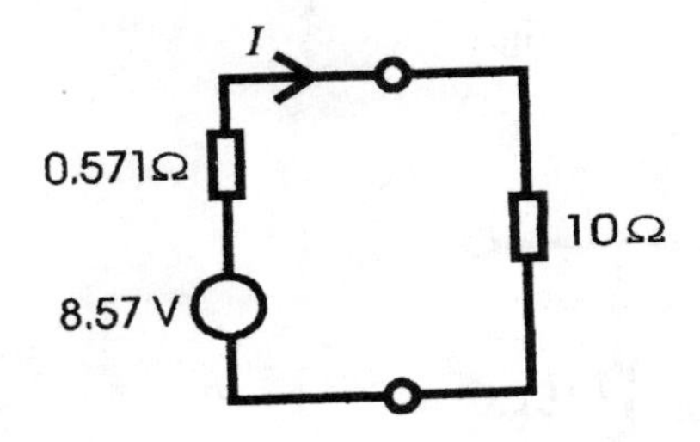

Fig. 117

$$I = \frac{8.57}{10.571} = 0.81 \text{ A.}$$

WORKED EXAMPLE 21

Determine the current through the load, I, and the power dissipated in the load.

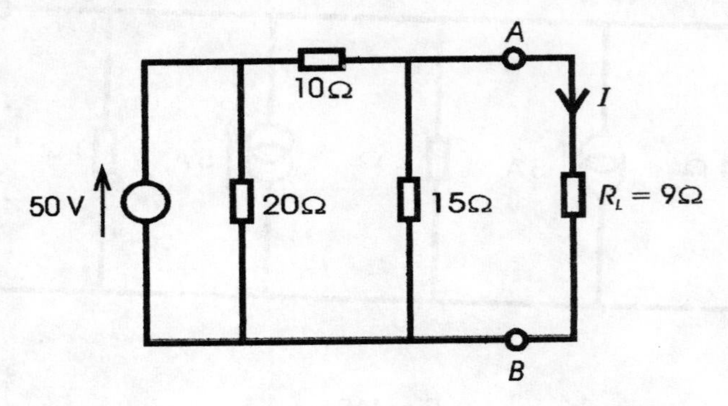

Fig. 118

SOLUTION 21

Remove the load.

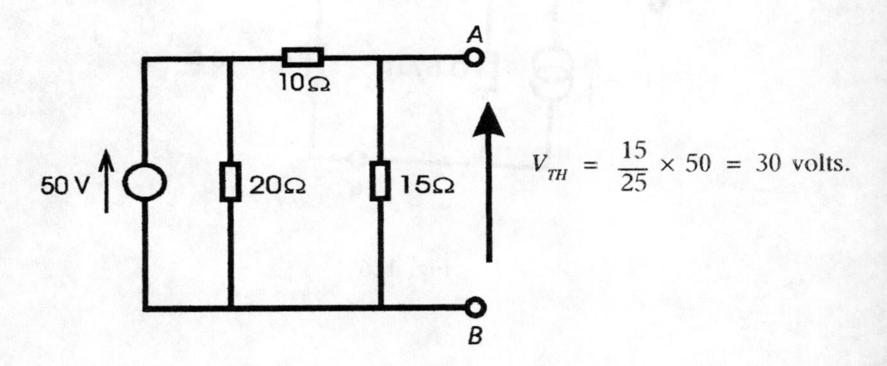

$$V_{TH} = \frac{15}{25} \times 50 = 30 \text{ volts.}$$

Fig. 119

Suppress the e.m.f.

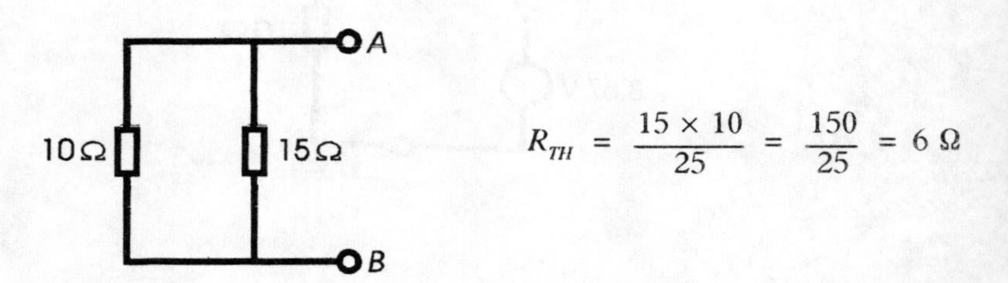

$$R_{TH} = \frac{15 \times 10}{25} = \frac{150}{25} = 6\ \Omega$$

Fig. 120

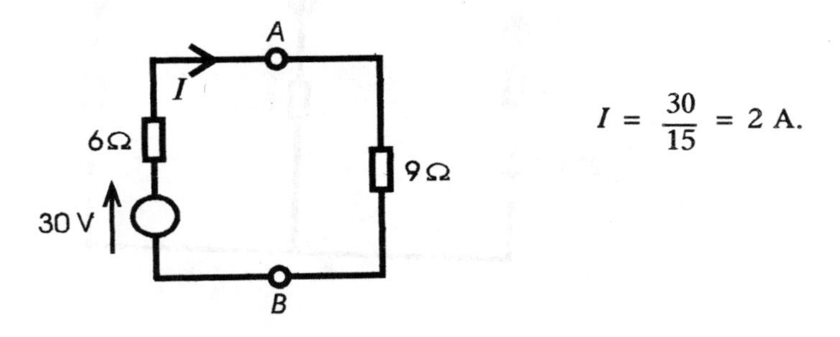

$$I = \frac{30}{15} = 2\text{ A.}$$

Fig. 121

Power in the load = I^2R_L = 4 × 9 = 36 W

SUPERPOSITION THEOREM

WORKED EXAMPLE 22

Two batteries are connected across a load resistor as shown in Fig. 122.

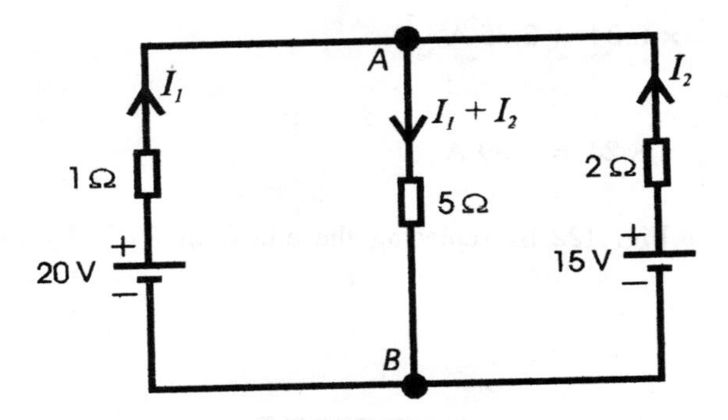

Fig. 122

Use the superposition theorem to determine the load current I.

SOLUTION 22

Redraw the circuit of Fig. 122 by replacing the e.m.f. of 15 V by its internal resistance of 2 Ω.

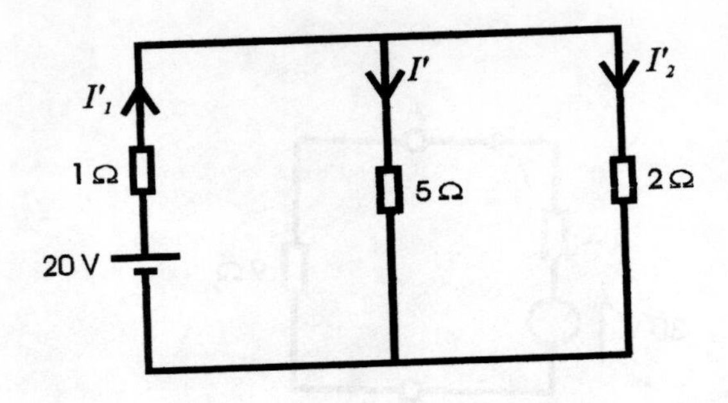

Fig. 123

The total current, I_1', of Fig. 123 is given by

$$I_1' = \frac{20}{1 + \dfrac{5 \times 2}{5 + 2}} = \frac{20 \times 7}{17} = 8.24 \text{ A.}$$

Using the current divider to find I' and I_2' we have,

$$I' = \frac{2}{7} \times I_1' = \frac{2}{7} \times 8.24 = 2.35 \text{ A}$$

$$I_2' = \frac{5}{7} \times I_1' = \frac{5}{7} \times 8.24 = 5.89 \text{ A.}$$

Redraw the circuit of Fig. 122 by replacing the e.m.f. of 20 V by its internal resistance of 1 Ω.

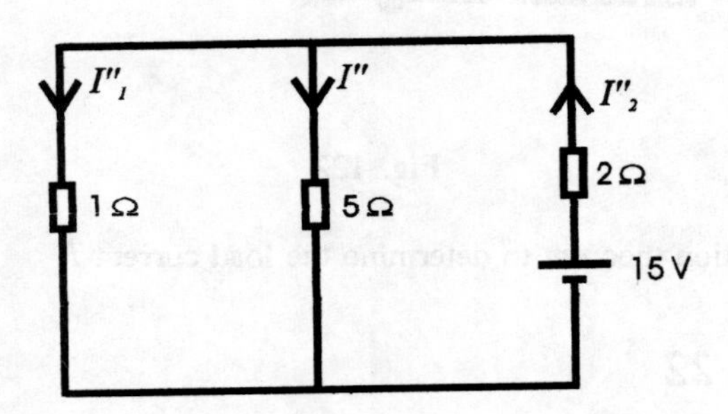

Fig. 124

The total current, I_2'', of Fig. 124 is given by

$$I_2'' = \frac{15}{2 + \frac{1 \times 5}{1 + 5}} = \frac{15 \times 6}{17} = 5.29 \text{ A}$$

$$I'' = \frac{1}{6} \times I_2'' = \frac{1}{6} \times 5.29 = 0.88 \text{ A}$$

$$I_1'' = \frac{5}{6} \times I_2'' = \frac{5}{6} \times 5.29 = 4.41 \text{ A.}$$

To find I we superimpose the currents in that branch, taking into account the direction of the currents.

$$I = I' + I''$$

$$I = 2.35 + 0.88 = 3.24 \text{ A}$$

$$I = 3.24 \text{ A.}$$

Similarly, we can find the currents I_1 and I_2.

$$I_1 = I_1' - I_1''$$

$$I_1 = 8.24 - 4.41 = 3.83 \text{ A}$$

$$I_1 = 3.83 \text{ A}$$

$$I_2 = -I_2' + I_2'' = -5.89 + 5.29$$

$$I_2 = -0.60 \text{ A}$$

WORKED EXAMPLE 23

Use the superposition theorem to determine the branch currents in the circuit of Fig. 125.

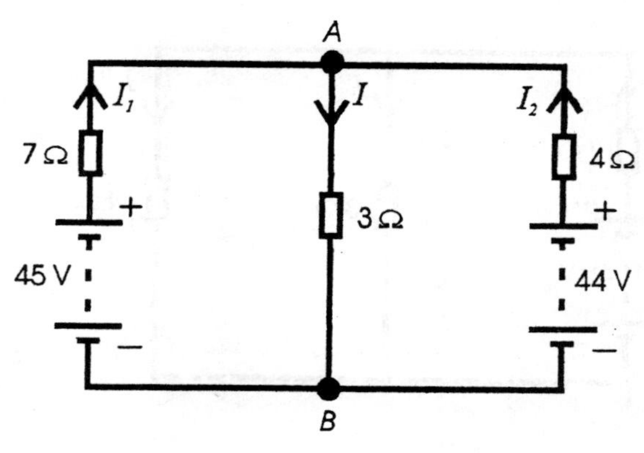

Fig. 125

SOLUTION 23

Replace one of the e.m.f.s by its internal resistance, we have;

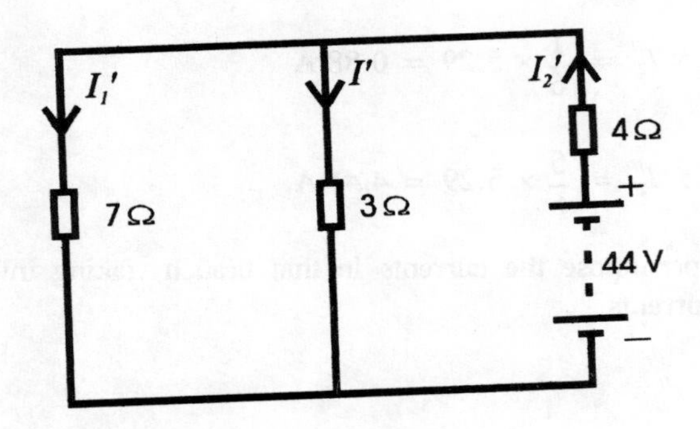

Fig. 126

$$I_2' = \frac{44}{\frac{3 \times 7}{3 + 7} + 4} = \frac{44 \times 10}{61} = 7.21 \text{ A}$$

$$I' = \frac{7}{10} \times 7.21 = 5.05 \text{ A}$$

$$I_1' = \frac{3}{10} \times 7.21 = 2.16 \text{ A}.$$

Replace the other e.m.f. by its internal resistance, we have

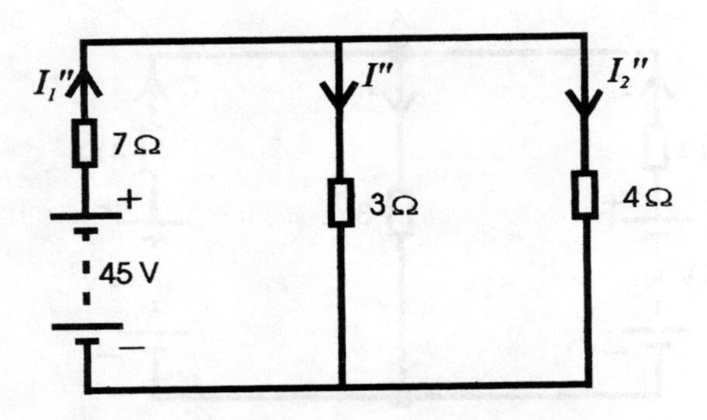

Fig. 127

$$I_1'' = \frac{45}{\frac{3 \times 4}{7} + 7} = \frac{45 \times 7}{61} = 5.16 \text{ A}$$

$$I'' = \frac{4}{7} \times 5.16 = 2.95 \text{ A}$$

$$I_2'' = \frac{3}{7} \times 5.16 = 2.21 \text{ A.}$$

Superimposing the currents, we have

$$I_1 = -I_1' + I_1'' = -2.16 + 5.16 = 3.0 \text{ A}$$

$$I = I' + I'' = 5.05 + 2.95 = 8.0 \text{ A}$$

$$I_2 = I_2' - I_2'' = 7.21 - 2.21 = 5.0 \text{ A.}$$

MAXIMUM POWER TRANSFER THEOREM

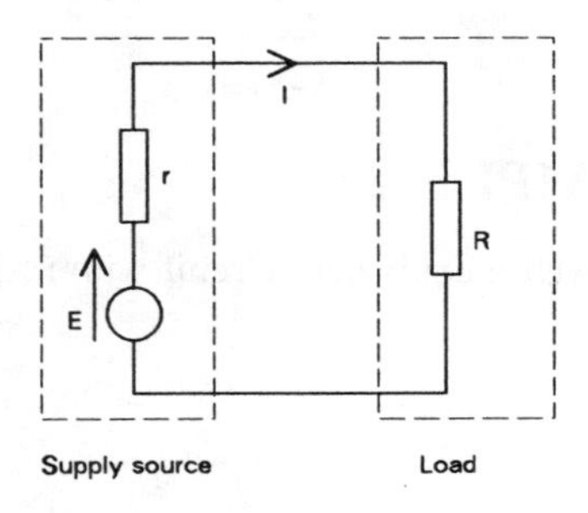

Fig. 128

The power transferred from a supply source to a load is at its maximum when the resistance of the load is equal to the internal resistance (or the output resistance) of the source.

Proof by calculus

$$I = \frac{E}{r + R} \qquad P = I^2R = \frac{E^2}{(r + R)^2}R = \frac{E^2R}{(r + R)^2}$$

Using the quotient rule

$$\frac{dP}{dR} = \frac{E^2(r + R)^2 - E^2R\,2(r + R)}{(r + R)^4}$$

$$= \frac{E^2(r + R)^2 - E^2(r + R)\,2R}{(r + R)^4}$$

$$= \frac{E^2(r + R) - 2E^2R}{(r + R)^3}$$

For maximum or minimum $\frac{dP}{dR} = 0$

$$E^2(r + R) = 2E^2R$$

$$r + R = 2R$$

$\boxed{R = r}$ condition for maximum power

$$P_{max} = \frac{E^2 r}{(2r)^2} = \frac{E^2}{4r}.$$

WORKED EXAMPLE 24

(a) Determine the Thévenin equivalent circuit with respect to terminals AB in Fig. 129.

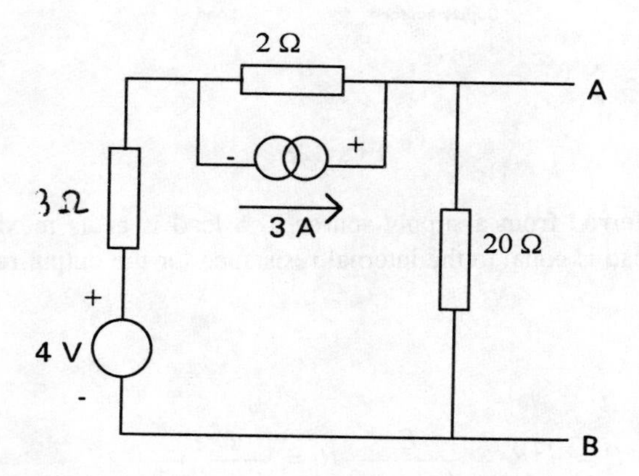

Fig. 129

(b) What value of resistive load will take maximum power from terminals AB and what is the value of this power?

SOLUTION 24

The equivalent circuits of Fig. 129 succeedingly are as shown in Fig. 130, 131, 132, 133 and 134.

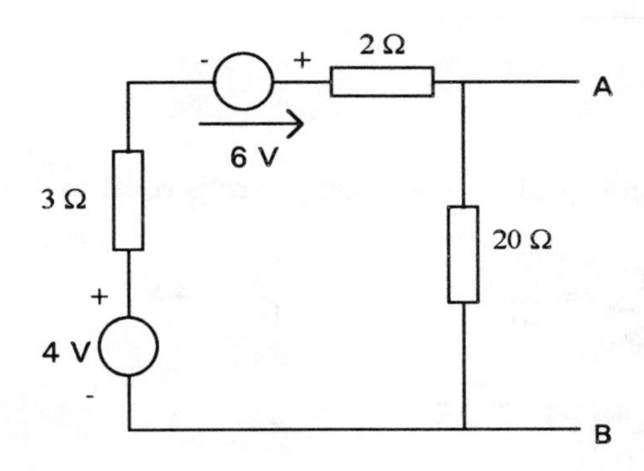

Fig. 130

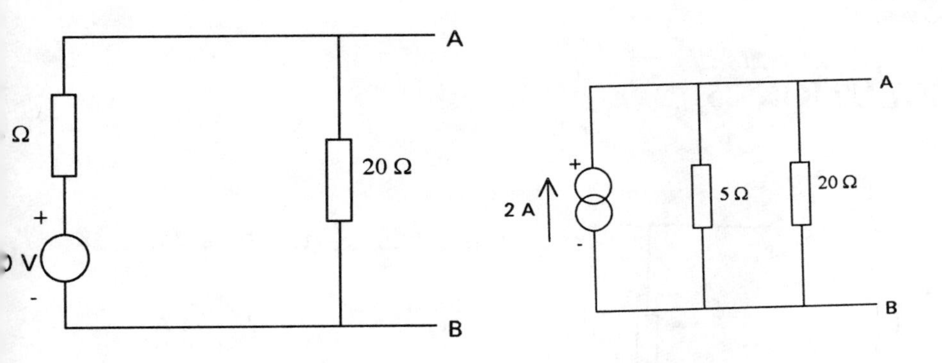

Fig. 131 Fig. 132

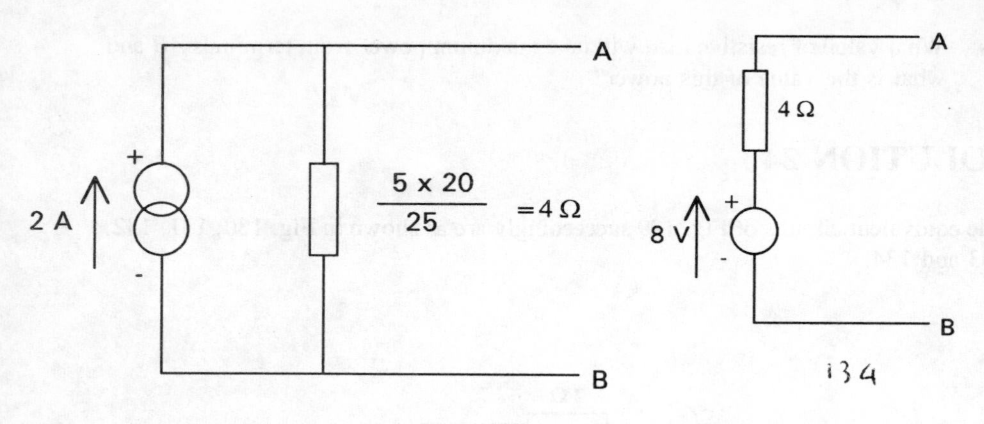

Fig. 133 Fig. 134

(b) The value of resistive load for maximum power is equal to 4 Ω.

$$P_{max} = \frac{E^2}{4r} = \frac{8^2}{4 \times 4} = \frac{64}{16} = 4 \text{ W.}$$

WORKED EXAMPLE 25

When a resistive load of 4 Ω is connected across a supply source, the power dissipated in it is 64 W. When a resistive load of 5 Ω is connected, the power dissipated is 61.25 W. Determine the value of resistance which will dissipate the maximum power when connected to the supply source, and calculate this maximum power.

SOLUTION 25

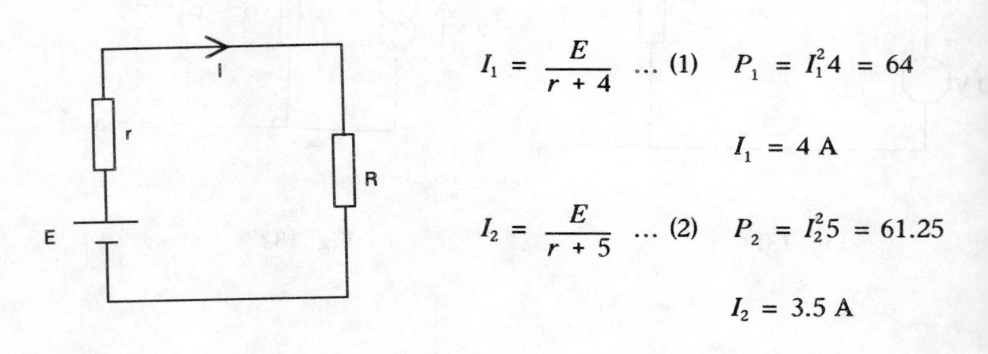

$$I_1 = \frac{E}{r+4} \quad \ldots (1) \qquad P_1 = I_1^2 4 = 64$$

$$I_1 = 4 \text{ A}$$

$$I_2 = \frac{E}{r+5} \quad \ldots (2) \qquad P_2 = I_2^2 5 = 61.25$$

$$I_2 = 3.5 \text{ A}$$

Fig. 135

Dividing (1) and (2) we have

$$\frac{I_1}{I_2} = \frac{r + 5}{r + 4} = \frac{4}{3.5} = \frac{r + 5}{r + 4} \Rightarrow 4r + 16 = 3.5r + 16.5$$

$$0.5r = 0.5 \Rightarrow r = 1\ \Omega$$

$$E = I_1\left(r + 4\right) = 4\ (1 + 4) = 20 \text{ volts.}$$

The resistance to be connected across the supply source for maximum power will be 1 Ω and the maximum power

$$P_{max} = \frac{20^2}{4 \times 1} = \frac{400}{4 \times 1} = 100 \text{ W.}$$

WORKED EXAMPLE 26

Apply Norton's theorem to calculate the load current I in the circuit shown in Fig. 136.

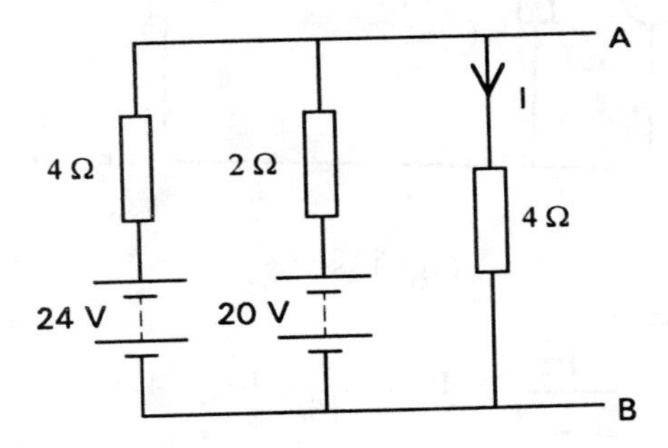

Fig. 136

Replace the load resistor by an other resistor for maximum power dissipation in it and calculate this power.

SOLUTION 26

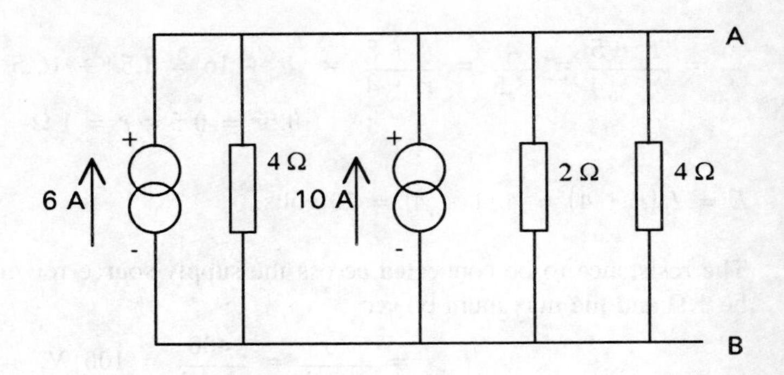

Fig. 137

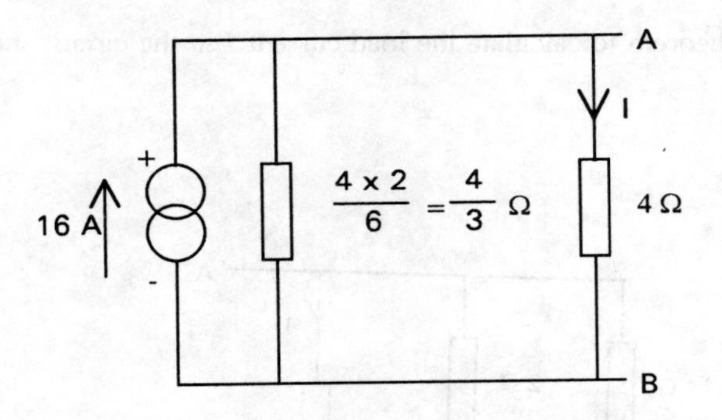

Fig. 138

Using the current divider

$$I = \frac{4/3}{4 + \frac{4}{3}} \times 16 = \frac{4/3}{16/3} 16 = 4 \text{ A}.$$

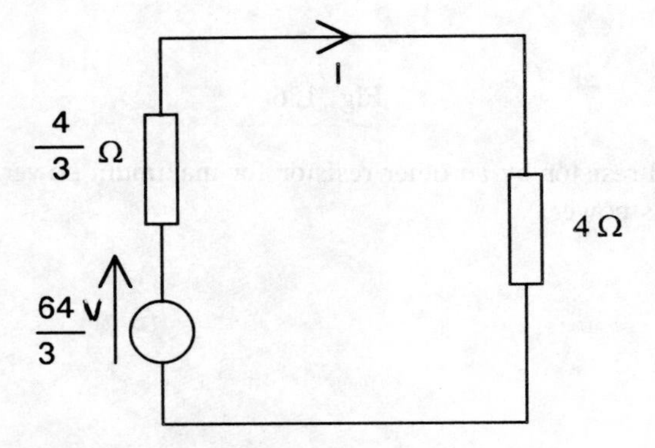

Fig. 139

$$I = \frac{64/3}{4 + \frac{4}{3}} = \frac{64/3}{16/3} = 4 \text{ A}$$

$4\ \Omega$ is replace by $\frac{4}{3}\ \Omega$ resistor

$$P_{max} = \frac{E^2}{4r} = \frac{(64/3)^2}{4 \times \frac{4}{3}} = \frac{64 \times 64 \times 3}{9 \times 16} = 85.3 \text{ watts.}$$

TRANSFORMER MATCHING

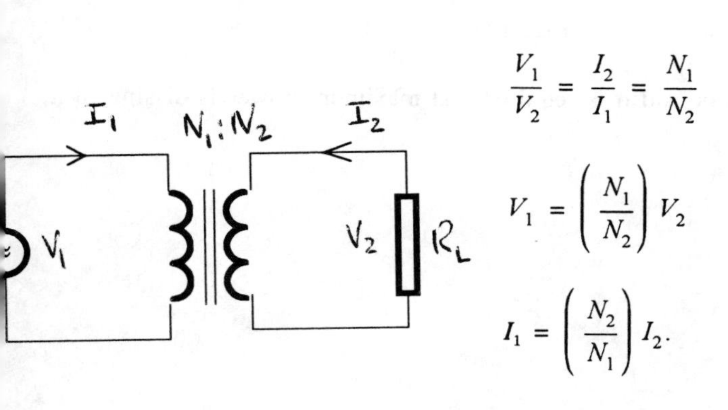

$$\frac{V_1}{V_2} = \frac{I_2}{I_1} = \frac{N_1}{N_2}$$

$$V_1 = \left(\frac{N_1}{N_2}\right) V_2$$

$$I_1 = \left(\frac{N_2}{N_1}\right) I_2.$$

Fig. 140

The input a.c. resistance, $R_{in} = \frac{V_1}{I_1} = \frac{(N_1/N_2)V_2}{(N_2/N_1)I_2} = \left(\frac{N_1}{N_2}\right)^2 \frac{V_2}{I_2}$

$$\boxed{R_{in} = \left(\frac{N_1}{N_2}\right)^2 R_L}$$

WORKED EXAMPLE 27

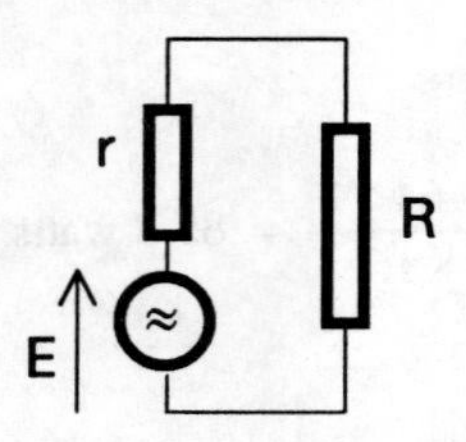

Fig. 141

R_L is a fixed load resistor and it is required that maximum power is dissipated in it.

SOLUTION 27

For this to occur $R_L = r$.

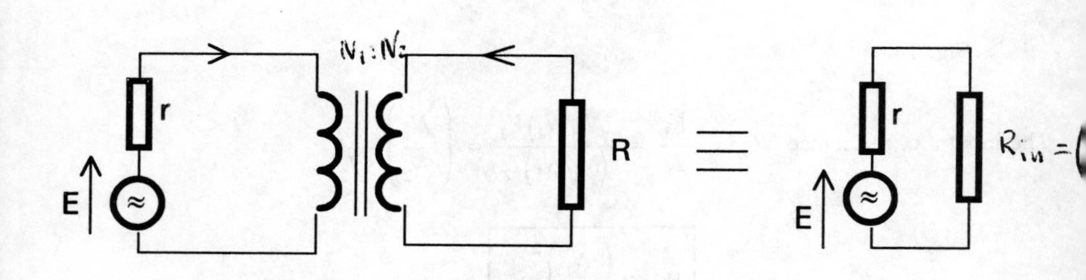

Fig. 142

$R_{\text{in}} = r$ for maximum power transfer.

$$\text{Hence} \quad \left(\frac{N_1}{N_2}\right)^2 = 1 \Rightarrow N_1 = N_2$$

WORKED EXAMPLE 28

An 8 Ω loudspeaker is to be matched to an amplifier of output resistance 20 KΩ. Calculate the turns ratio of the transformer required to achieve this matching.

SOLUTION 28

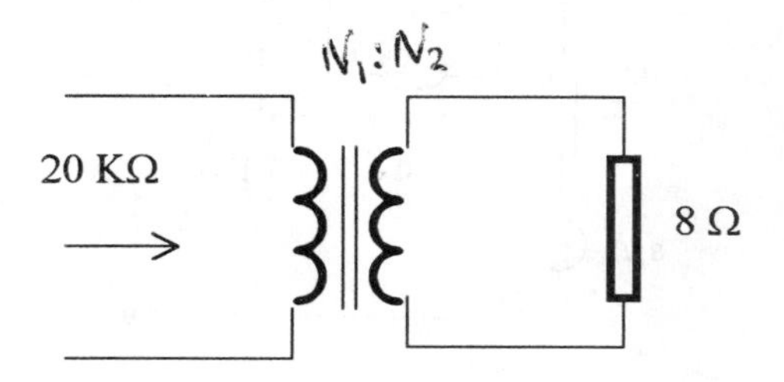

Fig. 143

$$20000 = \left(\frac{N_1}{N_2}\right)^2 8$$

$$\left(\frac{N_1}{N_2}\right)^2 = \frac{20000}{8} = 2500$$

$$\frac{N_1}{N_2} = 50$$

WORKED EXAMPLE 29

The output stage of an amplifier has an output resistance of 4.9 KΩ. Determine the optimum transformation ratio of a transformer which would match a load resistance of 4 Ω to the output resistance of the amplifier.

SOLUTION 29

$$R_{\text{in}} = \left(\frac{N_1}{N_2}\right)^2 4 = 4900$$

$$\frac{N_1}{N_2} = \sqrt{\frac{4900}{4}} = 35.$$

EXERCISES 2

1. (a) Determine the Thévenin equivalent circuit with respect to terminals AB in Fig. 144

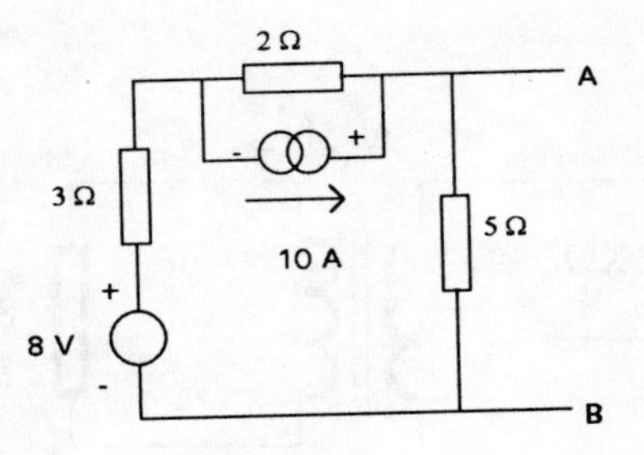

Fig. 144

(b) What value of resistive load will take maximum power from terminals AB and what is the value of this power?

2.

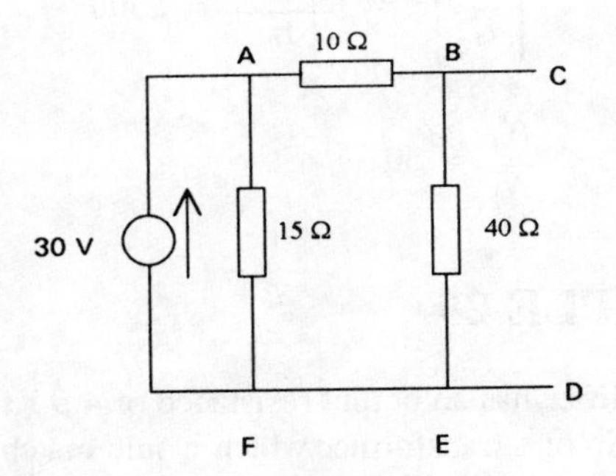

Fig. 145

(a) Determine and draw the Thévenin equivalent circuit for the circuit in Fig. 145 with respect to the terminals C and D.

(b) From the result in (a) or otherwise, determine and draw the Norton equivalent circuit.

(c) Calculate the load resistance to be connected between C and D for maximum power transfer.

(d) Calculate the values of the resistors and draw the equivalent STAR network to replace the DELTA network $ABEF$.

(e) Use the equivalent circuit of (d) to show that $V_{TH} = 24$ V, $R_{TH} = 8\ \Omega$.

3. Apply Norton's theorem to calculate the current I in the circuit shown in Fig. 146.

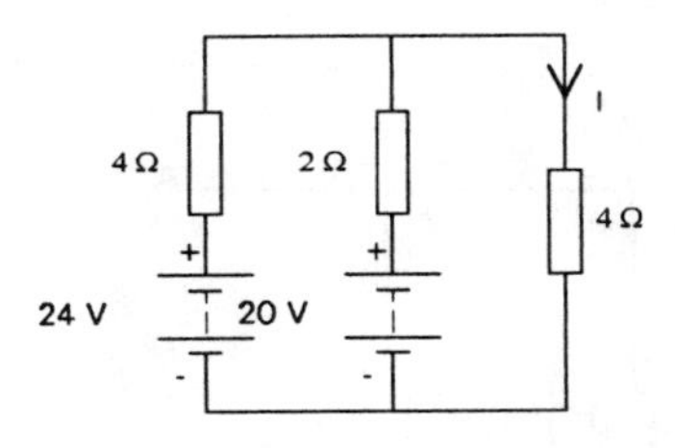

Fig. 146

4. Calculate the current in the 5 Ω resistor of Fig. 147 using Thévenin's theorem.

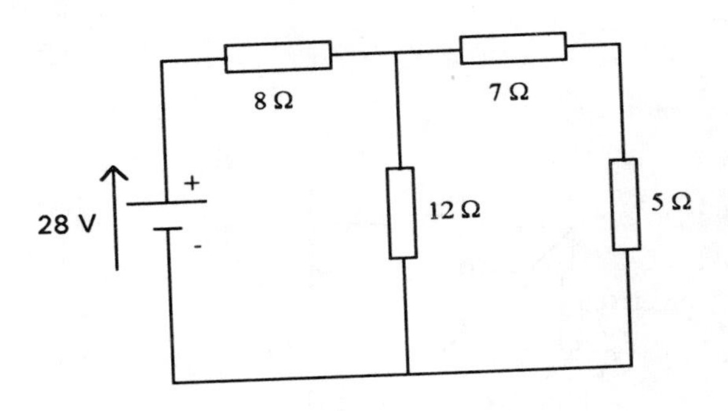

Fig. 147

5. Determine R_{TH}, V_{TH} & I_N.

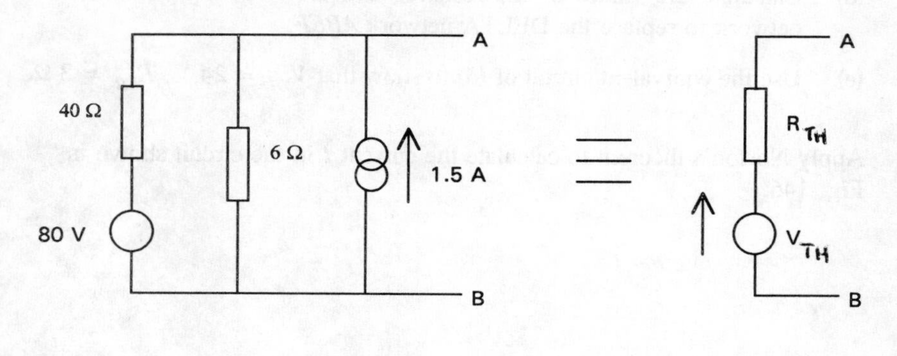

Fig. 148

Fig. 149

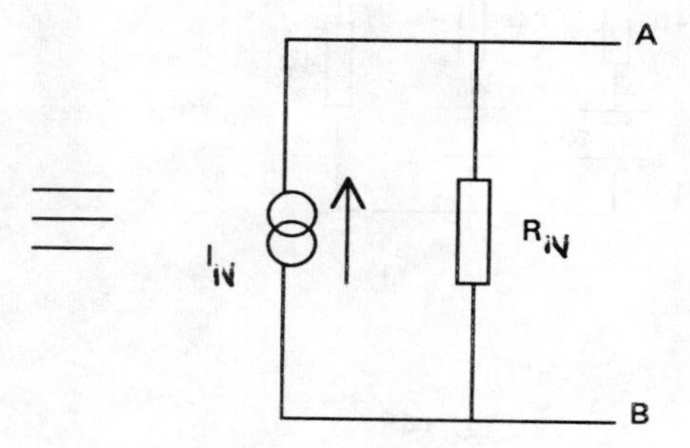

Fig. 150

6. Calculate the current through and the voltage across the meter *D*.

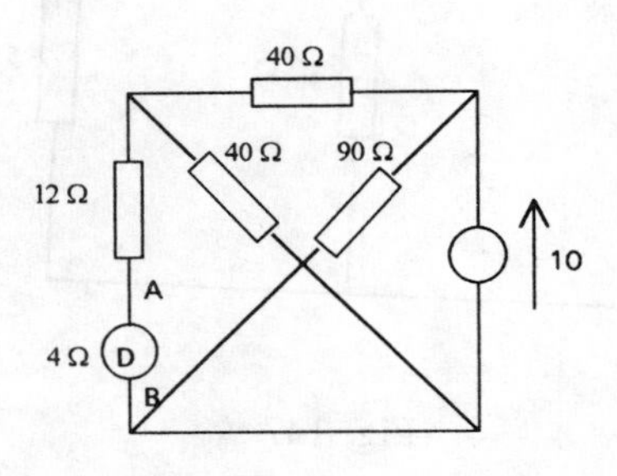

Fig. 151

7. Obtain the Thévenin equivalent circuits for the networks shown in Fig. 152, 153 and 154.

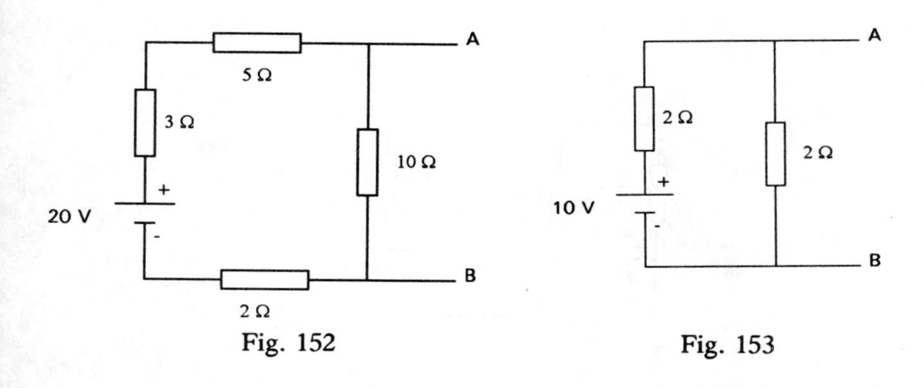

Fig. 152

Fig. 153

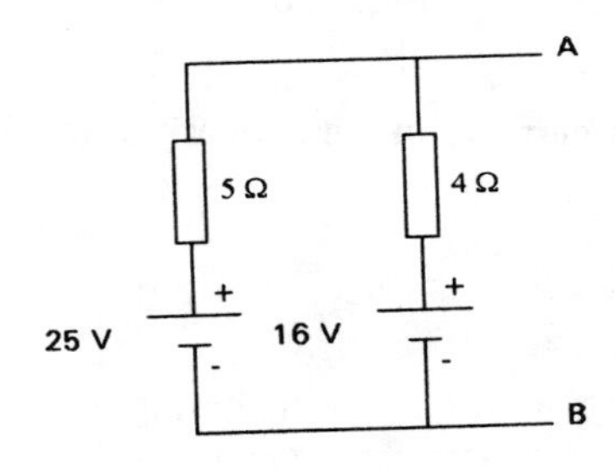

Fig. 154

8. Determine the current in the 5 Ω resistor using Thévenin theorem.

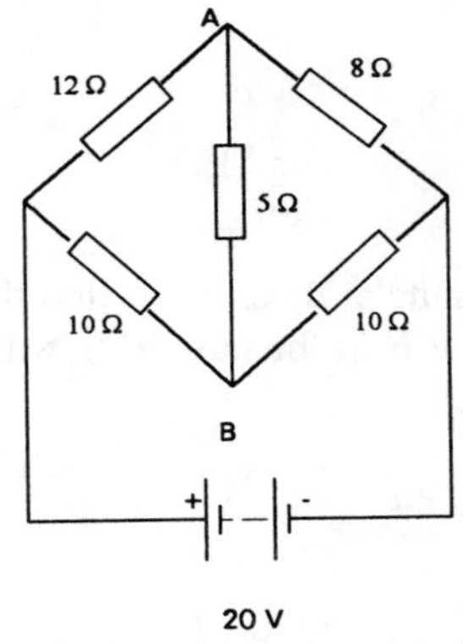

Fig. 155

9. Obtain the Thévenin equivalent of the network shown in Fig. 156 and hence deduce the Norton equivalent.

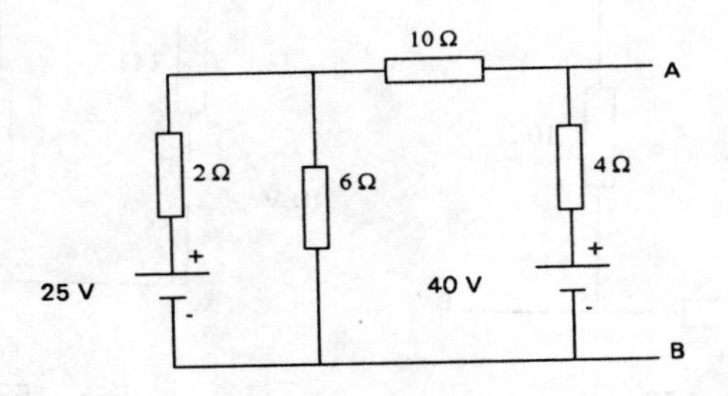

Fig. 156

10. Determine the branch currents in the network shown using the superposition principle.

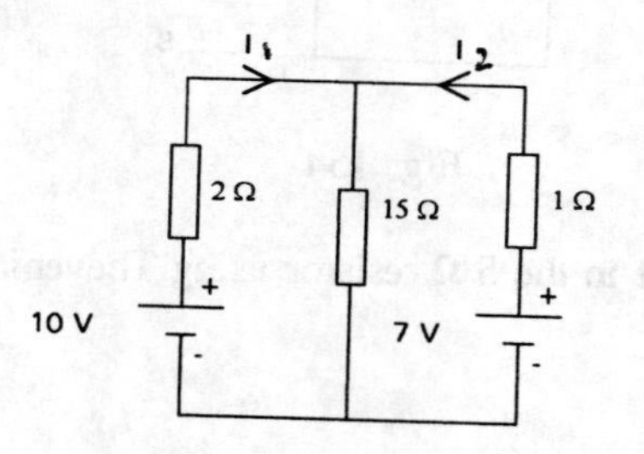

Fig. 157.

11. Use Thévenin's (Helmholtz's) theorem to find the current in 3.19 Ω resistor connected between *A* and *B* in the circuit shown.

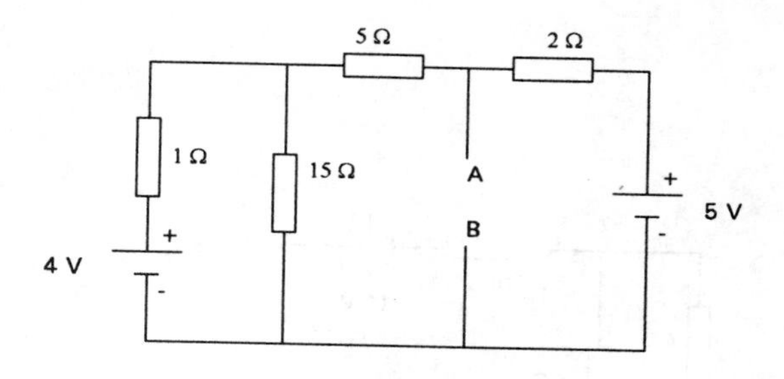

Fig. 158

12. Calculate the value of the current in the 10 Ω resistor in the network shown by means of the theorem.

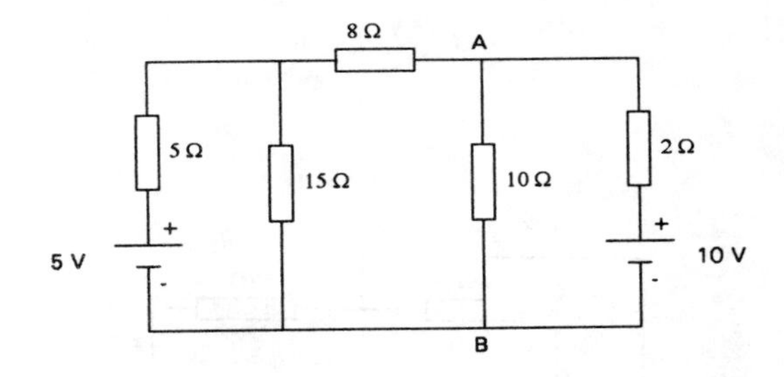

Fig. 159

13. A network of resistances and sources of d.c. e.m.f.'s has two output terminals. The open-circuit voltage at the terminals is 200 V. The current flowing when the terminals are short-circuited is 15 A and 10 A when connected through a resistor of R Ω. Determine the components of the equivalent circuit feeding the terminals and the load resistor. What value of load resistance would give maximum power output?

14. **Obtain the Norton equivalent circuit for the network.**

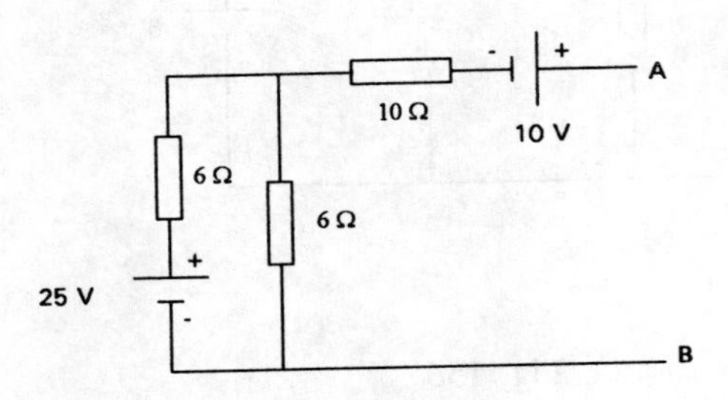

Fig. 160

15. **Obtain the Thévenin equivalent circuit at the terminals *AB* of the network.**

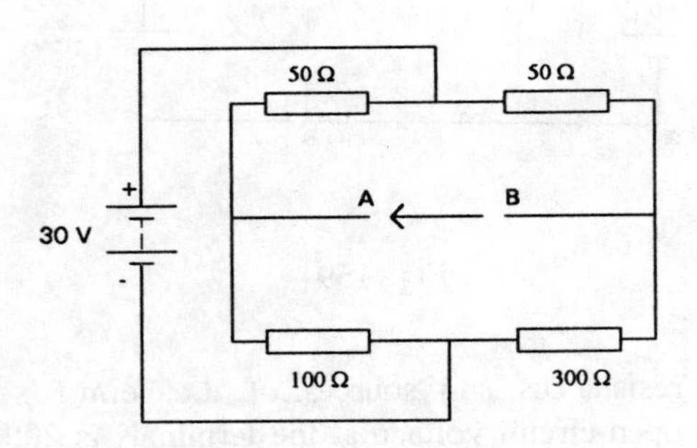

Fig. 161

16. Use Thévenin theorem to calculate the current through the ammeter.

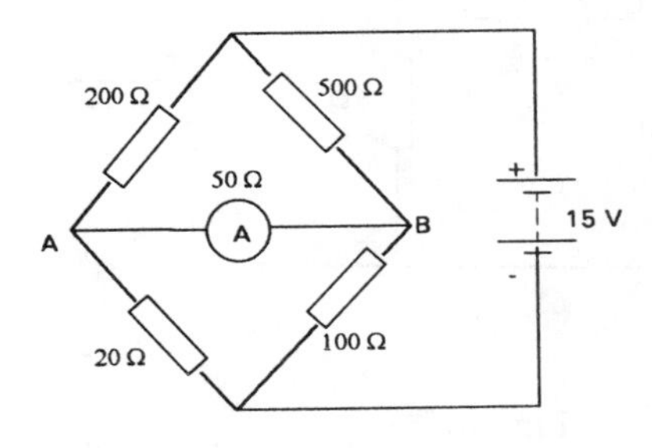

Fig. 162

17.

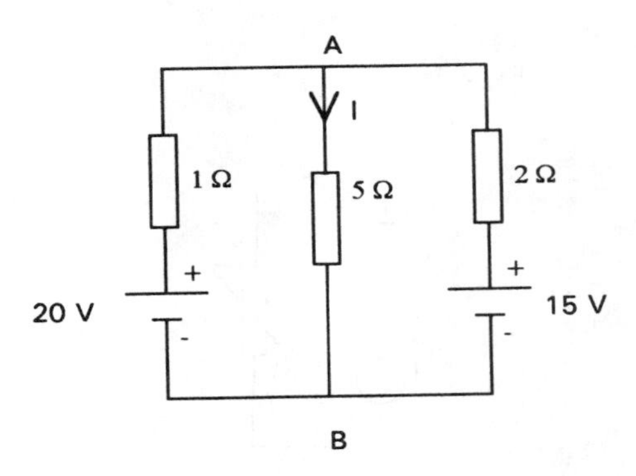

Fig. 163

Calculate I by applying
(i) Kirchhoff's laws
(ii) Thévenin's Theorem
(iii) Norton's Theorem
(iv) Superposition Theorem.

18. In the above question remove the 5 Ω resistor and determine the new value of resistance to be replaced which when connected between the terminals AB dissipates maximum power. Calculate the maximum power dissipated in this resistor.

19.

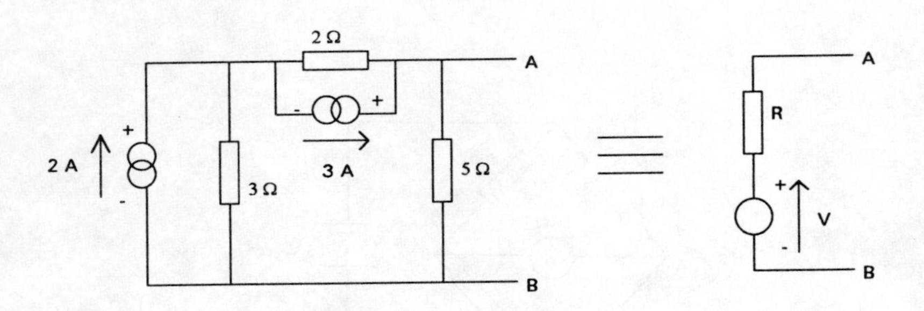

Fig. 164 Fig. 165

(i) Determine R_{TH} and V_{TH}.

(ii) What value of resistive load will take maximum power from terminals AB and what is the value of this power.

20.

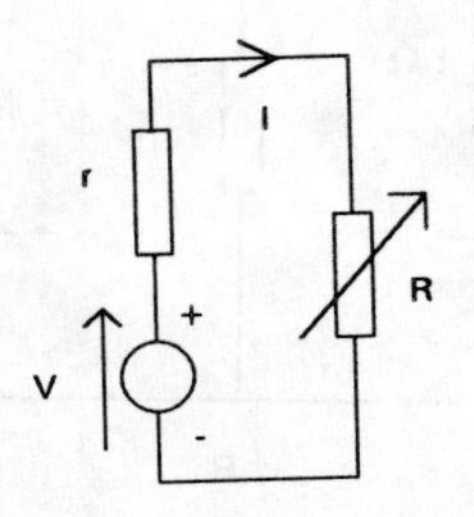

Fig. 166

(i) Show by means of calculus that the value of R required to dissipated maximum power in it is given by $\boxed{R = r}$

(ii) If $V_{TH} = 10$ volts and $r = 5\ \Omega$, calculate the power dissipated in R if $R = 0, 1, 2, 3, 4, 5, 6, 7, 9, 9, 10$.

Draw a table of P against R and indicate the maximum power.

21. (i) Determine the open circuit voltage, V_{TH}, across AB.

(ii) Determine the short circuit current, I_N, through AB when it is shorted. Hence find R_{TH} between AB.

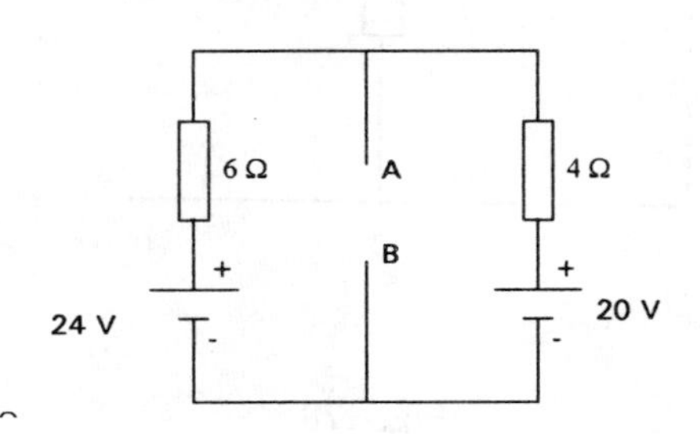

Fig. 167

22.

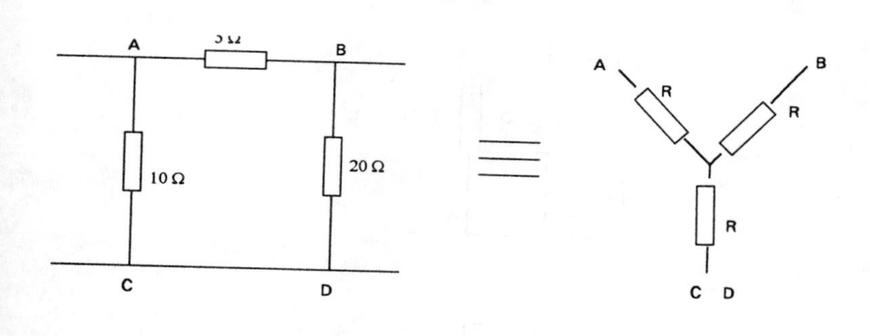

Fig. 168 Fig. 169

Determine R_A, R_B and R_C for the T or Y network which equivalent to the Δ network on the left hand side.

The T network is connected between the source and the load as shown in Fig. 170.

Determine R for maximum power dissipation in R and the value of this power.

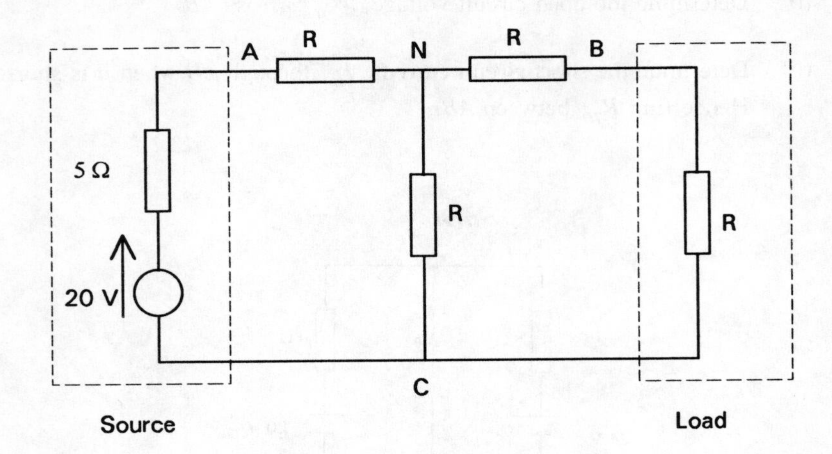

Fig. 170

23. Calculate the current in the 10 Ω resistor using Thévenin theorem. Check your answer with the current divider method.

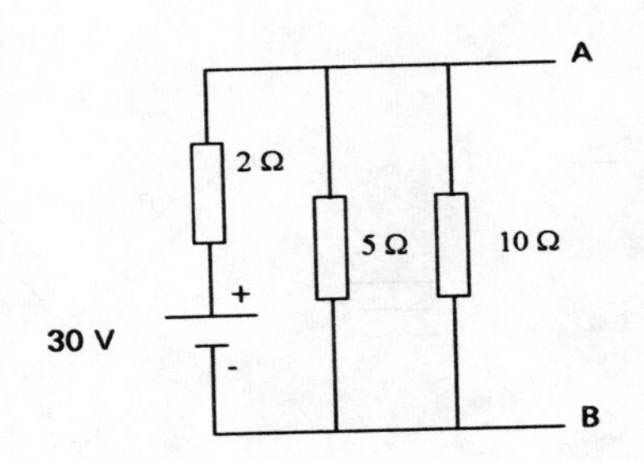

Fig. 171

24.

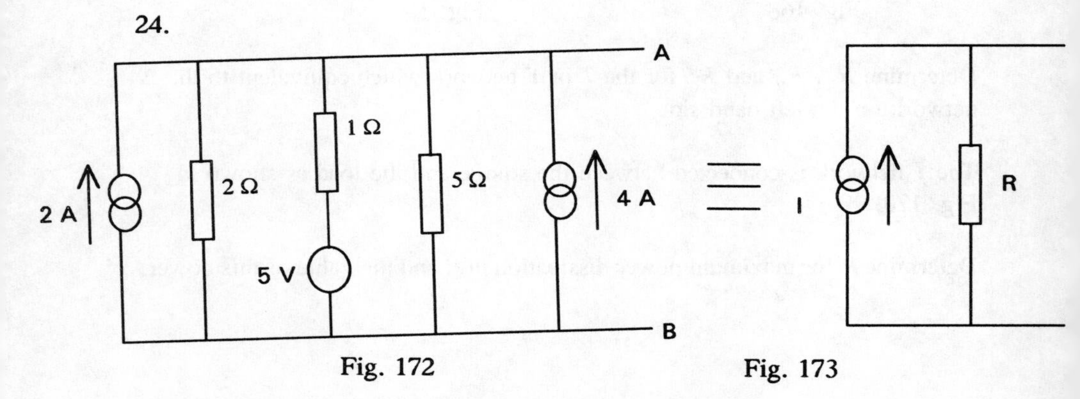

Fig. 172

Fig. 173

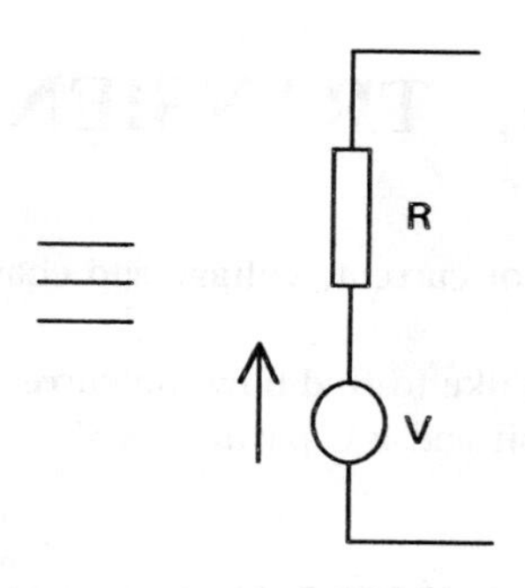

Fig. 174

Determine I_N, R_N, R_{TH} and V_{TH}.

25. Determine the optimum value of load resistance for maximum transfer of power if it is connected to an electronic amplifier at output resistance 288 Ω through a transformer with 6:1 step down ratio.

3. TRANSIENTS

Transient is the variation of current, voltage and charge with respect to time.

In this chapter, we would like to find how the current through a coil and capacitor and a voltage across a coil and a capacitor vary with time when a d.c. supply is impressed.

CHARGING UP A CAPACITOR

Consider a pure non-inductive resistor which is connected in series with a pure capacitor as shown in Fig. 175 in series with a d.c. supply and a switch.

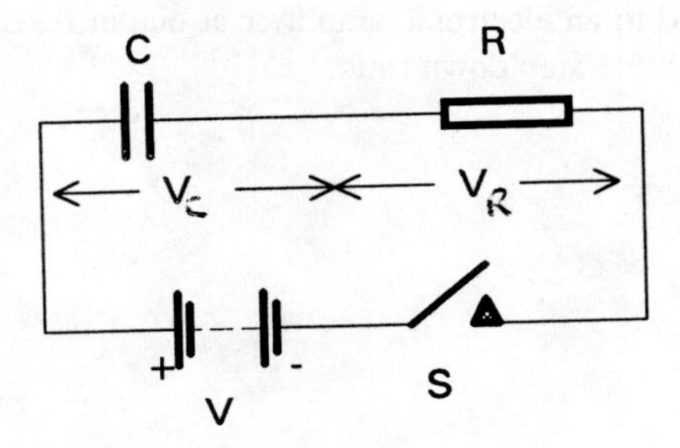

Fig. 175

At t, time, after closing the switch S_1 we have

$$V = v_C + v_R$$

$$V = \frac{1}{C}\int i\,dt + iR.$$

Note that, $i = C\frac{dv}{dt}$, the current through the capacitor is proportional to the rate of change of voltage, $\frac{dv}{dt}$, the constant of proportionality is C, the capacitance of the capacitor

$$C = \frac{i}{\frac{dv}{dt}} = \frac{\text{instantaneous current}}{\text{the rate of change of voltage across the capacitor}}$$

Re-arranging the relationship $i = C\frac{dv}{dt}$

$$dv = \frac{i\,dt}{C}$$

and integrating with respect to time, we have

$$\int dv = \frac{1}{C}\int i\,dt$$

$$v_C = \frac{1}{C}\int i\,dt$$

where v_C is the instantaneous voltage across the capacitor. When $t = 0$, at the instant of closing the switch S_1, the voltage across the capacitor is zero.

At $t = 0_-$, S_1 is open and $t = 0_+$, S_1 is closed.

At $t = 0_+$, the current in the circuit is maximum.

$I = \frac{V}{R}$ since the voltage across C is zero. At $t = \infty$, that is, after a long time,

$I = 0$, and $v_C = V$, remember that $Q = CV$.

$$V = \frac{1}{C}\int i\,dt + iR.$$

Differentiating with respect to time, t.

$$\frac{d}{dt}(V) = \frac{i}{C} + \frac{di}{dt}R$$

$$0 = \frac{i}{C} + \frac{di}{dt}R \; \ldots (1)$$

the derivative of a constant (V) is zero, integration is the reverse of differentiation,

therefore the derivative of the integral $\frac{1}{C}\int i\,dt$ is $\frac{1}{C} \times$ function of i (which is a function of time).

Separating the variables of equation (1)

$$\frac{i}{C} = -\frac{di}{dt}R$$

$$-\frac{dt}{CR} = \frac{di}{i} \quad \dots (2)$$

Observe that we have, on one side of the equation, the variable t, and on the other side, the variable i, CR is a constant and it is equal to $\tau = CR =$ time constant. Integrating both side of equation (2)

$$-\frac{t}{CR} + \text{an arbitrary constant} = \ln i + \text{an arbitrary constant.}$$

The arbitrary constants on the left and on the right of this equations are lumped together to one constant

$$-\frac{t}{CR} = \ln i + \text{constant}$$

when $t = 0$, $i = I$

$$0 = \ln I + \text{constant}$$

$$\text{constant} = -\ln I$$

$$-\frac{t}{CR} = \ln i - \ln I$$

$$\ln \frac{i}{I} = -\frac{t}{CR}$$

$$e^{-\frac{t}{RC}} = \frac{i}{I} \text{ since } \tau = RC$$

$$\boxed{i = Ie^{-\frac{t}{\tau}}} \quad \dots (3)$$

The voltage across C, after closing the switch is given

$$v_C = \frac{1}{C}\int i dt$$

$$v_C = \frac{1}{C}\int Ie^{-\frac{t}{\tau}}\, dt$$

$$v_C = \frac{I}{C}\frac{e^{-\frac{t}{\tau}}}{(-1/\tau)} + \text{constant}$$

$$v_C = -IRe^{-\frac{t}{\tau}} + \text{constant}$$

$$v_C = -Ve^{-\frac{t}{\tau}} + \text{constant}$$

when $t = 0$, $v_C = 0$, constant $= V$, the boundary condition

$$v_C = -Ve^{-\frac{t}{\tau}} + V$$

$$\boxed{v_C = V\left(1 - e^{-\frac{t}{\tau}}\right)}.$$

The current through the capacitor decays, the voltage across the capacitor grows, and hence the charge on the capacitor grows. The transients are shown in Fig. 176, Fig. 177 and Fig. 178.

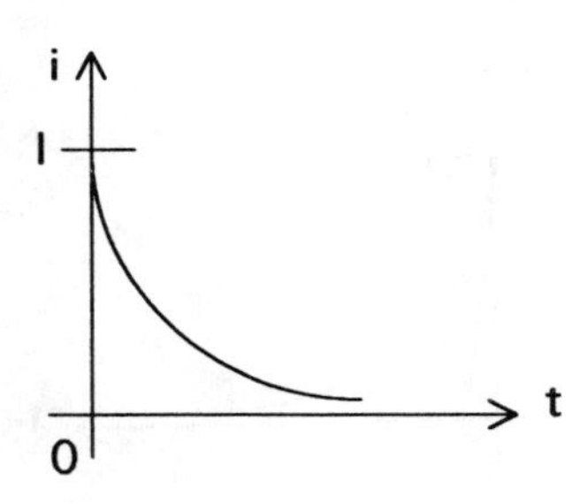

Fig. 176

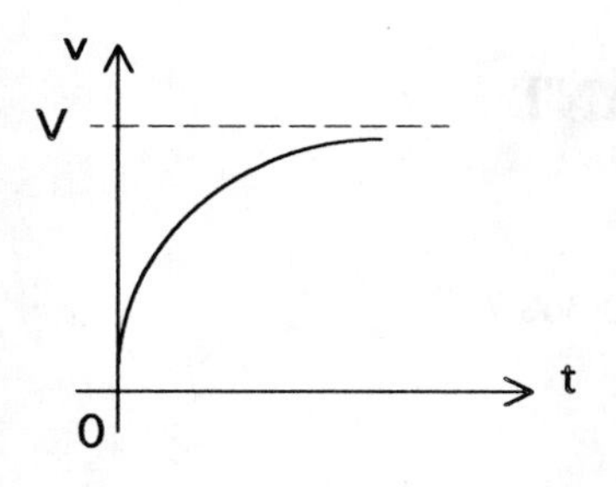

Fig. 177

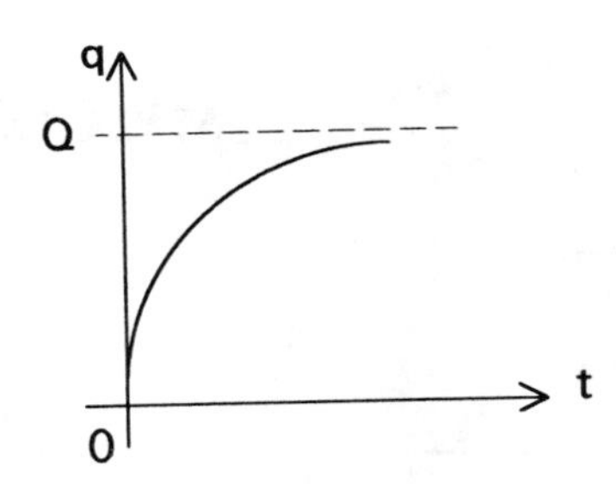

Fig. 178

The variations or transients of i, v and q are exponential.

$$Q = CV$$

$$q = Cv_C$$

$$q = CV\left(1 - e^{-\frac{t}{\tau}}\right)$$

$$\boxed{q = Q\left(1 - e^{-\frac{t}{\tau}}\right)}$$

Differentiating equation (3)

$$\frac{\mathrm{d}i}{\mathrm{d}t} = -\frac{I}{\tau}e^{-\frac{t}{\tau}}$$

at $t = 0$, $\boxed{\left(\frac{\mathrm{d}i}{\mathrm{d}t}\right)_{t=0} = -\frac{I}{\tau}}$ the initial rate of current decay

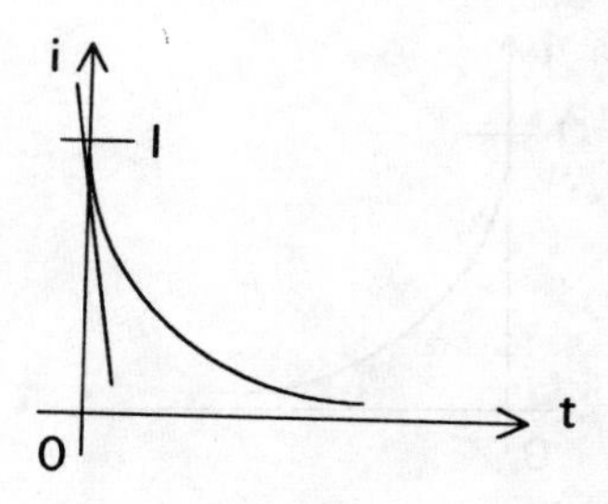

Fig. 179

TIME CONSTANT

$\boxed{i = Ie^{-\frac{t}{\tau}}}$ if $t = \tau$, $i = Ie^{-1} = 0.368\ I$

$\boxed{v_C = V\left(1 - e^{-\frac{t}{\tau}}\right)}$ if $t = \tau$, $v_C = V\left(1 - e^{-1}\right) = 0.632\ \mathrm{V}$

$$v_R = Ve^{-\frac{t}{\tau}}$$ if $t = \tau$, $v_R = Ve^{-1} = 0.368\ V$

$$q = Q\left(1 - e^{-\frac{t}{\tau}}\right)$$ if $t = \tau$, $q = Q(1 - e^{-1}) = 0.632\ Q$.

Consider the decay curve of the current.

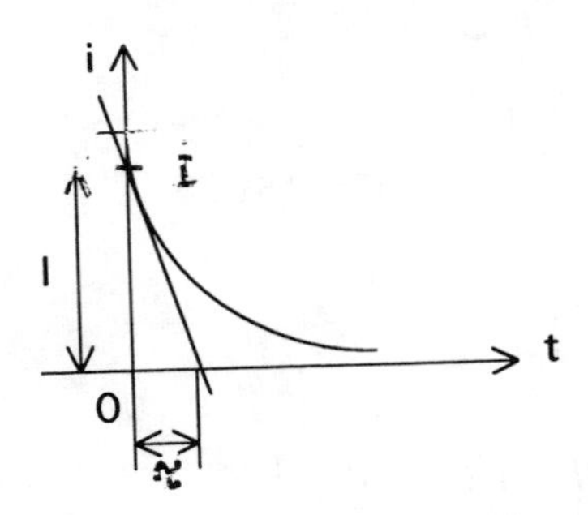

Fig. 180

The gradient at A, that is, at $t = 0$

$$\left(\frac{di}{dt}\right)_{t=0} = -\frac{I}{\tau}.$$

Observe this gradient is negative, since the angle of the tangent at A has an angle greater than 90° with the horizontal.

The initial rate of change of current, $\left(\frac{di}{dt}\right)_{t=0}$. τ = time constant, I = steady current, then $\tau = -\dfrac{\text{steady current, I}}{\text{the initial rate of change of current}}$

<u>Definition of time constant</u>

Considering the growth curves, time constant, τ, is the time taken for the voltage across the capacitor to reach to about 63.2 % of the final steady voltage, V, or the time taken for the charge on the capacitor to reach to 63.2 % of the final steady charge, Q.

Considering the decay curves, the time constant, τ, is the time taken for the current through the capacitor to drop to about 36.8 % of the initial steady current, I, or the time taken for the voltage across the resistor to drop to 36.8 % of the initial steady voltage, V.

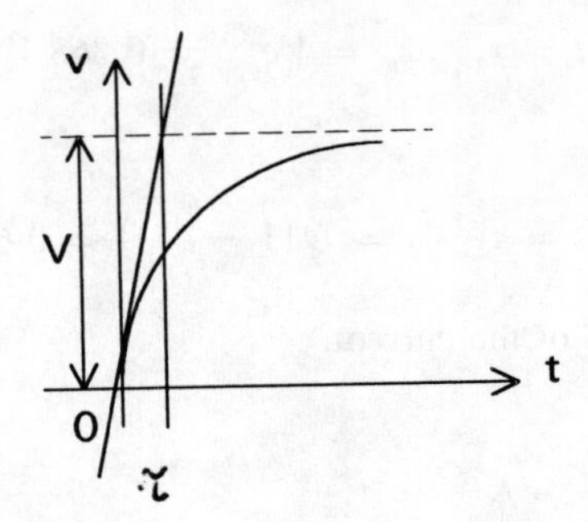

Fig. 181

$$v_C = V\left(1 - e^{-\frac{t}{\tau}}\right)$$

$$\frac{\mathrm{d}v_C}{\mathrm{d}t} = -Ve^{-\frac{t}{\tau}}\left(-\frac{1}{\tau}\right) = \frac{V}{\tau}e^{-\frac{t}{\tau}}$$

If $t = 0$, $\boxed{\left(\frac{\mathrm{d}v_C}{\mathrm{d}t}\right)_{t=0} = \frac{V}{\tau}}$.

The initial rate of growth of voltage.

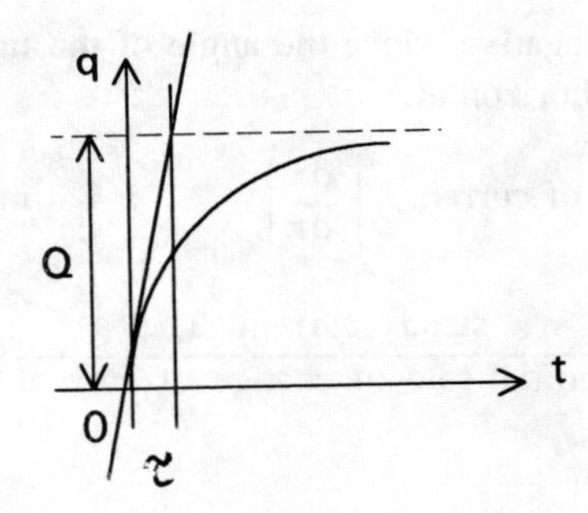

Fig. 182

$$q = Q\left(1 - e^{-\frac{t}{\tau}}\right)$$

$$\frac{\mathrm{d}q}{\mathrm{d}t} = \frac{Q}{\tau}e^{-\frac{t}{\tau}}$$

If $t = 0$

$$\boxed{\left(\frac{dq}{dt}\right)_{t=0} = \frac{Q}{\tau}}.$$

The initial rate of growth of charge.

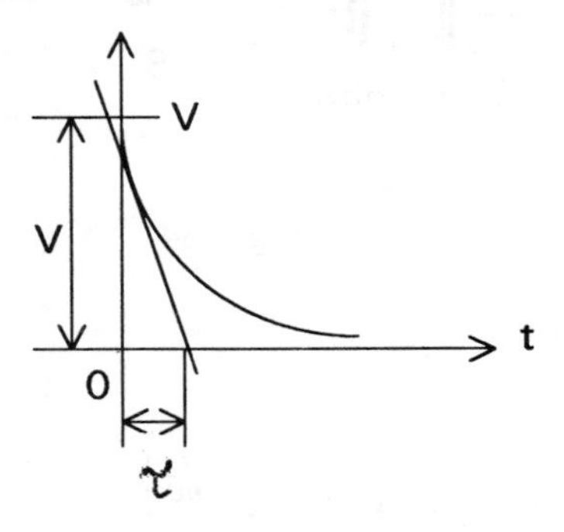

Fig. 183

$$v_R = Ve^{-\frac{t}{\tau}}$$

$$\frac{dv_R}{dt} = -\frac{V}{\tau}e^{-\frac{t}{\tau}}$$

At $t = 0$

$$\boxed{\left(\frac{dv_R}{dt}\right)_{t=0} = -\frac{V}{\tau}}.$$

The initial rate of change of voltage across R.

WORKED EXAMPLE 30

After closing the switch, S, find the time taken for a p.d. of 5 V to be developed across (i) the 10 KΩ resistor

(ii) the 10 μF capacitor.

The capacitor is assumed to be uncharged initially.

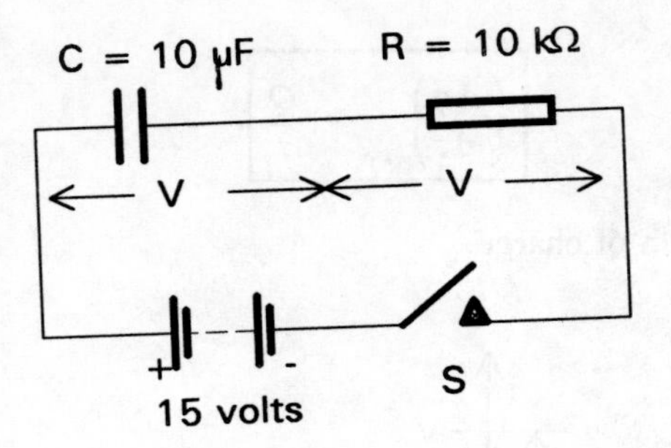

Fig. 184

SOLUTION 30

$$v_R = Ve^{-\frac{t}{RC}} \text{ and } v_C = V\left(1 - e^{-\frac{t}{RC}}\right)$$

(i) $v_R = 5$ volts, $RC = \tau = 10^1 \times 10^3 \times 10^1 \times 10^{-6} = 10^{-1} = 0.1$ s

$$5 = 15e^{-\frac{t}{0.1}}$$

$$\frac{1}{3} = e^{-\frac{t}{0.1}} = e^{-10t}$$

$$e^{10t} = 3.$$

Taking logarithms on both sides of the equation

$$\log_e e^{10t} = \log_e 3$$

$$10\,t = \ln 3$$

$$t = \frac{1}{10} \ln 3 = 0.11 \text{ s}$$

(ii) $v_C = 5$ V

$$5 = 15\left(1 - e^{-\frac{t}{0.1}}\right)$$

$$\frac{1}{3} = 1 - e^{-\frac{t}{0.1}}$$

$$e^{-10t} = 1 - \frac{1}{3} = \frac{2}{3}$$

$$e^{10t} = \frac{3}{2}.$$

Taking logarithms on both sides of the equation

$$10t = \ln \frac{3}{2}$$

$$t = \frac{1}{10} \ln \frac{3}{2} = 0.041 \text{ s or } 41 \text{ ms.}$$

WORKED EXAMPLE 31

A capacitor of capacitance 5 μF is connected in series with a resistor of resistance of 250 KΩ to a d.c. supply of 85 V.

(a) Calculate the time constant of the circuit.

(b) Derive a formula for the initial rate of rise of voltage and hence calculate its value.

(c) What is the rate of rise of voltage at $t = 1$ s?

(d) Show that when $t = 1.25$ s, the voltage across the capacitor is 63.2 % of 85 V.

SOLUTION 31

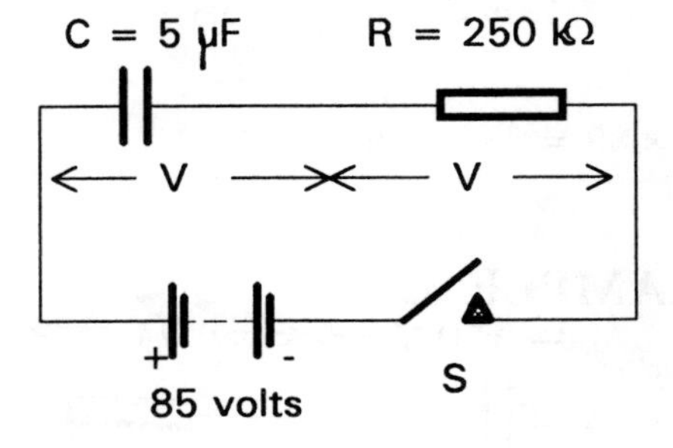

Fig. 185

(a) $\tau = RC = 250 \times 10^3 \times 5 \times 10^{-6} = 1250 \times 10^{-3} = 1.25$ s

(b) $v_C = V\left(1 - e^{-\frac{t}{\tau}}\right)$

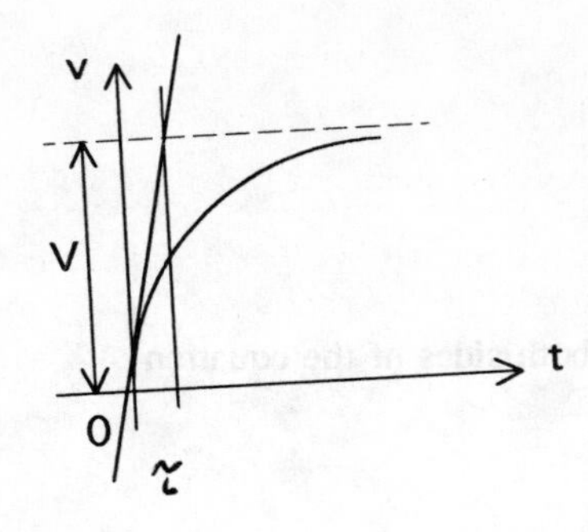

Fig. 186

$$\frac{\mathrm{d}v_C}{\mathrm{d}t} = \frac{V}{\tau} e^{-\frac{t}{\tau}}$$

$$\text{when } t = 0, \left(\frac{\mathrm{d}v_C}{\mathrm{d}t}\right)_{t=0} = \frac{V}{\tau} = \frac{85}{1.25} = 68 \text{ V/s}$$

(c) $\frac{\mathrm{d}v_C}{\mathrm{d}t} = \frac{V}{\tau} e^{-\frac{t}{\tau}} = 68 e^{-\frac{1}{1.25}} = 30.6 \text{ V/s}$

(d) When $t = \tau = 1.25$ s

$$v_C = V\left(1 - e^{-\frac{t}{\tau}}\right) = 85\left(1 - e^{-1}\right)$$

$$v_C = 85\,(1 - 0.3678794) = 0.632 \times 85$$

$$= \frac{63.2}{100} \times 85 = 53.7 \text{ volts.}$$

WORKED EXAMPLE 32

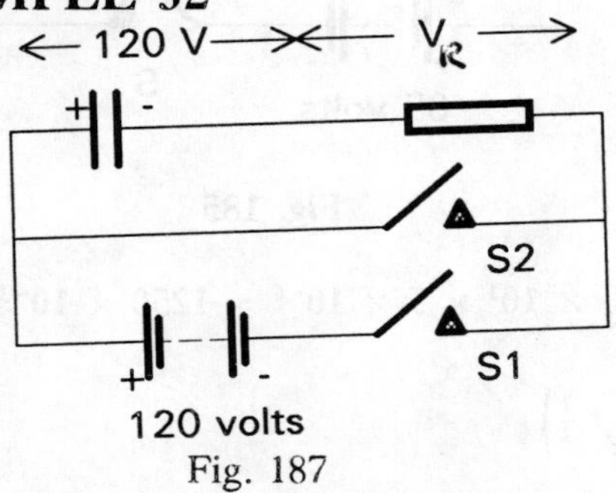

Fig. 187

A capacitor of capacitance 2.5 μF is fully charged through a series resistor of resistance 400 KΩ when connected across a d.c. supply of 120 V.

Draw a circuit for this situation and label the p.d.s across C and R.

The switch S_1 is now opened and switch S_2 is closed at the same time, assuming no energy is lost in opening and closing the switches.

Write down the p.d.s across C and R after t seconds, and show that the sum of these voltages is zero.

SOLUTION 32

$$v_C = Ve^{-\frac{t}{\tau}}$$

$$v_R = -Ve^{-\frac{t}{\tau}}$$

Find the time taken for a p.d. of 1 V to appear across C

$$v_C = 1 = 120e^{-\frac{t}{RC}}$$

$$\frac{1}{120} = e^{-\frac{t}{400 \times 10^3 \times 2.5 \times 10^{-6}}}$$

$$e^{-t} = \frac{1}{120}$$

$$e^t = 120$$

$$t = \ln 120 = 4.79 \text{ s.}$$

WORKED EXAMPLE 33

An uncharged capacitor is connected in series with a 1 MΩ resistor to a 100 V d.c. supply. If the initial rate of rise of voltage across the capacitor is 35 V/s, determine:- (i) the time constant;
(ii) the capacitance of the capacitor.

SOLUTION 33

(i) $$v_C = V\left(1 - e^{-\frac{t}{RC}}\right) = V\left(1 - e^{-\frac{t}{\tau}}\right) = 100\left(1 - e^{-\frac{t}{\tau}}\right)$$

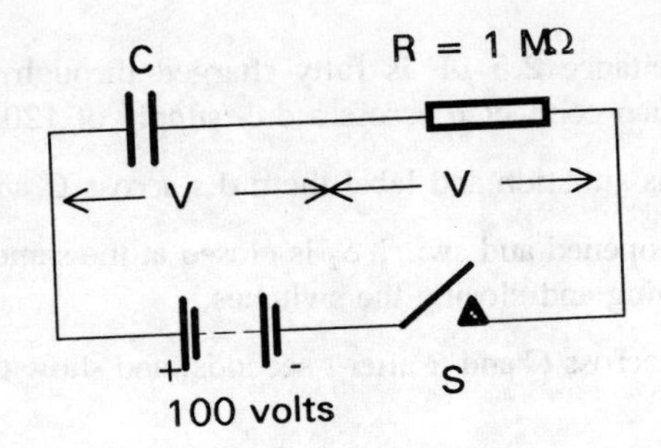

Fig. 188

$$\frac{dv_C}{dt} = \frac{100}{\tau} e^{-\frac{t}{\tau}}$$

If $t = 0$, (the initial rate of rise of voltage across the capacitor).

(i) $$\left(\frac{dv_C}{dt}\right)_{t=0} = \frac{100}{\tau} = 35 \text{ V/s}$$

$$\tau = \frac{100}{35} = \frac{20}{7} = 2.86 \text{ s}$$

(ii) $\tau = CR = 2.86$

$$C = \frac{2.86}{1 \times 10^6} = 2.86 \ \mu\text{F}.$$

WORKED EXAMPLE 34

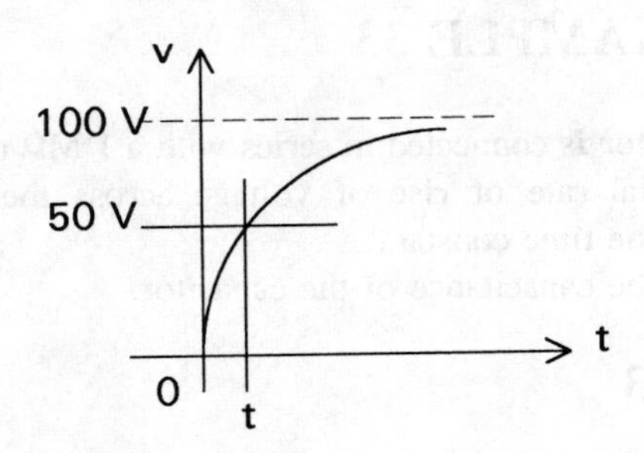

Fig. 189

The growth curve of an RC circuit with a d.c. battery of 100 V is shown.

If $R = 10\ \text{K}\Omega$, $C = 100\ \mu\text{F}$, $v = V\left(1 - e^{-\frac{t}{\tau}}\right)$.

(i) Calculate t.

(ii) If $t = 0.5$ ms, calculate v_C.

SOLUTION 34

(i) $v = V\left(1 - e^{-\frac{t}{\tau}}\right)$

$50 = 100\left(1 - e^{-\frac{t}{CR}}\right)$

$\tau = CR$

$= 10 \times 10^3 \times 100 \times 10^{-6}$

$= 1$

$50 = 100\left(1 - e^{-\frac{t}{T}}\right)$

Divide both sides by 100

$$\frac{50}{100} = \frac{100\left(1 - e^{-\frac{t}{T}}\right)}{100}$$

$0.5 = 1 - e^{-\frac{t}{T}}$

$0.5 - 1 = 1 - 1 - e^{-\frac{t}{T}}$

$-0.5 = -e^{-\frac{t}{T}}$ $0.5 = e^{-\frac{t}{T}}$ $e^t = 2$

$t = \ln 2$

$t = 0.693$ s

(ii) $t = 0.5$ s, $v = ?$

$v = V\left(1 - e^{-\frac{t}{\tau}}\right)$

$v = 100\left(1 - e^{-0.5}\right) = 39.34$ V.

WORKED EXAMPLE 35

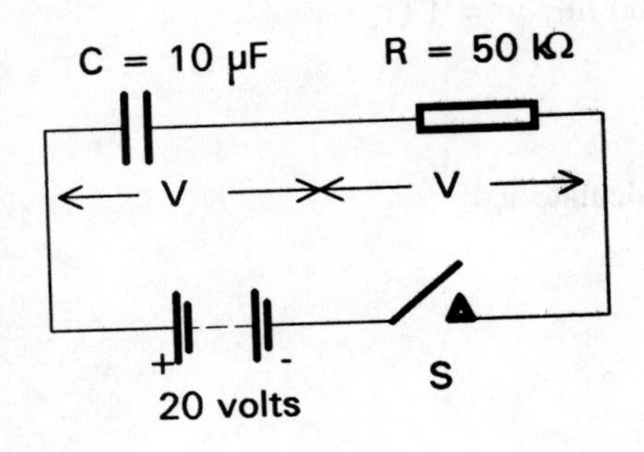

Fig. 190

Calculate:- (a) The steady current after closing the switch **S**,

(b) the time constant,

(c) the currents at t = 0, 0.1, 0.2, 0.3, 0.5, 0.7, 1.0.

(d) Draw the graph i against t and from the graph determine the current at t = 0.8 s

SOLUTION 35

(a) $I = \frac{V}{R} = \frac{20}{50 \times 10^3} = 0.4 \text{ mA}$

(b) $\tau = RC = 50 \text{ K}\Omega \times 10 \text{ }\mu\text{F} = 0.5 \text{ s}$

t (s)	0	0.1	0.2	0.3	0.5	0.7	1.0
i (mA)	0.4	0.33	0.268	0.2195	0.147	0.099	0.054

Specimen calculation

$$i = Ie^{-\frac{t}{\tau}}$$

$$i = 0.4 \times 10^{-3} e^{-\frac{t}{0.5}}$$

If t = 0.1 s

$$i = 0.4 \times 10^{-3} e^{\frac{-0.1}{0.5}} = 0.4 \times 10^{-3} e^{-0.2}$$

$$= 0.33 \text{ mA.}$$

(d) The graph is drawn as shown when $t = 0.8$ s, $i = 0.08$ mA.

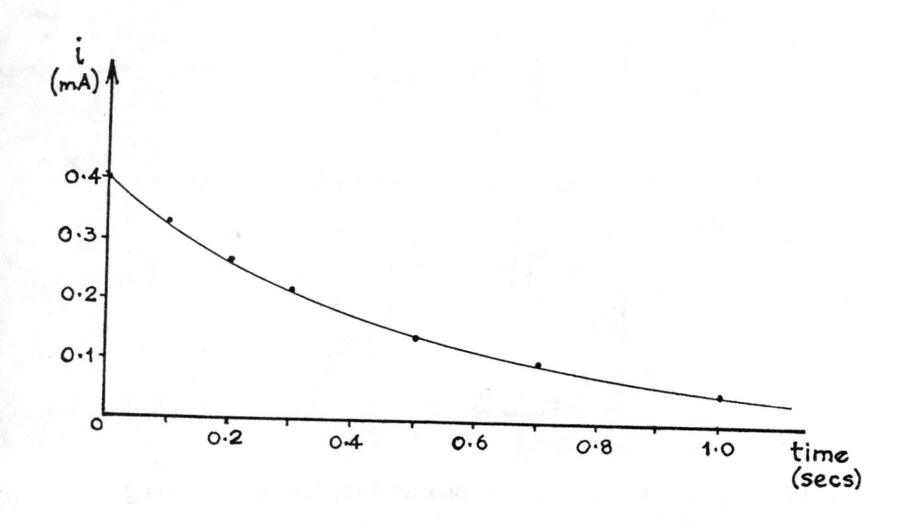

Fig. 191

DISCHARGING OF A CAPACITOR

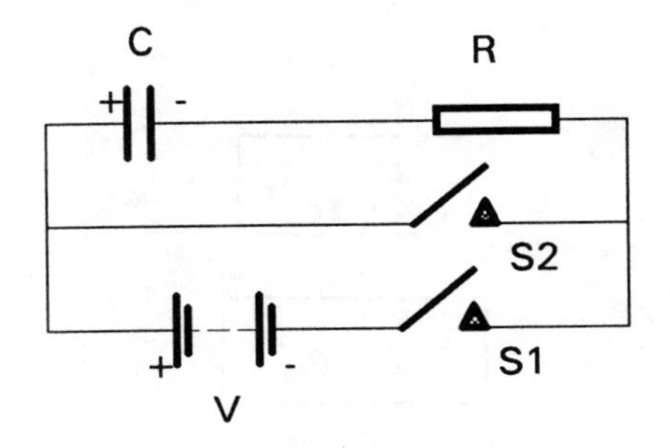

Fig. 192

In discharging

S_1 is opened and S_2 is closed. The current is flowing in the opposite direction

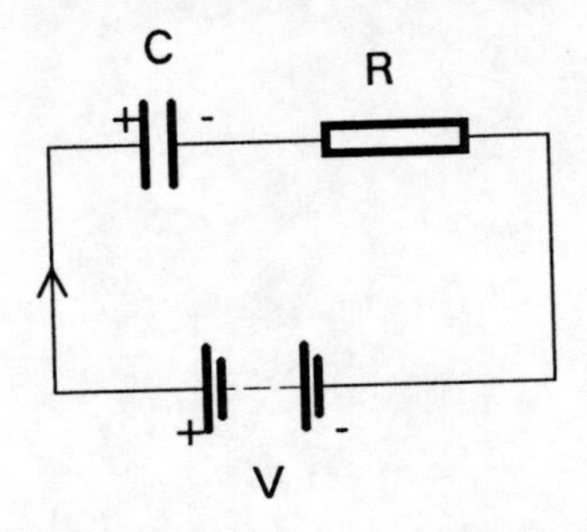

Fig. 193 The current flows clockwise when charging

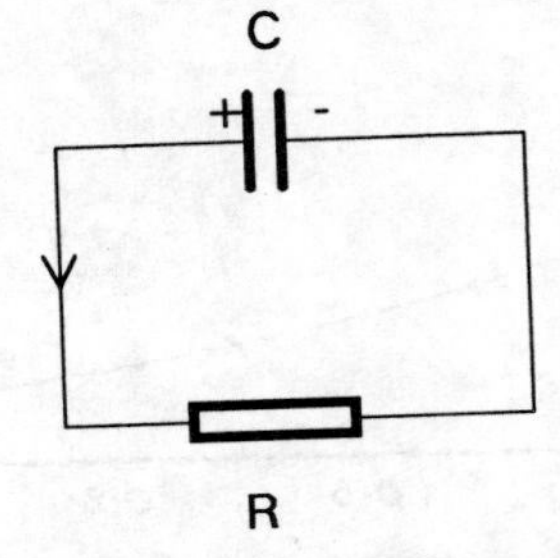

Fig. 194 The current flows in the opposite direction when discharging

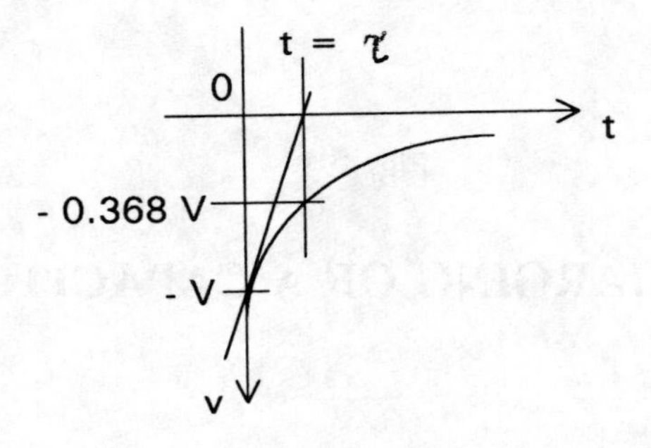

Fig. 195

$$i = -Ie^{-\frac{t}{\tau}}$$

$$v_R = -Ve^{-\frac{t}{\tau}}$$

$$v_c = Ve^{-\frac{t}{\tau}}$$

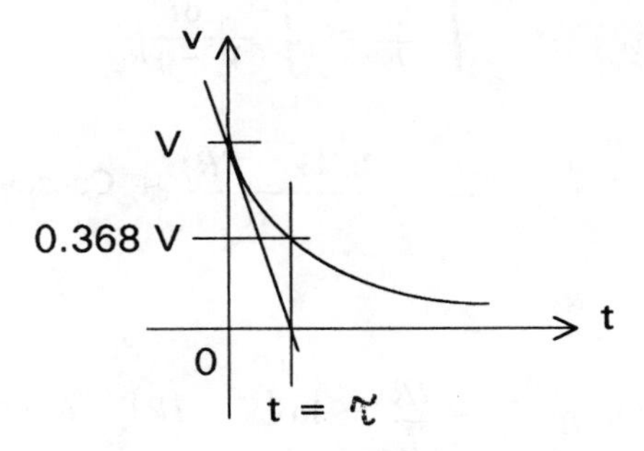

Fig. 196

PURE INDUCTOR IN SERIES WITH A RESISTOR

Magnetizing the coil.

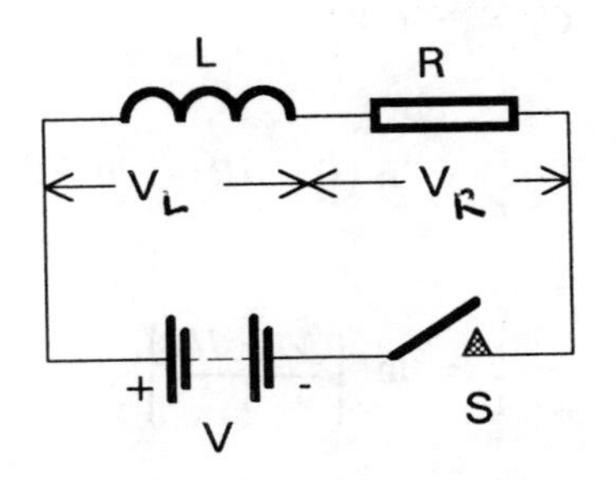

Fig. 197

After closing the switch, t time.

Applying Kirchhoff's voltage law, we have

$$V = v_L + v_R$$

$$V = L\frac{di}{dt} + iR$$

$$\text{At } t = 0,\ i = 0$$

$$V - iR = L\frac{di}{dt}$$

$$\int \frac{dt}{L} = \int \frac{di}{V - iR}$$

$$\frac{t}{L} = \frac{\ln(V - iR)}{-R} + \text{Const.}$$

multiply each term by $-R$

$$-\frac{tR}{L} = \ln(V - iR) - R\text{const.}$$

let $\tau = \dfrac{L}{R}$

$$-\frac{t}{\tau} = \ln(V - iR) + C_1$$

when $t = 0$, $i = 0$

$$C_1 = -\ln V$$

$$-\frac{t}{\tau} = \ln(V - iR) - \ln V$$

$$-\frac{t}{\tau} = \ln\left(\frac{V - iR}{V}\right)$$

$$e^{-\frac{t}{\tau}} = \frac{V - iR}{V}$$

$$V - iR = Ve^{-\frac{t}{\tau}}$$

$$V\left(1 - e^{-\frac{t}{\tau}}\right) = iR$$

$$i = \frac{V}{R}\left(1 - e^{-\frac{t}{\tau}}\right)$$

$$\boxed{i = I\left(1 - e^{-\frac{t}{\tau}}\right)}$$

the growth of current

$$\frac{di}{dt} = I\left(-e^{-\frac{t}{\tau}}\right)\left(-\frac{1}{\tau}\right) = \frac{I}{\tau}e^{-\frac{t}{\tau}}$$

$$= \frac{IR}{L}e^{-\frac{t}{\tau}}$$

$$L\frac{di}{dt} = IRe^{-\frac{t}{\tau}} = v_L$$

$$\boxed{v_L = Ve^{-\frac{t}{\tau}}}$$ the decay of voltage

$$v_R = iR = IR\left(1 - e^{-\frac{t}{\tau}}\right)$$

$$\boxed{v_R = V\left(1 - e^{-\frac{t}{\tau}}\right)}$$ the growth of voltage

$$V = v_L + v_R = Ve^{-\frac{t}{\tau}} + V\left(1 - e^{-\frac{t}{\tau}}\right)$$

Kirchhoff's voltage law is verified.

EXERCISES 3

1. A 25 μF capacitor is connected in series with a 4 KΩ resistor to a 120 volt d.c. supply.

   Construct curves of the supply current and the voltage across the capacitor, from the time when the uncharged capacitor is connected to the supply until the capacitor is fully charge.

   From the curves, determine (i) the time taken for the supply current to reach 7.5 mA, and

   (ii) the voltage across the capacitor at that time.

2. A capacitor of capacitance 2.5 μF is connected in series with resistor of resistance 400 KΩ to a d.c. supply of 120 V. Calculate the time constant of the circuit.

   Construct graphically the curve of voltage across the capacitor from zero time, when the capacitor is fully discharged, using a time scale of 0 to 2.3 s.

   From your graph read: (i) the voltage across the capacitor at a time of 1.3 s;

   (ii) the time needed for the voltage to reach 80 V.

3. (a) A fully discharged capacitor is connected in series with a 1 megaohm resistor to a 100 V d.c. supply. If the initial rate of rise of voltage across the capacitor is 27.5 V/s, determine:-

   (i) the time constant,

   (ii) the capacitance of the capacitor

   (b) A 100 V d.c. supply is switched on to a field circuit.

   (i) If the initial rate of rise of current is 1.76 A/s, determine the inductance of the circuit.

   If the time constant of the circuit is 3.125 s, determine:-

   (ii) the resistance of the circuit,

   (iii) the final current.

4. A coil of inductance 4 H and resistance 100 Ω is suddenly switched across a 50 V, d.c. supply.

   Calculate:- (i) the time constant of the circuit

   (ii) the current after 10 milliseconds.

5. (a) With reference to Fig. 198 and Fig. 199, calculate

   (i) the value of current at the instant the switch is closed

   (ii) the time constant of each circuit, and

   (iii) the final value of current in each case

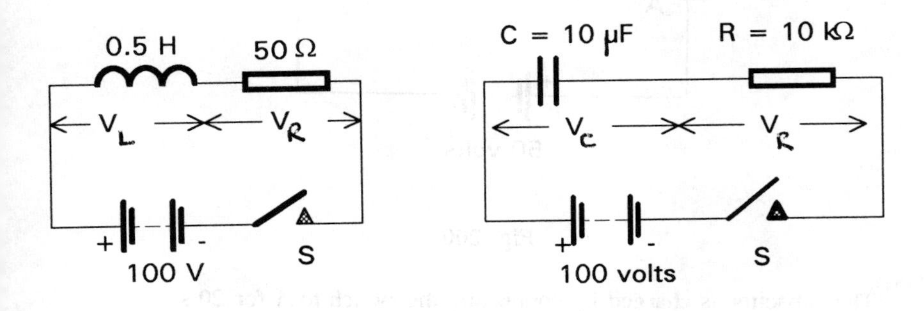

Fig. 198

Fig. 199

(b) Sketch a curve showing how the voltage across the inductance varies with time.

(c) Sketch a curve showing how the voltage across the capacitor varies with time.

(Ans. (i) 0 A, 10 mA (ii) 10 ms, 100 ms (iii) 2 A, 0 A)

6. A coil of resistance 50 ohms and inductance 0.5 H is connected to a 250 V d.c. supply.

Calculate:-
(a) the time constant of the circuit,
(b) the final value of the current
(c) the initial rate of rise of current
(d) the energy stored in the magnetic field when the current has reached its final value
(e) the value of the current after a period equal to one time constant.

(Ans. (a) 10 ms (b) 5 A (c) 500 A/s (d) 6.25 J (e) 3.160 A)

7.

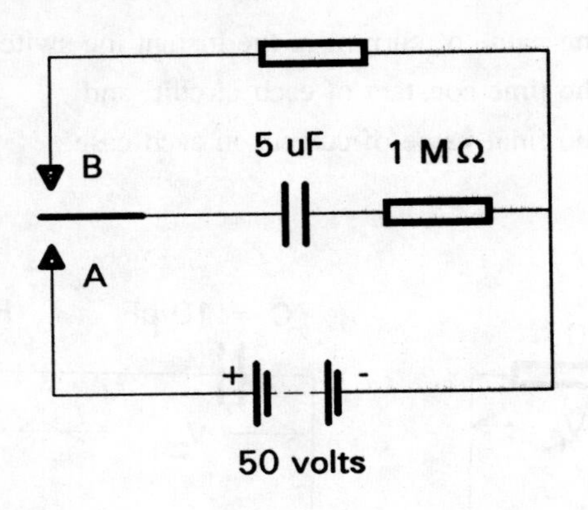

Fig. 200

The capacitor is charged by connecting the switch to *A* for 20 s.

The *RC* circuit is then disconnected from *A* and connected to *B*. Sketch waveforms showing how the capacitor voltage and current vary with time.

4. THREE-PHASE SYSTEMS

Three sets of supply voltages having a fixed phase angle difference of 120° between Pairs of voltages.

The total cost of the conductors in the cables is less than the equivalent single phase. The National Grid distribution system uses a three-phase system for economic reason.

STAR CONNECTION

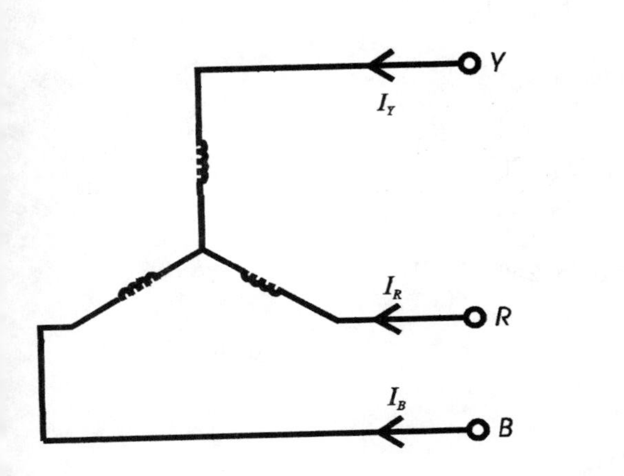

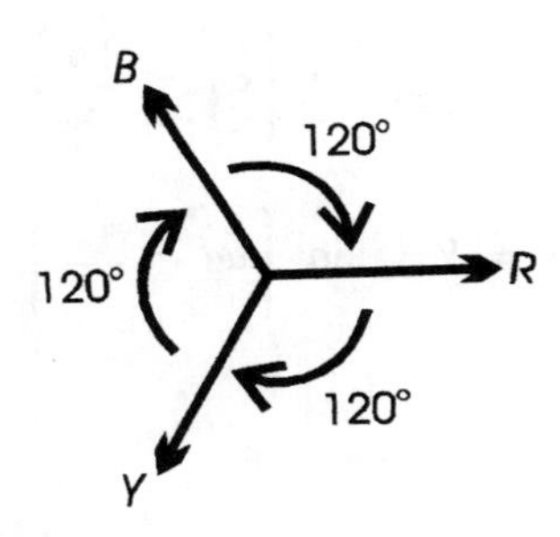

Fig. 201

The voltage between a pair of lines is called the line voltage.

The voltage across each coil is called the Phase Voltage or line to neutral voltage.

What is the relationship between phase and line voltage?

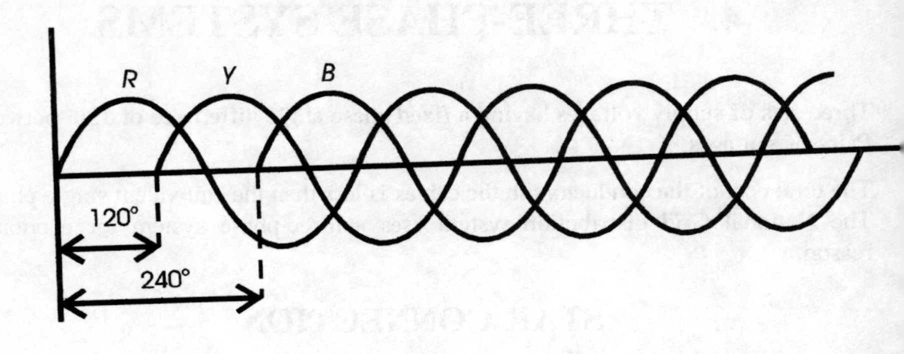

Fig. 202

$R = V_{max} \sin \omega t$ reference

$Y = V_{max} \sin \left(\omega t - \frac{2\pi}{3} \right)$ lags by 120°

$B = V_{max} \sin \left(\omega t - \frac{4\pi}{3} \right)$ lags by 240°

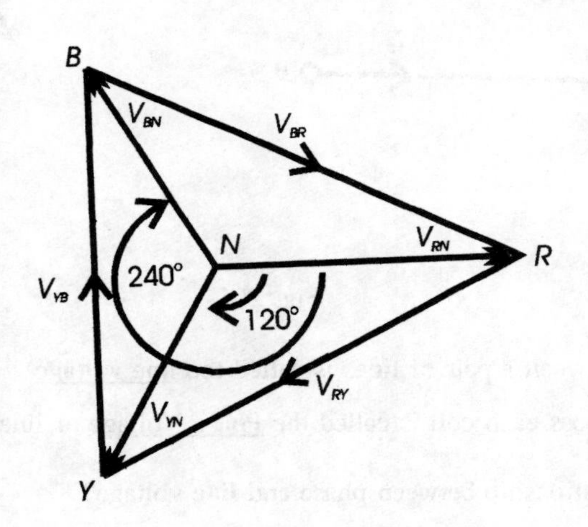

Fig. 203

V_{RY}, V_{YB}, V_{BR} are line voltages.

V_{RN}, V_{YN}, V_{BN} are phase voltages.

$$V_{RN} + V_{RY} - V_{YN} = 0$$

$$V_{RY} = V_{YN} - V_{RN}$$

$$V_{YB} = V_{BN} - V_{YN}$$

$$V_{BR} = V_{RN} - V_{BN}$$

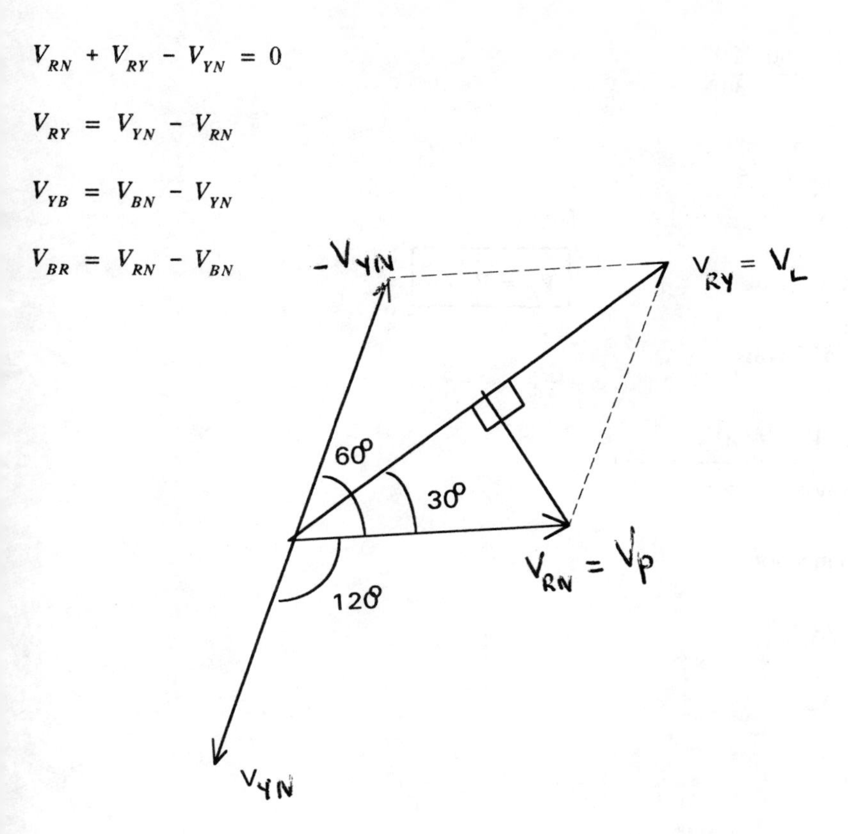

Fig. 204 Relationship between phase and line voltages in a star connected system.

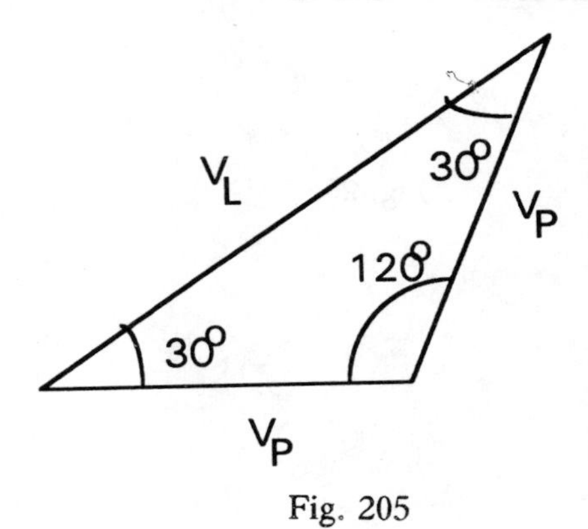

Fig. 205

$$\frac{V_L}{\sin 120°} = \frac{V_P}{\sin 30°}$$

$$V_L = V_P \frac{\sin 120°}{\sin 30°}$$

$$= V_P \frac{\sqrt{3}/2}{1/2}$$

$$\boxed{V_L = V_P\sqrt{3}}$$

If V_L = 415 volts

$$V_P = \frac{V_L}{\sqrt{3}} = \frac{415}{\sqrt{3}} = 240 \text{ V.}$$

For Δ connection

$$I_L = \sqrt{3}\, I_P$$

$$V_L = V_P$$

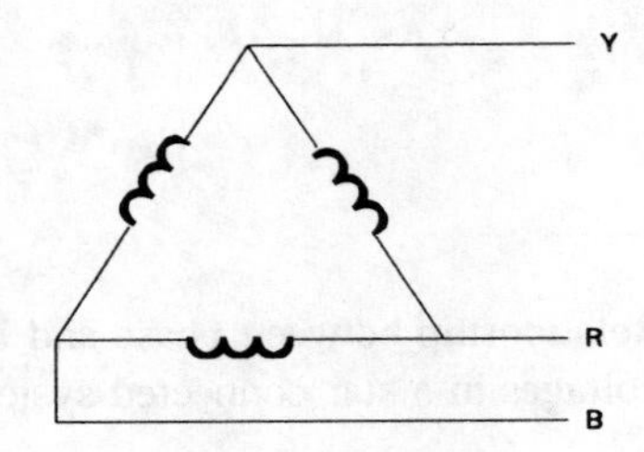

Fig. 206

For either Y or Δ

$$P = 3V_P I_P \cos\phi$$

$$P = \sqrt{3}\, V_L I_L \cos\phi$$

WORKED EXAMPLE 36

Three coils, each of 20 mH inductance and 10 ohms resistance, are connected (a) in star, and (b) in delta, to a 415 V, 50 Hz, 3-phase supply.

Calculate, for both cases, the line current and the total power dissipated by the full load.

SOLUTION 36

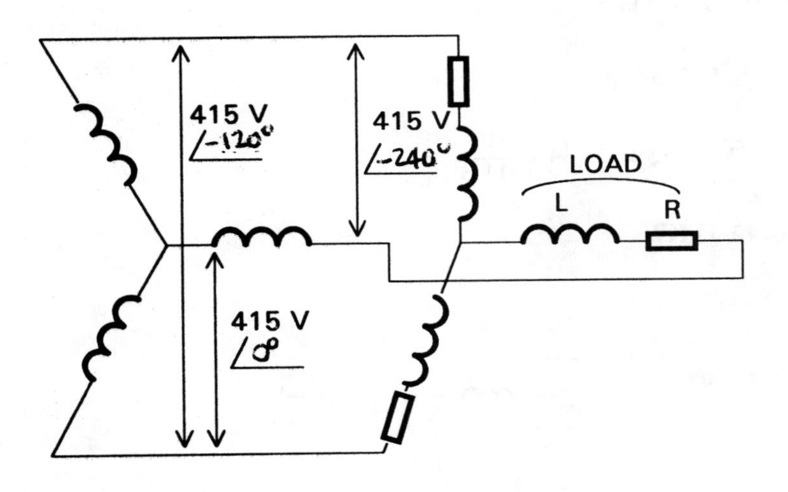

Fig. 207

$$Z = \sqrt{R^2 + X_L^2} = \sqrt{(10)^2 + (6.28)^2} = 11.8\ \Omega$$

$$X_L = 2\pi 50 \times 20 \times 10^{-3} = 6.28\ \Omega$$

$$V_P = \frac{V_L}{\sqrt{3}} = \frac{415}{\sqrt{3}} = 240 \text{ volts}$$

$$I_P = I_L = \frac{V_P}{Z} = \frac{240}{11.8} = 20.34 \text{ A}$$

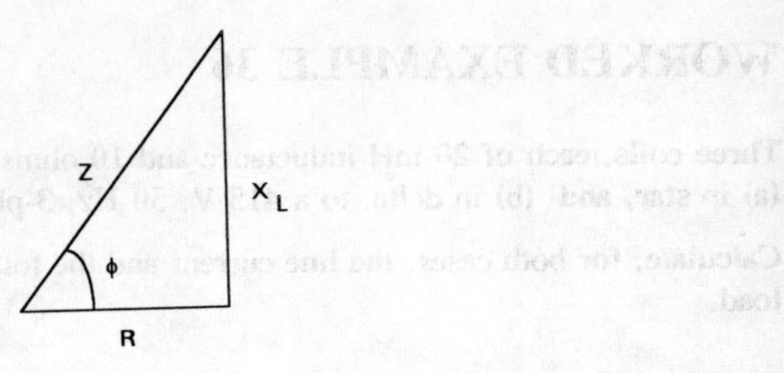

Fig. 208

$$\cos\,\phi \;=\; \frac{R}{Z} \;=\; \frac{10}{11.8}$$

$$P \;=\; \sqrt{3}\,V_L I_L \cos\,\phi$$

$$=\; \sqrt{3} \times 415 \times 20.34 \times \frac{10}{11.8}$$

$$P \;=\; 12.39 \text{ KW}$$

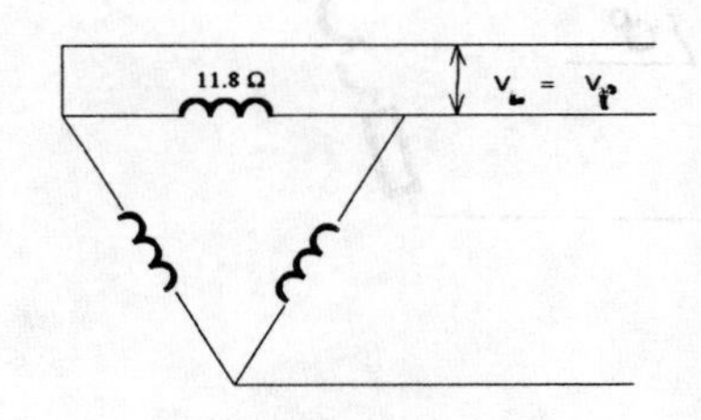

Fig. 209

$$I_P \;=\; \frac{415}{11.8} \;=\; 35.17 \text{ A}$$

$$I_L \;=\; \sqrt{3} \times 35.17 \;=\; 60.9 \text{ A}$$

$$P \;=\; 3V_P I_P \cos\,\phi$$

$$P \;=\; 37.11 \text{ KW}$$

WORKED EXAMPLE 37

A 415 V, 3-phase supply delivers a current of 12.6 A at a power factor of 0.86 lagging to a delta-connected motor.

Calculate:- (i) the total input power to the motor,

(ii) the impedance, resistance and reactance per phase of the motor winding.

SOLUTION 37

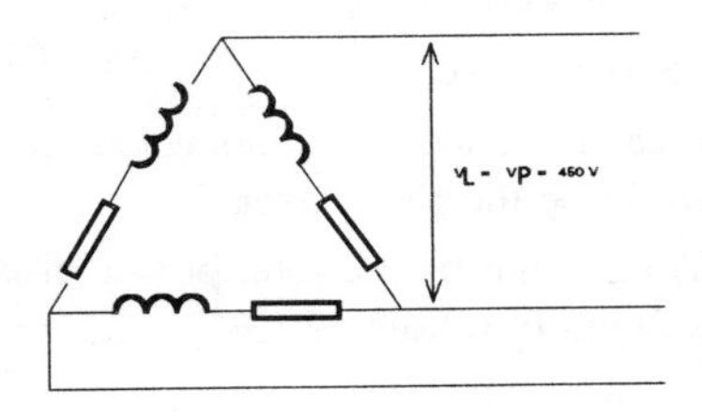

Fig. 210

(i) $P = \sqrt{3}\, V_L I_L \cos\phi$

$$= \sqrt{3} \times 415 \times 12.6 \times 0.86 = 7789 \text{ W}$$

(ii) $P = 3 I_P^2 R_P \qquad R_P = \dfrac{P}{3 I_P^2} = \dfrac{7789}{3 \times (7.27)^2} = 49.1\ \Omega$ where

$$I_P = \frac{I_L}{\sqrt{3}} = \frac{12.6}{\sqrt{3}} = 7.27 \text{ A}$$

$$Z = \frac{V_P}{I_P} = \frac{415}{7.27} = 57.1\ \Omega$$

$$Z = \sqrt{R^2 + X_L^2}$$

$$X_L = \sqrt{Z^2 - R^2}$$

$$= \sqrt{(57.1)^2 - (49.1)^2}$$

$$= 29.2\ \Omega$$

WORKED EXAMPLE 38

Each branch of a delta-connected load consists of a 30 ohms resistor in series with an inductive reactance of 40 ohms. The line voltage is 450 volts.

Calculate:- (a) the line current, and

(b) the total power

(c) If each branch was connected in a star configuration, what would be the line current?

(d) What would be the total power dissipation for the condition as outlined in (c) above.

SOLUTION 38

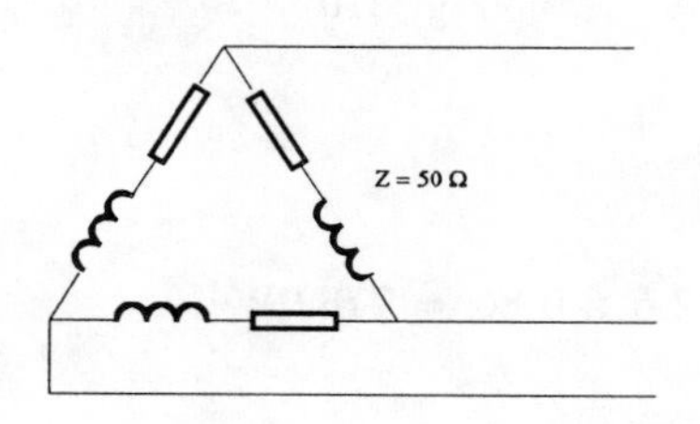

Fig. 211

(a) $I_P = \dfrac{V_P}{Z} = \dfrac{450}{50} = 9\ \text{A}$

$$Z = \sqrt{R^2 + X_L^2} = \sqrt{(30)^2 + (40)^2} = 50\ \Omega$$

$$I_L = \sqrt{3}\, I_P = 9\sqrt{3}\ \text{A.}$$

(b)

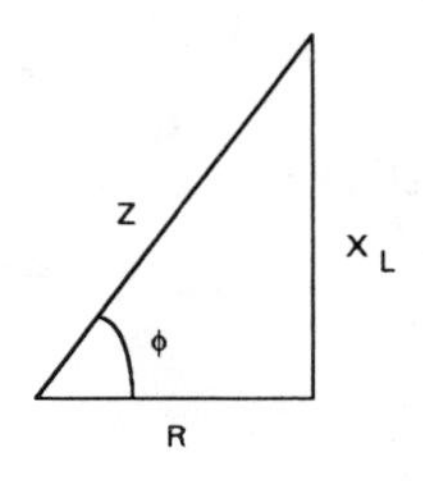

Fig. 212

$$\cos\phi = \frac{R}{Z} = \frac{30}{50}$$

$$P = 3V_P I_P \cos\phi$$

$$= 3 \times 450 \times 9 \times \frac{30}{50} = 7.29 \text{ KW}$$

(c) $V_L = 450$ V

$$V_P = \frac{450}{\sqrt{3}} = 260 \text{ V}$$

$$I_L = I_P = \frac{V_P}{Z} = \frac{260}{50} = 5.2 \text{ A}$$

(d) $P = \sqrt{3}\, V_L I_L \cos\phi$

$$P = \sqrt{3} \times 450 \times 5.2 \times \frac{30}{50}$$

$$P = 2.43 \text{ KW}$$

WORKED EXAMPLE 39

Three coils having resistance 12 Ω and inductive reactance 5 Ω are connected

(i) in star

(ii) in delta.

If V_L of a 3ϕ 440 V supply, calculate in both cases

(a) the line and phase currents and

(b) the line and phase voltages at the load.

SOLUTION 39

$$Z = \sqrt{R^2 + X_L^2} = \sqrt{12^2 + 5^2} = 13\ \Omega$$

(i) Star

$$V_L = 440\ \text{V}$$

$$V_P = \frac{V_L}{\sqrt{3}} = 254\ \text{V}$$

$$I_P = \frac{V_P}{Z} = \frac{254}{13} = 19.5\ \text{A}$$

$$I_L = I_P = 19.5\ \text{A}$$

(ii) Delta

$$V_L = V_P = 440\ \text{V}$$

$$I_P = \frac{V_P}{Z} = \frac{440}{13} = 33.9\ \text{A}$$

$$I_L = \sqrt{3}\, I_P = \sqrt{3}\ 33.9 = 58.6\ \text{A}$$

WORKED EXAMPLE 40

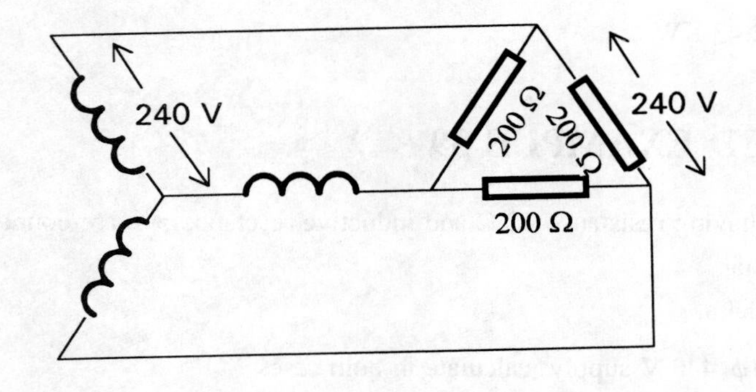

Fig. 213

A three-phase, star-connected alternator of 240 V phase voltage is connected to a balanced delta-connected load as shown in Fig. 213.

Calculate:-
(i) The line voltage,
(ii) the phase current
(iii) the line current
(iv) the power per phase
(v) the total power.

SOLUTION 40

(i) $V_L = \sqrt{3}\, V_P = \sqrt{3} \times 240 = 415.7\text{ V}$

(ii) $I_P = \dfrac{240\sqrt{3}}{200} = 2.08\text{ A}$

(iii) $\sqrt{3} \times 2.08 = 3.6\text{ A}$

(iv) $P = V_P I_P \cos\phi$

$= 415.7 \times 2.08 \times 1 = 865\text{ W}$

(v) $P = 3 V_P I_P \cos\phi$

$= 3 \times 415.7 \times 2.08 \times 1 = 2594\text{ W}$

EXERCISES 4

1. Three inductors, each of reactance 30 Ω and resistance 40 Ω, are connected in a star-connection to a 240 V, three-phase supply. Calculate:-

   (i) The phase current

   (ii) the line current

   (iii) the phase voltage

   (iv) the line voltage

   (v) the power taken from the supply.

2. Three identical coils are connected to a 440 V, 50 Hz, three-phase supply and the power dissipated is 5 KW at a power factor of 0.75 lagging. Calculate the resistance and inductance of each coil if they are connected:-

   (a) in star,

   (b) in delta.

3. Three inductors each of resistance 5 Ω and reactance 12 Ω are connected in delta when connected to a three-phase supply they consume 1.2 KW.

   Calculate:- (a) the line and phase currents,

   (b) the power factor,

   (c) the supply voltage.

4. A balanced three-phase load takes a line current of 50 A at a line voltage of 440 V. If the power consumed is 30 KW, calculate the power factor of the load.

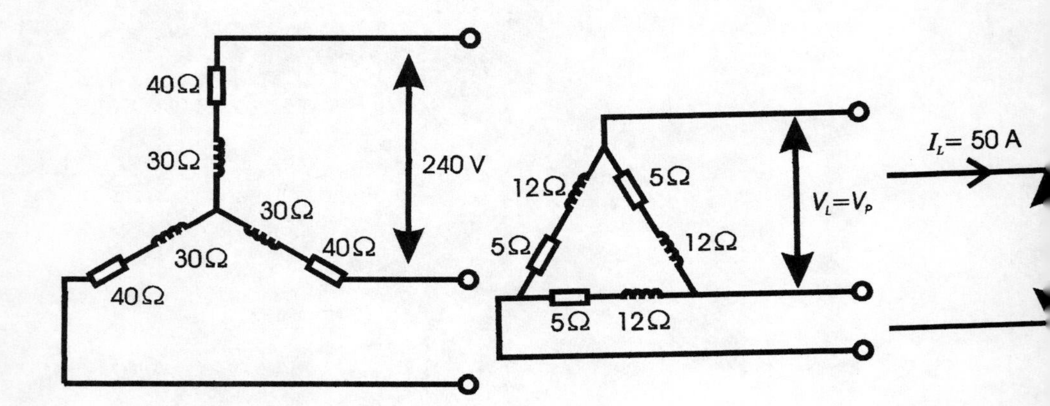

Fig. 214 Fig. 215 Fig. 216

5. DC MACHINES

MAIN POINTS CONCERNED WITH DC MACHINES

When the input to an electrical machine is electrical energy, (seen as applying a voltage to the electrical terminals of the machine), and the output is mechanical, (seen as a rotating shaft), the machine is called an electric motor.

Thus an electric MOTOR CONVERTS ELECTRICAL ENERGY INTO MECHANICAL ENERGY.

When the input to an electrical machine is mechanical energy, (seen as, say a diesel engine, or other PRIME MOVER, coupled to and the output is electrical energy, (seen as a voltage appearing at the electrical terminals of the machine), the machine is called generator.

Thus a GENERATOR CONVERTS MECHANICAL ENERGY TO ELECTRICAL ENERGY.

DC machines use the facts that a conductor carrying current in a magnetic field will experience a force ($F = BlI$) and a conductor moving in a magnetic field will have an e.m.f. induced in it ($e = Blv$).

THE ACTION OF A COMMUTATOR

In an electric motor, conductors rotate in a uniform magnetic field. A single-loop conductor, mounted between permanent magnets is shown in Fig. 217. A voltage is applied at points A and B in Fig. 217.

A force, F, acts on the loop due to the interaction of the magnetic fields of the permanent magnets and the magnetic field created by the current flowing in the loop.

This force is proportional to the flux density, B, the effective length of the conductor, l, and the current flowing, I. i.e. $F = BlI$ newtons.

The force is made up of two parts, one acting vertically downwards due to the current flowing from C to D and the other acting vertically upwards due to the current flowing from E to F (from Fleming's left-hand rule).

If the loop is free to rotate then when it has rotated through 180°, the conductors are as shown in Fig. 218, i.e. from D to C and from F to E.

This apparent reversal in the direction of current flow is achieved by a process called COMMUTATION.

With reference to Fig. 217, when a direct voltage is applied at A and B, then as the single-loop conductor rotates, current flow will always be away from the commutator for the part of the conductor adjacent to the *N*-pole and towards the commutator for the part of the conductor adjacent to the *S*-pole.

Thus the forces act to give continuous rotation in an anti-clockwise direction.

The arrangement shown in Fig. 217 is called a 'two-segment' commutator and the voltage is applied to the rotating segments by stationary BRUSHES (usually carbon blocks), which slide on the commutator material (usually copper), when rotation takes place.

In practice, there are many conductors on the rotating part of a d.c. machine and these are attached to many commutator segments. A schematic diagram of a multi-segment commutator is shown in Fig. 219 (b)

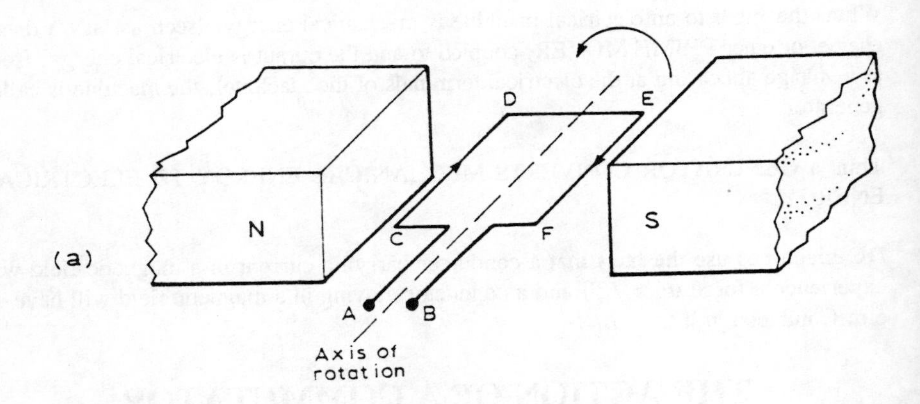

Fig. 217

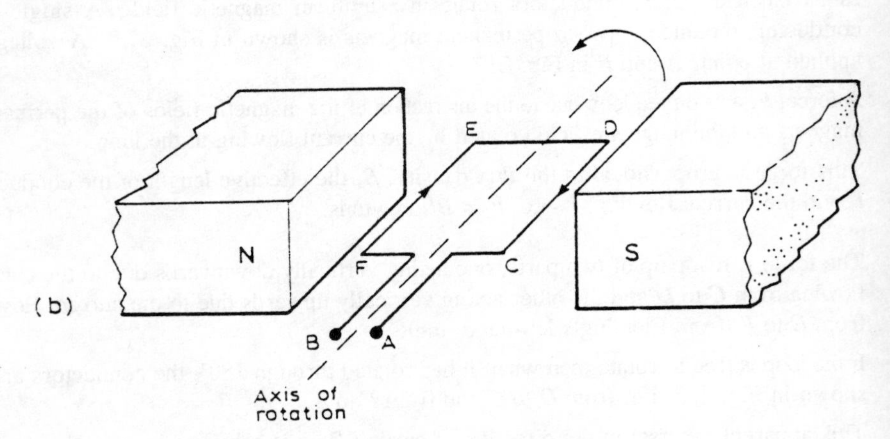

Fig. 218

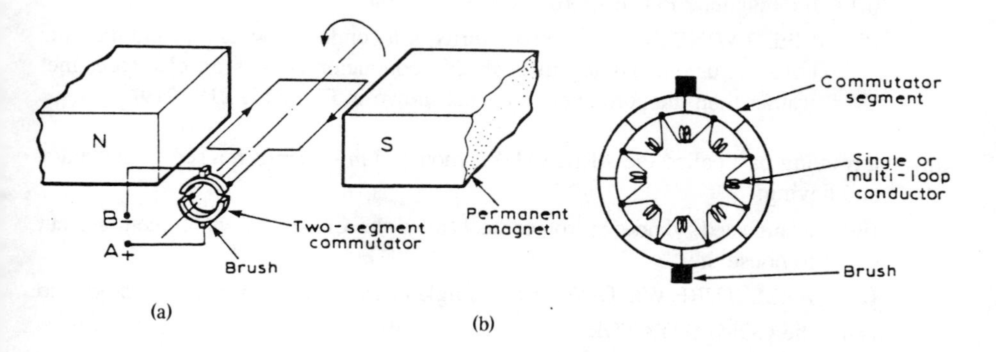

Fig. 219

DC MACHINE CONSTRUCTION

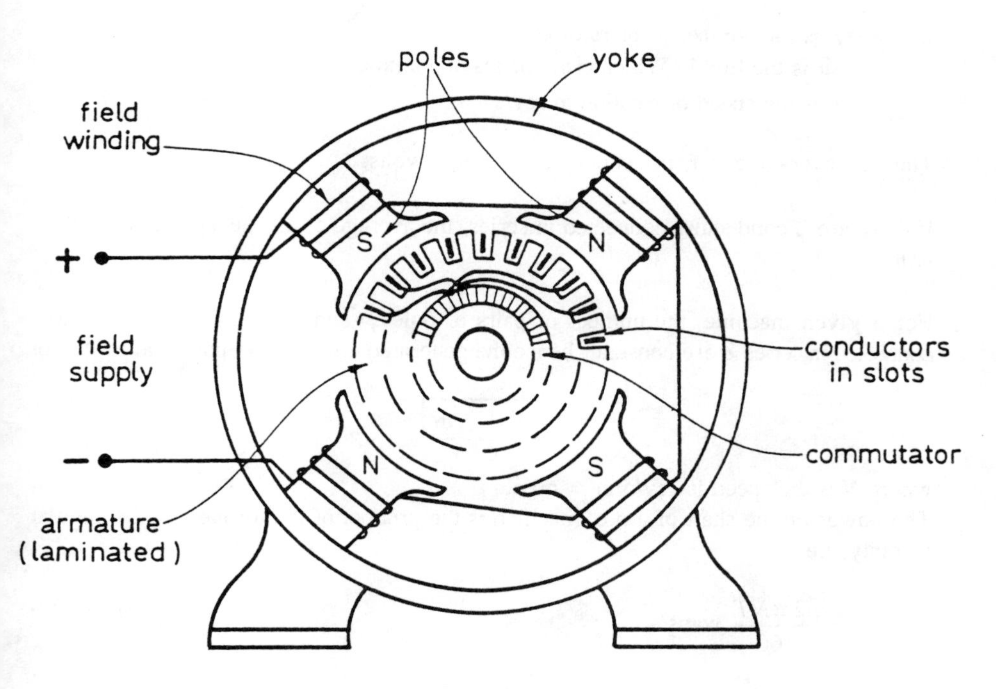

Fig. 220

The basic parts of any d.c. machine are shown in Fig. 220, above, and comprise:-

(a) a stationary part called the STATOR having:-
(i) a steel ring called the YOKE, to which are attached

(ii) the magnetic POLES, around which are the

(iii) FIELD WINDINGS, i.e. many turns of a conductor wound round the pole core. Current passing through this conductor creates an electromagnet (rather than the permanent magnets shown in Fig. 217, 218, 219).

(b) a rotating part called the ARMATURE mounted in bearings housed in the stator and having:-

(iv) a laminated cylinder of iron or steel called the CORE, in which teeth are cut to house the

(v) ARMATURE WINDING, i.e. a single or multi-loop conductor system and

(vi) the COMMUTATOR.

EMF AND TORQUE

The average e.m.f. induced in a single conductor on the armature of a d.c. machine

is given by $\dfrac{\text{flux cut/rev}}{\text{time of 1 rev}} = \dfrac{2p\phi}{1/n}$

where p is the number of pairs of poles
ϕ is the flux in Wb entering or leaving a pole and
n is the speed of rotation in rev/s

Thus the average e.m.f. per conductor is $2p\phi n$ volts.

If there are Z conductors connected in series, the average e.m.f. generated is $2p\phi nZ$ volts.

For a given machine, the number of pairs of poles p and the number of conductors connected in series Z are constant, hence the generated e.m.f. is proportional to ϕn or

$$\boxed{E \propto N\phi}$$

where N is the speed in rev/min = 60 n.
The power on the shaft of a d.c. machine is the product of the torque and the angular velocity, i.e.

$$P = \frac{2\pi NT}{60} \text{ watts.}$$

The power developed by the armature is EI_a watts,

where E is the generated e.m.f. in volts and
I_a is the armature current in amperes.

If losses are neglected then

$$\frac{2\pi NT}{60} = EI_a.$$

But $E \propto N\phi$

Hence $\frac{2\pi NT}{60} \propto N\phi I_a$

$$\text{i.e. } \boxed{T \propto I_a\phi}$$

DC MOTORS

The coils on each pole of the DC machine are all connected in series with each other and referred to as the FIELD CIRCUIT.

THE ARMATURE CIRCUIT consists of the armature, commutator and brushes in series.

The field circuit may be connected in parallel (or SHUNT) with the armature, in SERIES; or there may be two field circuits, one in shunt and the other in series - this is called COMPOUND excitation.

Classified according to their method of excitation. They may be Shunt, Series or Compound types. Compound motors - a combination of shunt and series - are not dealt with in this unit.

DC MOTOR CHARACTERISTIC

Graphs may be plotted showing relationships between:-

(1) Speed and armature current, $\frac{N}{I_a}$

(2) Torque and armature current, $\frac{T}{I_a}$.

(3) Speed and torque, $\frac{N}{T}$.

SHUNT MOTOR

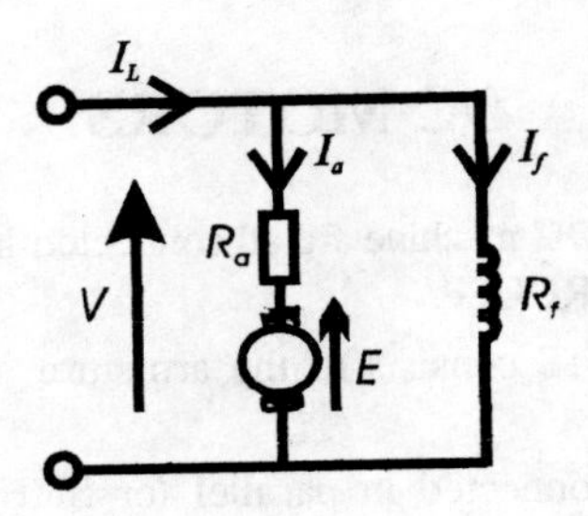

Fig. 221

$$E = V - I_a R_a$$

$$I_f = \frac{V}{R_f}$$

$$I_L = I_a + I_f$$

R_a = armature-circuit resistance

R_f = field-circuit resistance

For constant V; I_f and hence ϕ are constant. As armature current I_a increases E falls slightly. It will fall by 5 per cent between no-load and full load. Hence speed falls slightly.

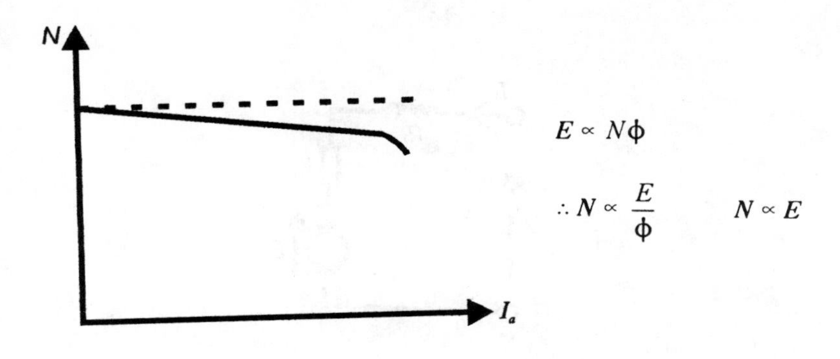

Fig. 222

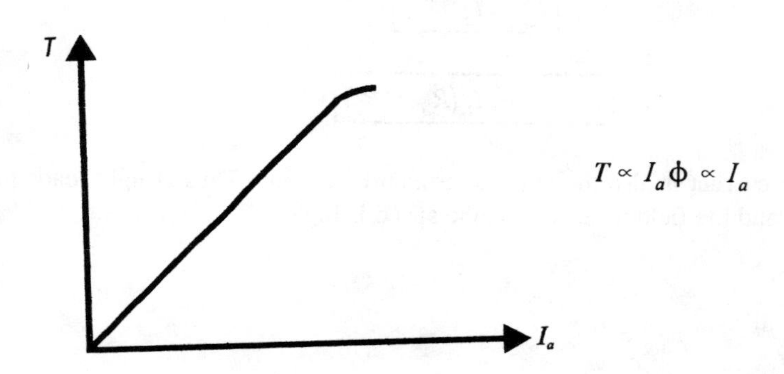

Fig. 223

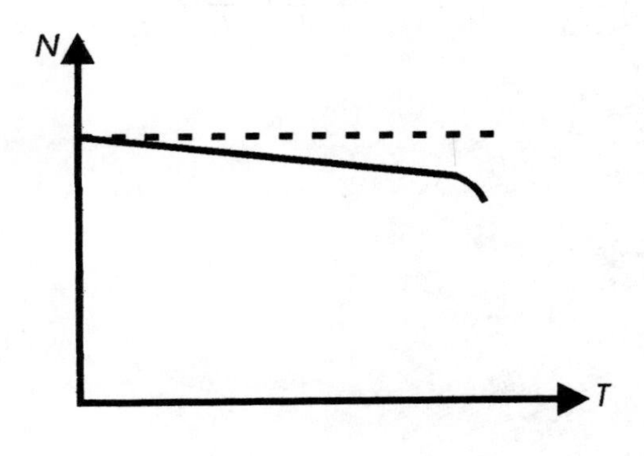

Fig. 224

Used in application where reasonably constant speed is needed.

SERIES MOTOR

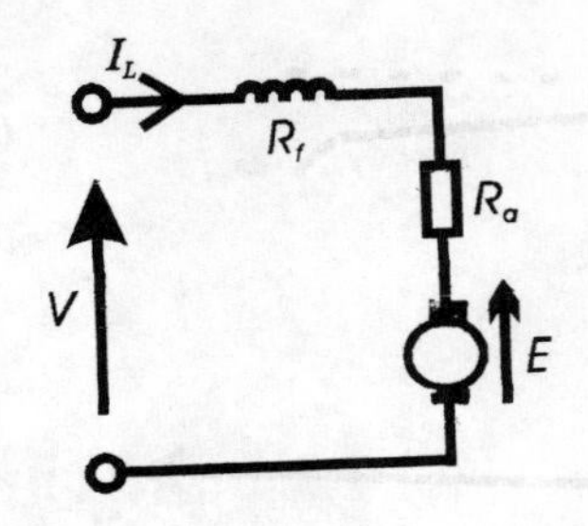

Fig. 225

$$\boxed{I_L = I_a = I_f}$$

$$\boxed{E = V - I_L(R_a + R_f)}$$

The field-circuit current is also the line and armature current. Thus at light loads the current is small and the field weak so that the speed is high.

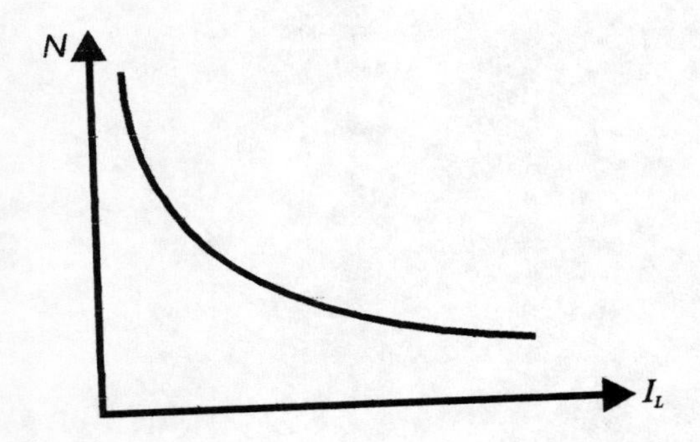

$$N \propto \frac{E}{\phi} \propto \frac{E}{I_L} \propto \frac{1}{I_L}$$

Fig. 226

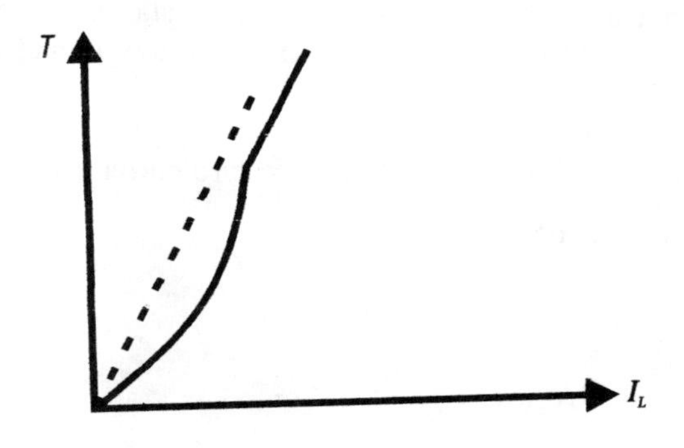

Fig. 227

$T \propto \phi I_a \propto I_L^2$ when I_L is large, magnetic circuit becomes saturated and $T \propto I_L$

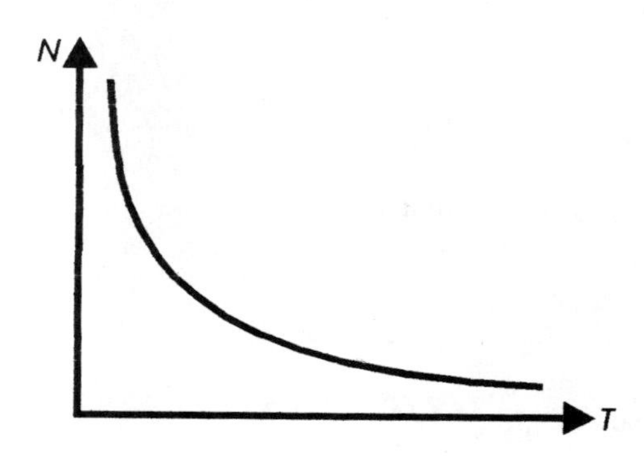

Fig. 228

The series motor gives high torque at low speeds i.e. rapid acceleration and is therefore advantageous in traction and crane applications.

On the other hand the speed is very high when the torque is low. In fact under no-load conditions (torque virtually zero) the speed could be dangerously high. For this reason the series, motor must be permanently, mechanically complete to the load, not by belt or clutch.

WORKED EXAMPLE 41

A shunt motor is connected to a 460 V mains and runs at a speed of 700 rev/min. The armature current is 60 A and the armature resistance 0.1 ohm. The flux per pole is decreased by 20 per cent and the armature torque and the supply voltage remain unchanged.

Calculate:- (a) the final value of the armature current; and

(b) the new speed.

SOLUTION 41

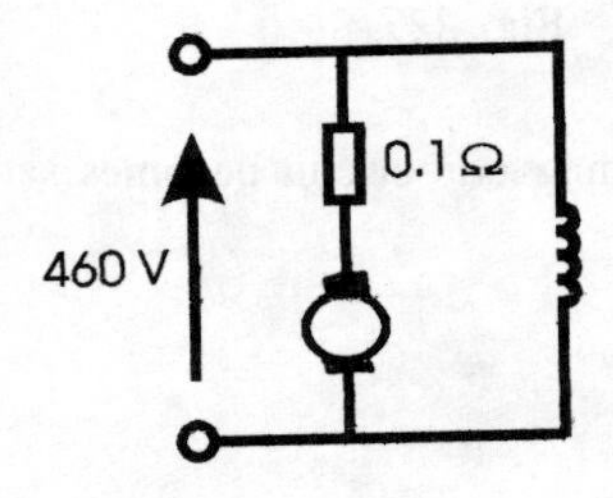

$N_1 = 700$ rev/min

$I_{A1} = 60$ A

$\phi_2 = 0.8\phi_1$

Fig. 229

(a) $E_1 = 460 - 60 \times 0.1 = 454$ V

$T_1 = T_2 \qquad \therefore\ I_{A1}\phi_1 = I_{A2}\phi_2$

$$I_{A2} = \frac{\phi_1}{\phi_2} . I_{A1} = \frac{\phi_1}{0.8\phi_1} \times 60 = 75 \text{ A}$$

(b) $E_2 = 460 - 75 \times 0.1 = 452.5$ V

$$N \propto \frac{E}{\phi} \qquad \therefore\ \frac{N_2}{N_1} = \frac{E_2}{E_1} . \frac{\phi_1}{\phi_2}$$

$$N_2 = \frac{452.5}{454} \times \frac{\phi_1}{0.8\phi_1} \times 700 = 872 \text{ rev/min}$$

WORKED EXAMPLE 42

A series motor has a combined armature and field resistance of 1 ohm. The supply voltage is 440 V. When the motor takes a current of 20 A the speed is 1260 rev/min and the torque is 325 Nm. The current is increased to 40 A and the pole flux for this current increases by 50 per cent.

Calculate:- (a) the new speed; and

(b) the new torque.

SOLUTION 42

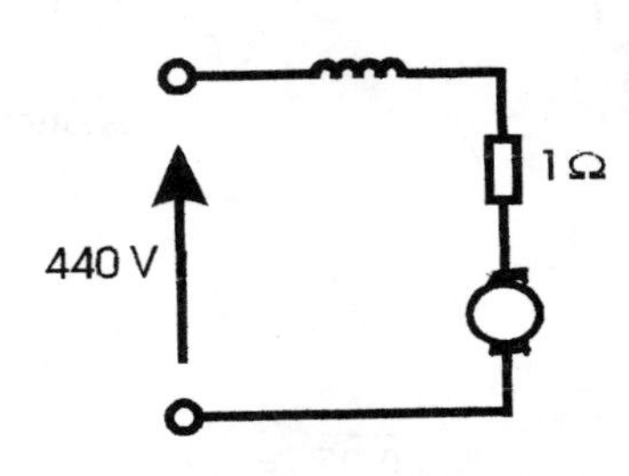

$I_1 = 20$ A

$N_1 = 1260$ rev/min

$T_1 = 325$ Nm

$I_2 = 40$ A

$\phi_2 = 1.5\phi_1$

Fig. 230

(a) $E_1 = 440 - 20 \times 1 = 420$ V

$E_2 = 440 - 40 \times 1 = 400$ V

$$N_2 = \frac{E_2}{E_1} \cdot \frac{\phi_1}{\phi_2} \cdot N_1 = \frac{400}{420} \times \frac{\phi}{1.5\phi_1} \times 1260 \text{ rev/min} = 800 \text{ rev/min.}$$

(b) $$\frac{T_2}{T_1} = \frac{I_2\phi_2}{I_1\phi_1}$$

$$T_2 = \frac{40}{20} \times \frac{1.5\phi_1}{\phi_1} \times 325 = 975 \text{ Nm.}$$

WORKED EXAMPLE 43

A 110 V series motor has a combined armature-circuit and field resistance of 0.15 ohm. Full-load torque is produced when the armature speed is 1200 rev/min and the input current is 24 A. When the motor is producing one-quarter full-load torque.

Calculate:- (a) the armature current; and
(b) the speed.

SOLUTION 43

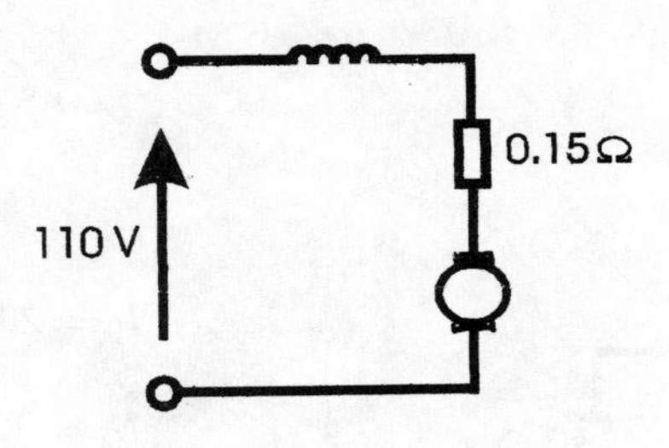

$N_1 = 1200$ rev/min

$I_1 = 24$ A

$T_2 = 0.25\,T_1$

Assume $\phi \propto I$

Fig. 231

(a) $T \propto I\phi \propto I^2$ $\qquad I_2 = I_1\sqrt{\dfrac{T_2}{T_1}} = 24\sqrt{0.25} = 12$ A

(b) $E_1 = 110 - 24 \times 0.15 = 106.4$ V

$E_2 = 110 - 12 \times 0.15 = 108.2$ V

$E \propto N\phi \propto NI$

$$N_2 = \frac{E_2}{E_1}\cdot\frac{I_1}{I_2}\cdot N_1 = \frac{108.2}{106.4} \times \frac{24}{12} \times 1200 \text{ rev/min} = 2441 \text{ rev/min.}$$

WORKED EXAMPLE 44

When on normal full load, a 500 V, d.c. shunt motor runs at 800 rev/min and takes an armature current of 42 A. The flux per pole is reduced to 75 per cent of its normal value by suitably increasing the field-circuit resistance.

Calculate the speed of the motor if the total torque exerted on the armature is:-
(a) unchanged; or
(b) reduced by 20 per cent.

The armature-circuit resistance is 0.6 ohm and the total voltage loss at the brushes is 2 V.

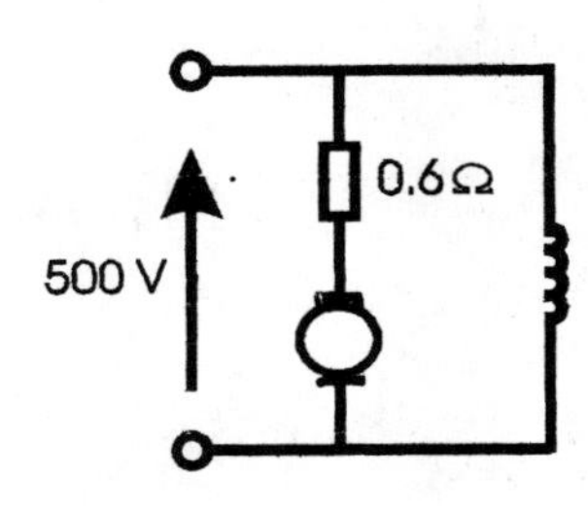

$N_1 = 800$ rev/min

$I_{A1} = 42$ A

$\phi_2 = 0.75\,\phi_1$

$V_B^2 = 2$ V

Fig. 232

(a) $E_1 = 500 - 42 \times 0.6 - 2 = 472.8$ V

$T_2 = T_1 \qquad \therefore I_{A2}\,\phi_2 = I_{A1}\,\phi_1$

$$I_{A2} = \frac{\phi_1}{0.75\,\phi_1} \times 42 = 56 \text{ A}$$

$E_2 = 500 - 56 \times 0.6 - 2 = 464.4$ V

$$N_2 = \frac{E_2}{E_1} \cdot \frac{\phi_1}{\phi_2} \cdot N_1 = \frac{464.4}{472.8} \times \frac{\phi_1}{0.75\,\phi_1} \times 800 \text{ rev/min} = 1048 \text{ rev/min}$$

(b) $T_2 = 0.8T_1 \qquad I_{A2}\,\phi_2 = 0.8 I_{A1}\,\phi_1$

$$I_{A2} = \frac{\phi_1}{0.75\,\phi_1} \times 42 \times 0.8 = 44.8 \text{ A}$$

$E_2 = 500 - 448 \times 0.6 - 2 = 471.12$ V

$$N_2 = \frac{471.12}{472.8} \times \frac{1}{0.75} \times 800 = 1063 \text{ rev/min}$$

DC GENERATORS

As with motors, classified according to field excitation.

SHUNT GENERATOR

Field coils with many turns of fine wire.

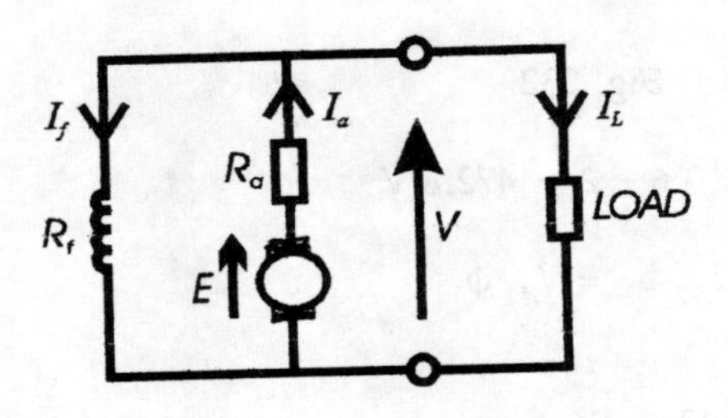

Fig. 233

$$I_f = \frac{V}{R_f}$$

$$I_a = I_f + I_L$$

$$E = V + I_a R_a$$

WORKED EXAMPLE 45

A shunt generator supplies a load of 15 KW at 200 V through feeders of resistance 0.08 ohm. Resistance of shunt-field winding 80 ohms; resistance of armature-circuit 0.04 ohm.

Calculate:- (a) the terminal voltage; and
(b) the e.m.f. generated.

SOLUTION 45

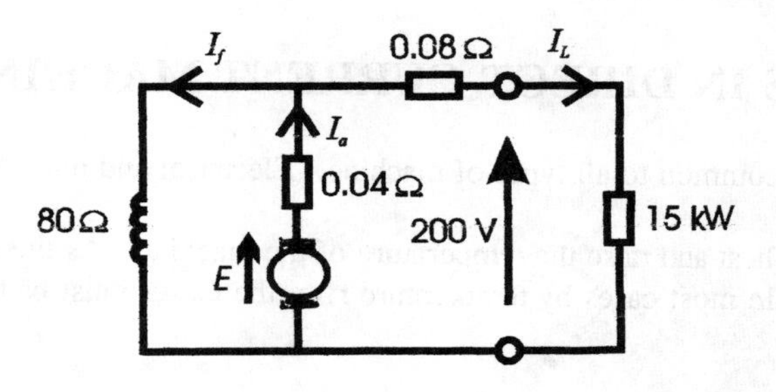

Fig. 234

$$I_L = \frac{15 \times 10^3}{200} = 75 \text{ A}$$

$$V = 200 + 75 \times 0.08 = 206 \text{ V}$$

$$I_F = \frac{206}{80} = 2.575 \text{ A}$$

$$I_A = 75 + 2.575 = 77.575$$

$$E = 206 + 77.575 \times 0.04 = 209.1 \text{ V.}$$

SERIES GENERATOR

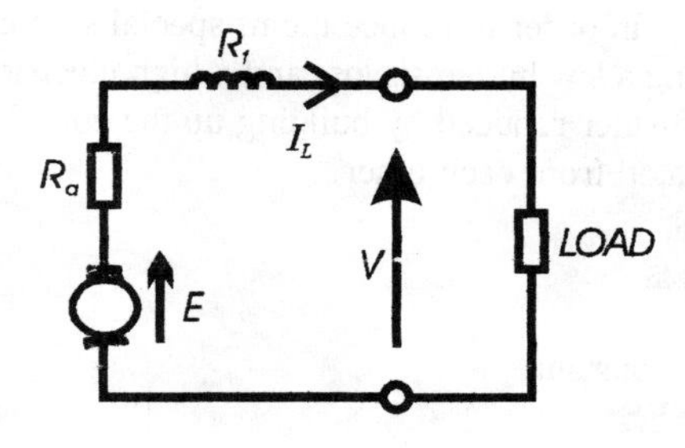

Fig. 235

Field coils wound with a few turns of wire of a large cross-sectional area.

Hardly ever used except for special purposes, e.g. boosters.

Boosters are series generators connected in series with d.c. feeders. They inject a voltage into the feeder to compensate for the volt drop. Both volt drop and injected voltage are proportional to current in the feeder, so the compensation is adequate.

LOSSES IN DIRECT-CURRENT MACHINES

Mechanical losses-common to all types of machine. Electrical and magnetic losses.

All losses appear as heat and raise the temperature of the machine. As the output of the machine is limited in most cases by temperature rise, the losses must be kept as small as possible.

Losses can be roughly divided into two categories:-

(1) constant losses - independent of load, and

(2) variable losses - dependent on load.

(1) CONSTANT LOSSES

(a) Friction and windage losses.

Bearing friction, friction between brushes and commutator, friction and eddy losses set up by the armature in the surrounding air. In ball-bearing machines, the bearing friction is very small compared with windage and brush friction, particularly in high-speed, heavy-current machines.

Windage loss is often increased by the addition of a fan for cooling purposes.

Both losses increase with speed, (particularly windage).

(b) Iron or core losses

Hysteresis and eddy-current losses.

These not only lower the efficiency but also raise the temperature of the magnetic core. In order to reduce them, special silicon steels(e.g. Stalloy) are used, having a low hysteresis loss and a high electrical resistivity. Eddy currents are further reduced by building up the core from thin laminations (sheets) insulated from each other.

(c) Shunt-field loss

Is reasonably constant.

(2) VARIABLE LOSSES

(a) Armature-circuit copper losses

Resistance losses in armature and series-field windings.

Proportional to (armature current)^2.

Can be calculated from measured or estimated values at the working temperature (75° C).

(b) Brush-contact loss

Resistance loss in brush itself is negligible.

Most of loss occurs at contact between brush and commutator.

Not a pure resistance loss and volt drop at the contact varies little with load. Hence loss is almost proportional to current.

EFFICIENCY

$$\text{Efficiency, } \eta = \frac{\text{output}}{\text{input}}$$

which may also be written:-

$$\eta = \frac{\text{input} - \text{losses}}{\text{input}} = 1 - \frac{\text{losses}}{\text{input}} = \frac{\text{output}}{\text{output} + \text{losses}}$$

Rotating machines in general operate relatively efficiently, except at light loads.

The full-load efficiency of average motors is in the neighbourhood of:-

0.74 per unit for 1 KW motors,
0.90 per unit for 50 KW motors,
0.95 per unit for 500 KW motors, and
0.98 per unit for 4 MW motors.

It can be shown that maximum efficiency occurs at the load for which the constant losses and the variable losses are equal. This point usually occurs at the average OPERATING load - about $\frac{2}{3}$ to $\frac{3}{4}$ of RATED load.

The values of maximum efficiency and full-load efficiency are a function of the amount of iron and copper used. They are thus an economic balance between the incremental cost of losses and material. As a result, the efficiency of large machines is higher than that of small ones.

WORKED EXAMPLE 46

A 200 V, d.c. motor develops a shaft torque of 15 Nm at 1200 rev/min. If the efficiency is 0.8 per unit, determine the current supplied to the motor.

SOLUTION 46

Output power of motor, $P_o = \dfrac{2\pi NT}{60} = \dfrac{2\pi \times 1200 \times 15}{60} = 1885$ W.

Now efficiency, $\eta = \dfrac{P_o}{P_i} \quad \therefore P_i = \dfrac{P_o}{\eta}$

i.e. input power, $P_i = \dfrac{1885}{0.8} = 2356$ W.

But $P_i = VI \quad \therefore I = \dfrac{P}{V}$

i.e. input current to motor, $I = \dfrac{2356}{200}$ A $= 11.8$ A

WORKED EXAMPLE 47

A 100 V d.c. generator supplies a current of 15 A when running at 1500 rev/min. If the torque on the shaft driving the generator is 12 Nm, determine:-

(a) the efficiency of the generator, and

(b) the power loss in the generator.

SOLUTION 47

(a) Output power of generator, $P_o = VI = 100 \times 15 = 1500$ W

Input power to generator, $P_i = \dfrac{2\pi NT}{60} = \dfrac{2\pi \times 1500 \times 12}{60} = 1885$ W

$\therefore$ Efficiency, $\eta = \dfrac{P_o}{P_i} = \dfrac{1500}{1885} = 0.796$ per unit.

(b) Power loss = input power − output power

= 1885 − 1500 = 385 W

WORKED EXAMPLE 48

A 250 V, series-wound motor is running at 500 rev/min and its shaft torque is 130 Nm. If its efficiency at this load is 0.88 per unit, calculate the current taken from the supply.

SOLUTION 48

Output power of motor, $P_o = \dfrac{2\pi NT}{60} = \dfrac{2\pi \times 500 \times 130}{60} = 6807$ W

Efficiency, $\eta = \dfrac{P_o}{P_i}$ $\therefore P_i = \dfrac{P_o}{\eta}$

i.e. input power to motor, $P_i = \dfrac{6807}{0.88} = 7735$ W

Now $P_i = VI$ $\therefore I = \dfrac{P_i}{V}$

i.e. input current to motor, $I = \dfrac{7735}{250} = 30.9$ A

WORKED EXAMPLE 49

A 220 V d.c. generator supplies a load of 37.5 A and runs at 1550 rev/min. Determine the shaft torque of the diesel engine driving the generator, if the generator efficiency is 0.78 per unit.

SOLUTION 49

Output power of generator, $P_o = VI = 220 \times 37.5 = 8250$ W

Efficiency, $\eta = \dfrac{P_o}{P_i}$ $\therefore P_i = \dfrac{P_o}{\eta}$

i.e. input power to generator, $P_i = \dfrac{8250}{0.78} = 10577$ W

Now $P_i = \dfrac{2\pi NT}{60}$ $\therefore T = \dfrac{60P_i}{2\pi N}$

i.e. shaft torque, $T = \dfrac{60 \times 10577}{2\pi \times 1550} = 65.2$ Nm.

THE DC MOTOR STARTER

If a d.c. motor whose armature is stationary is switched directly to its supply voltage, it is likely that the fuses protecting the motor will burn out. This is because the armature resistance is small, frequently being less then one ohm. Thus, additional resistance must be added to the armature circuit at the instant of closing the switch to start the motor.

As the speed of the motor increases, the armature conductors are cutting flux and a generated voltage, acting in opposition to the applied voltage, is produced, which limits the flow of armature current. Thus the value of the additional armature circuit resistance can then be reduced.

When at normal running speed, the generated e.m.f. is such that no additional resistance is required in the armature circuit.

To achieve this varying resistance in the armature circuit on starting, a d.c. motor starter is used, as shown on Fig. 236.

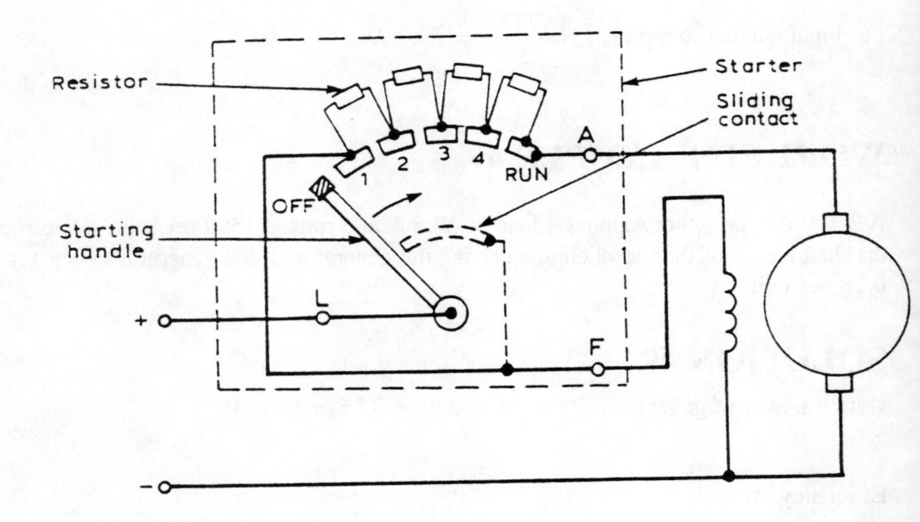

Fig. 236

The starting handle is moved slowly in a clockwise direction to start the motor. For a shunt-wound motor, the field winding is connected to stud 1 or to L via a sliding contact

The starting handle is moved slowly in a clockwise direction to start the motor. For a shunt-wound motor, the field winding is connected to stud 1 or to L via a sliding contact on the starting handle, to give maximum field current, hence maximum flux, hence maximum torque on starting, since $T \propto \phi I_a$.

A similar arrangement without the field connection is used for series motors.

WORKED EXAMPLE 50

A 300 V d.c. shunt motor has an armature resistance of 0.5 Ω. The starting current is to be limited to 75 A.

(a) Calculate the value of starting resistance required.

(b) When the armature has accelerated to a particular speed, the e.m.f. has risen to 225 V. What should the starting resistance be adjusted to so that the current is still 75 A?

SOLUTION 50

(a) $E = V - I(R_A + R)$ where R is starting resistance.

$$\therefore\ (R_A + R) = \frac{(V - e)}{I} = \frac{V}{I} \text{ at starting } (E = 0)$$

$$\text{i.e. } R = \frac{V}{I} - R_A = \frac{300}{75} - 0.5 = 3.5\ \Omega$$

(b) $$(R_A + R) = \frac{(V - E)}{I}$$

$$\therefore\ R = \frac{(V - E)}{I} - R_A$$

$$\text{i.e. } R = \frac{(300 - 225)}{75} - 0.5 = 0.5\ \Omega$$

EXERCISES 5

1. A series motor has a combined armature and field resistance of 1 Ω. The supply voltage is 440 V. When the motor takes a current of 20 A the speed is 1260 rev/min and the torque 325 Nm. The current is increased to 40 A and the pole flux for this current increases by 50%.

   Find (a) the new speed and

   (b) the new torque.

   (800 rev/min; 975 Nm)

2. A 110 V series motor has a combined armature circuit and field resistance of 0.15 Ω. Full load torque is produced when the armature speed is 1200 rev/min and the input current 24 A. When the motor is producing one-quarter full load torque find:-

   (a) the armature current,

   (b) the speed.

   Assume the pole flux is proportional to the field current.

   (12 A; 2440 rev/min)

3. The following test results refer to a d.c. series motor which has a total resistance of 0.87 Ω.

Supply p.d	125	125	125	125	volts
Current	13	16.7	20.6	25.0	amperes
Speed	1275	1050	900	790	rev/min
Back e.m.f					volts
Torque					Nm

   (a) Complete the above table and plot a curve of gross torque against current.

   (b) Obtain the kilowatt output at 900 rev/min if the rotational losses require a torque of 4.05 Nm at this speed.

   (113.69 V, 110.47 V, 107.1 V, 103.25 V: 11.07 Nm, 16.8 Nm, 23.4 Nm, 31.2 Nm; 1.824 kW)

4. When on normal full load, a 500 V d.c. shunt motor runs at 800 rev/min and takes an armature current of 42 A. The flux per pole is reduced to 75% of its normal value by suitably increasing the field circuit resistance.

(b) reduced by 20%.

The armature resistance is $0.6\,\Omega$ and the total voltage loss at the brushes is 2 V.

(1048 rev/min; 1063 rev/min)

5. A d.c. shunt motor has an armature-circuit resistance of 0.6 ohm and a shunt-field resistance of 250 ohms. It is connected to a 250 V supply. On no-load the input current is 2.5 A and the speed is 1500 rev/min. When fully loaded the input current is 10 A. Calculate the value of the counter-e.m.f. generated and the speed of the motor on full-load.

(244.6 V; 1473 rev/min)

6. The torque developed by the armature of a d.c. shunt motor is 95 Nm when the armature current is 50 A. Calculate the torque developed when the armature current is 25 A assuming:-
   (a) constant field current,
   (b) an increase in field current of 20 per cent.

(47.5 Nm, 57 Nm)

6. TRANSFORMERS

Ideal Transformer

$$\frac{N_1}{N_2} = \frac{E_1}{E_2} = \frac{V_1}{V_2} = \frac{I_2}{I_1}$$

$V_1 I_1 = V_2 I_2$ Primary volt-amperes = Secondary volt-amperes

$I_1 N_1 = I_2 N_2$. Primary ampere-turns = Secondary ampere-turns

The transformer as a matching device

Input resistance $R_1 = \dfrac{V_1}{I_1}$ $V_1 = I_1 R_1$

Output resistance $R_2 = \dfrac{V_2}{I_2}$ $V_2 = I_2 R_2$

For voltage and turns ratio

$$R_1 = R_2 \left(\frac{N_1}{N_2} \right)^2$$

$$R_2 = R_1 \left(\frac{N_2}{N_1} \right)^2$$

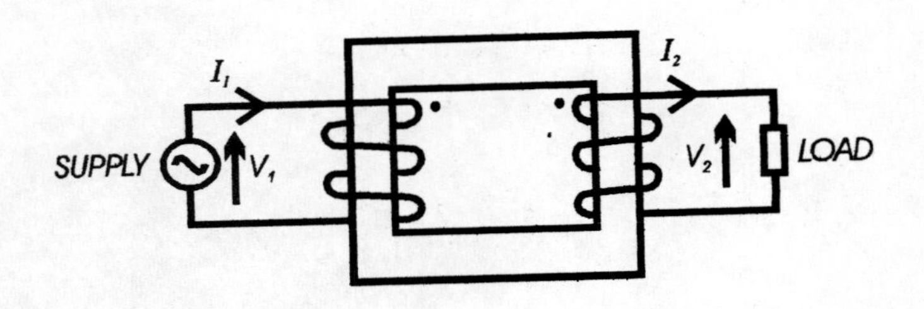

Fig. 237

Power = I^2R

Induced e.m.f. = $4.4\phi f N_1$ in Primary

Induced e.m.f. = $4.4\phi f N_2$ in Secondary

ϕ = Magnetic flux in Wb

Input power = Output power

$I_1 V_1 \cos\phi_1 = I_2 V_2 \cos\phi_2$

$\phi_1 \approx \phi_2$ the primary and secondary phase angles are nearly equal.

TRANSFORMERS LOSSES

IRON LOSSES

Ferro magnetic materials are used in order to obtain large self inductance in coils. Iron loss or core loss or no-load loss

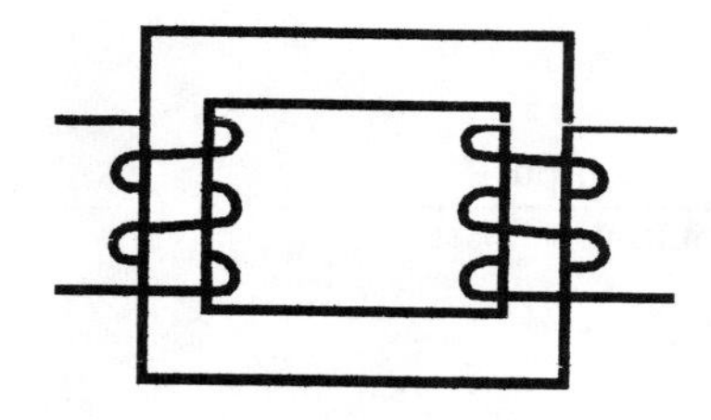

Fig. 238

These materials leads to two major types of losses called the Iron losses. Hysteresis Loss and Eddy Current Loss.

Iron losses are subdivided into hysteresis loss, P_h, and eddy current loss, P_e. Hysteresis loss is due to the power consumed when the magnetic domains in the core rotate each time the magnetising force reverses. See Electrical and Electronic Principles II.

$$P_h = kVfB_m^n \text{ (W)}$$

P_h is the power loss in W due to the hysteresis loop

B_m = maximum flux density (T)

n = number of steinmetz index (1.6 to 2.2)

n = 1.7 (typically), it lies between 1.6 to 2.2.

f = supply frequency (Hz) V = volume of iron (m^3)

k = hysteresis coefficient

Hysteresis loss is proportional to the *B-H* loop area.

When current flows in the windings which are electrically insulated, current tends to flow in the surface of the iron (eddy current).

P_e = Power due to eddy currents

$$P_e \propto (\text{eddy current})^2 \propto (fB_m)^2$$

$$P = P_h + P_e$$

The iron is preferred to be of laminated sheets insulated by paint, the eddy current is reduced. Silicon iron has high resistivity and the eddy current is reduced Ferrite cores are used in high frequency transformers.

Copper Losses

The windings are made of Copper and therefore $I_p^2 R_p$ and $I_s^2 R_s$

Transformer efficiency, η

$$(\text{eta})\ \eta = \frac{\text{output}}{\text{input}} = \frac{\text{output}}{\text{output} + \text{losses}}$$

Other losses

- Magnetic leakage
- Inter turn capacitance
- No-load losses

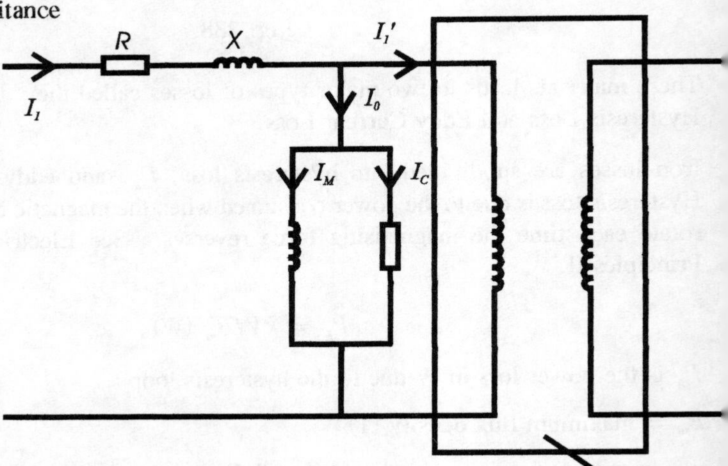

R X I_1' I_1 I_0 I_M I_C IDEAL

$$\text{Copper losses} = P_c = I_P^2 R_P + I_S^2 R_S$$

$$\text{Iron losses} = P_i = P_h + P_e.$$

Fig. 239

For an ideal transformer.

No load condition

The flux linking the primary and secondary windings is ϕ and it lags the primary and secondary voltages by 90°.

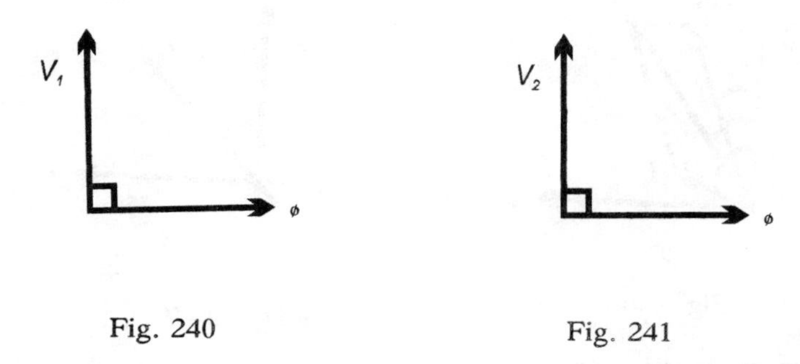

Fig. 240 Fig. 241

On load with lagging power factor.

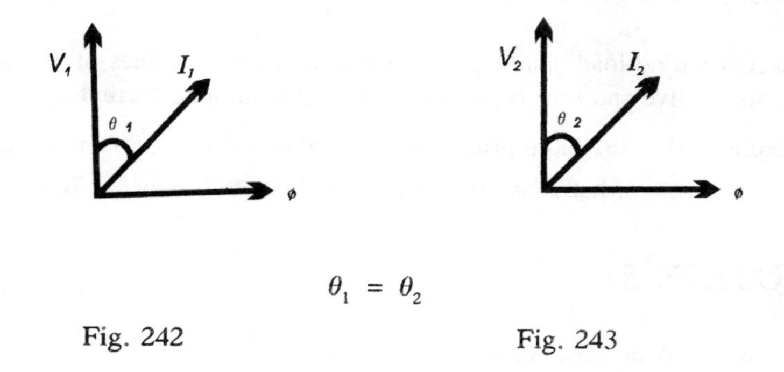

$$\theta_1 = \theta_2$$

Fig. 242 Fig. 243

For a practical transformer, where we take into account the losses we have

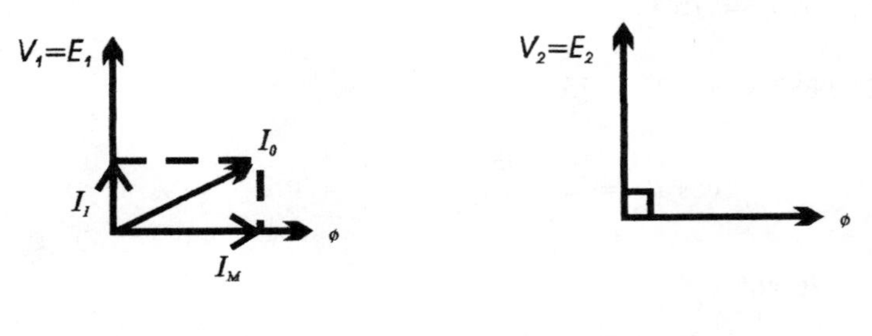

Fig. 244 Fig. 245

Where I_O = no-load current

I_C = core loss current

I_m = magnetizing current

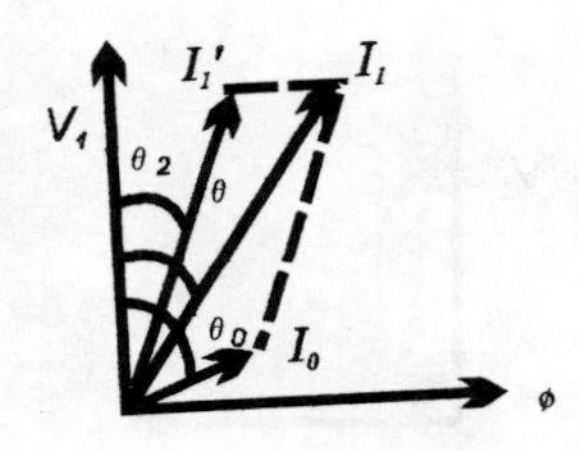

Fig. 246

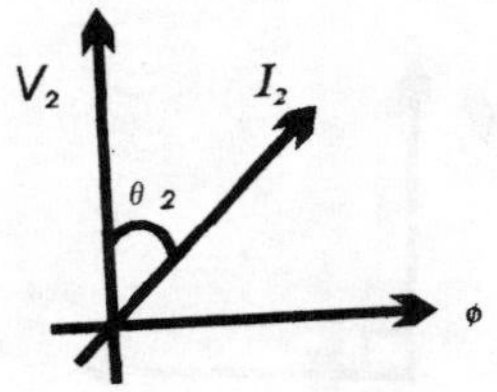

Fig.247

WORKED EXAMPLE 51

The following results were obtained from a no-load test on a single-phase transformer:- V_1 = 250 V, V_2 = 125 V; I_O = 1.5 A; P = 75 W.

(a) Sketch the no-load phasor diagram and indicate the values of no-load power factor, active and reactive components of the no-load current.

(b) Explain (i) the basic principle of operation of the transformer, and

(ii) what constitutes the no-load power loss of 75 W.

SOLUTION 51

(a) $I_O = 1.5$ A no load current

$P = 75$ W

$I_O V_1 \cos\theta_0 = 75$

$1.5 \times 250 \times \cos\theta_0 = 75$

$$\cos\theta_0 = \frac{75}{1.5 \times 250} = \frac{1}{5} = 0.2$$

$\theta_0 = 78.46°$

$$I_c = I_O \sin(90 - 78.46°) = I_O \sin 11.54° = 1.5 \times 0.2 = 0.3 \text{ A reactive}$$

$$I_m = I_O \cos(90 - 78.46°) = I_O \cos 11.54°$$

$$= 1.5 \times 0.98 = 1.47 \text{ A active.}$$

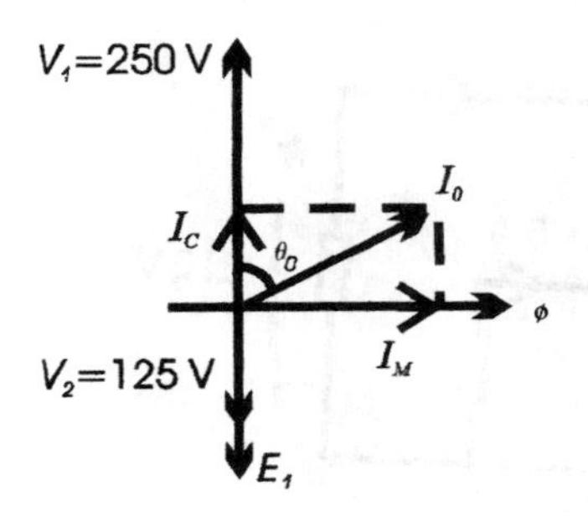

Fig.248 No-load Phasor diagram

(b) Two coils, called the primary and secondary windings, which are not connected to one another in any way, are wound on a complete soft iron core, when an alternating voltage is applied to the primary, the resulting current produces a large alternating magnetic flux which links the secondary and induces an e.m.f. in it. The value of this e.m.f. depends on the number of turns on the secondary.

WORKED EXAMPLE 52

Draw and label a phasor diagram for a transformer on no-load, showing flux, primary terminal voltage, no-load current and phase angle. Resolve the no-load current into its magnetising and core-loss components.

A transformer has 492 turns on the primary winding and 640 turns on the secondary winding, which has a centre tap.

(a) When a p.d. of 240 V a.c. is applied to the primary winding, calculate the secondary voltage between (i) the end terminals;

(ii) one end terminal and the centre tap.

(b) Neglecting the no-load current, determine the primary current if a current of 5.0 A is taken from (i) the terminals (a) (i) above

(ii) the terminals (a) (ii) above.

SOLUTION 52

See text.

(a)

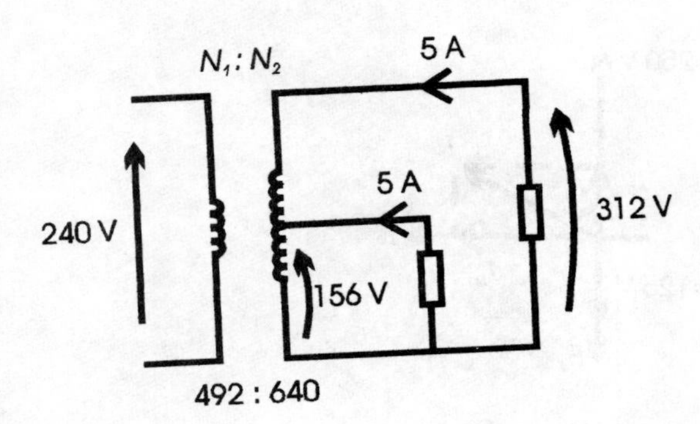

$$\frac{N_1}{N_2} = \frac{V_1}{V_2}$$

Fig.249

(i) $V_2 = 240 \cdot \frac{N_2}{N_1} = 240 \times \frac{640}{492} = 312$ volts.

(ii) 156 V.

(b) (i) $I_1 = \frac{312}{240} \times 5 = 6.5$ A

(ii) $I_1 = \frac{156}{240} \times 5 = 3.25$ A.

WORKED EXAMPLE 53

Under normal operating conditions power loss occurs in a transformer.

(i) What are these power losses?

(ii) What causes them?

(iii) How are their effect reduced?

How is the efficiency of a transformer measured?
An audio frequency transformer issued to match a source impedance of 18 Ω to a load resistance of 7200 Ω. Calculate the turns ratio for maximum power transfer.

SOLUTION 53

See text

$R_{in} = 18\ \Omega,\ R_{out} = 7200\ \Omega.$

$$R_1 = \left(\frac{N_1}{N_2}\right)^2 R_2$$

$$\left(\frac{N_1}{N_2}\right)^2 = \frac{18}{7200}$$

$$\frac{N_1}{N_2} = 0.05 \text{ or } \frac{N_2}{N_1} = 20.$$

WORKED EXAMPLE 54

An a.c. transformer has the following data:

Primary: 250 V, 5000 turns.

Secondary: 30000, 1000 turns
2 loads 1000 Ω, 100 Ω.

Determine: V_2, V_3, I_1, I_2, I_3 and the total power drawn from the supply.

SOLUTION 54

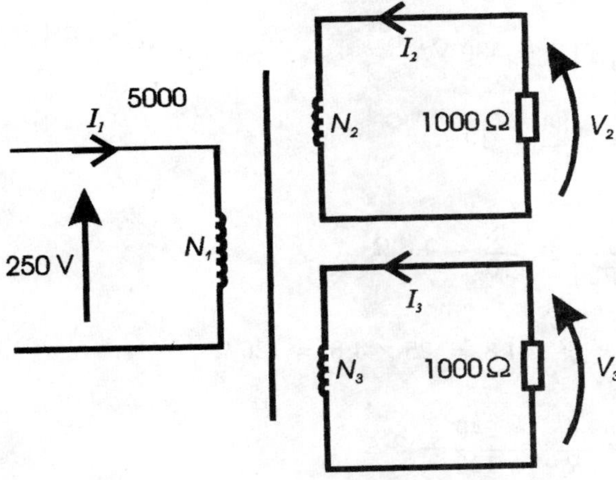

Fig.250

$$\frac{N_1}{V_1} = \frac{N_2}{V_2} = \frac{N_3}{V_3}$$

$$\frac{5000}{250} = \frac{30000}{V_2} = \frac{1000}{V_3} \Rightarrow V_3 = 50 \text{ volts}, V_2 = 1500 \text{ volts}$$

$$I_P N_P = I_S N_S \Rightarrow I_P = \frac{I_S N_S}{N_P} = 0.9 \text{ A}, I_3 = \frac{250}{500} 0.5 \text{ A}$$

$$\frac{5000}{250} = \frac{30000}{1000 I_2}$$

$$I_2 = 1.5 \text{ A}$$

$$P = I_1 V_1 = 0.9 \times 250 = 225 \text{ W}.$$

WORKED EXAMPLE 55

The turns ratio of a step down single-phase transformer is 5:1. The a.c. supply voltage is 240 V. Calculate the resistive load which can be connected. Neglect losses in the transformer. Transformer rating is 480 VA. Calculate the input resistance and hence the input current.

SOLUTION 55

$$\frac{V_2}{V_1} = \frac{N_2}{N_1} \Rightarrow V_2 = 240 \times \frac{1}{5} = 48 \text{ volts}$$

$$V_2 I_2(\text{FL}) = 480 \text{ VA}$$

$$I_2 = \frac{480}{48} = 10 \text{ A}$$

$$\frac{V_2}{I_2(\text{FL})} = \frac{48}{10} = 4.8 \ \Omega$$

$$R_1 = 5^2 \times 4.8 = 25 \times 4.8 = 120 \ \Omega$$

$$I_1 = \frac{V_1}{R_1} = \frac{240}{120} = 2 \text{ A}$$

WORKED EXAMPLE 56

Express the efficiency of a single phase transformer in terms of losses and input power.

A 3300/660 V, 6.6 KVA transformer has iron losses of 200 W and its primary and secondary resistances are 10 Ω and 0.4 Ω respectively.

Calculate the efficiency at unity power factor.

SOLUTION 56

$$\eta = \frac{\text{output power}}{\text{input power}} = \frac{\text{input power} - \text{losses}}{\text{input power}}$$

$$= 1 - \frac{\text{losses}}{\text{input power}}.$$

Transformer rating = 6.6 KVA = $V_2 I_2$ (F.L.)

$$I_2 \text{ (F.L.)} = \frac{6600}{660} = 10 \text{ A}$$

$$I_1 \text{ (F.L.)} = 10 \times \frac{660}{3300} = 2 \text{ A.}$$

Full-load copper losses $= I^2_{1\,(F.L.)} R_1 + I^2_{2\,(F.L.)} R_2$

$= 2^2 \times 10 + 10^2 \times 0.4$

$= 40 + 40 = 80$ W.

Total loss at full load = copper losses + iron losses

= 200 + 80 = 280 W

output power = $V_2 I_2$ (F.L.) $\cos \phi_2 = 660 \times 10 \times 1 = 6600$

$$\eta = \frac{6600}{6600 + 280} \times 100 = 95.9\ \%$$

EXERCISES 6

1. Calculate the no-load current in a single-phase transformer where the core loss current is 0.5 A and the magnetizing current is 1 A. Represent these currents in a phasor diagram and an equivalent circuit referred to the primary.

2. The following results were obtained from a no-load test on a single-phase transformer:-

$$V_1 = 240 \text{ V}, \; V_2 = 40 \text{ V}, \; I_0 = 2.0 \text{ A}$$

$$P = 50 \text{ W (the no-load power loss).}$$

Calculate:- (i) the no load power factor
(ii) the reactive and active components of the no-load current.

3. If R_L is the load resistor of a single-phase transformer and the turns ratio 10:1 show that the resistance viewed at the input terminals is $100R_L$.

If $R_L = 8400\ \Omega$, calculate R_{in}, if $\dfrac{N_1}{N_2} = \dfrac{1}{10}$.

4. A single-phase transformer has the following losses
(i) iron losses = 400 W
(ii) copper losses = 1000 W.

If the input power is 10000 W, calculate the efficiency of the power transformer.

5. The supply voltage of a single-phase transformer is 20 V and the load resistance is 5 Ω. If the resistance viewed at the input is 605 Ω.

Calculate:- (i) the turns ratio
(ii) the primary current
(iii) the load current
(iv) the primary voltage
(v) the secondary voltage
(vi) the power in the load.

Assume that the internal resistance of the supply is 605 Ω.

7. MEASURING INSTRUMENTS AND MEASUREMENTS

a. Makes dB measurements and explains the use of dB.

The Decibel.

$$\text{bel} = \log_{10} \frac{P_{out}}{P_{in}}$$

$$\text{decibel} = 10 \log_{10} \frac{P_{out}}{P_{in}}$$

$$\text{or dB} = 10 \log_{10} \frac{P_{out}}{P_{in}}$$

$$\frac{P_{out}}{P_{in}} = \text{power ratio}$$

If the power ratio is greater than unity, that is, there is a power gain, if the power ratio is less than unity, that is, there is a power loss, and if $P_{out} = P_{in}$ the power is zero decibel.

WORKED EXAMPLE 57

Calculate the power ratio in decibels.

(a) If the output power of an amplifier is 100 mW and the input power is 100 μW.

(b) If the output power of an attenuator is 1 W and the input power is 10 W.

(c) If the input and output powers of a system are equal.

SOLUTION 57

(a) $$\text{dB} = 10 \log_{10} \frac{100 \times 10^{-3}}{100 \times 10^{-6}} = 10 \log_{10} 1000 = 30$$

(b) $$\text{dB} = 10 \log_{10} \frac{1}{10} = 10 \times (-1) = -10$$

(c) $$\text{dB} = 10 \log_{10} \frac{P_{out}}{P_{in}} = 10 \log_{10} 1 = 0.$$

If $N = \log \frac{P_2}{P_1}$ bels where N are the bel,

$n = 10 \log \frac{P_2}{P_1}$ decibels where n are the decibels.

Power ratio	Power ratio (dB)
$\frac{P_{out}}{P_{in}}$	$10 \log_{10} \frac{P_{out}}{P_{in}}$
100	$10 \log_{10} 100 = 20$
1000	$10 \log_{10} 1000 = 30$
1000000	$10 \log_{10} 1000000 = 60$
1	$10 \log_{10} 1 = 0$
0.01	$10 \log_{10} 0.01 = -20$
0.001	$10 \log_{10} 0.001 = -30$
0.000001	$10 \log_{10} 0.000001 = -60$

The decibels can be positive (indicating gain), negative (indicating loss) and zero (indicating neither gain or loss).

Voltage or current ratio

$$dB = 10 \log \frac{P_{out}}{P_{in}} = 10 \log A_p$$

$$= 10 \log \frac{V_{out}^2 / R_{out}}{V_{in}^2 / R_{in}}$$

If $R_{in} = R_{out}$

$$dB = 10 \log \left(\frac{V_{out}}{V_{in}}\right)^2$$

$$dB = 20 \log \frac{V_{out}}{V_{in}}$$

$$\boxed{dB = 20 \log A_v}$$

$$dB = 10 \log \frac{P_{out}}{P_{in}}$$

$$dB = 10 \log \frac{I_{out}^2 \ R_{out}}{I_{in}^2 \ R_{in}}$$

If $R_{out} = R_{in}$

$$dB = 10 \log \left(\frac{I_{out}}{I_{in}}\right)^2$$

$$dB = 20 \log \frac{I_{out}}{I_{in}}$$

$$\boxed{dB = 20 \log A_i}$$

where A_i, A_v, A_p are the current, voltage and power gains.

WORKED EXAMPLE 58

The voltage gain of an amplifier is 40 dB. The input voltage is 50 mV. Calculate the output voltage, assuming that the output and input resistances are equal.

SOLUTION 58

$$40 = 20 \log \frac{V_{out}}{V_{in}}$$

$$2 = \log \frac{V_{out}}{V_{in}}$$

$$10^2 = \frac{V_{out}}{V_{in}}$$

$$V_{out} = 100 \times 50 \times 10^{-3} = 5 \text{ volts.}$$

WORKED EXAMPLE 59

A system of three cascaded amplifiers have power gains of 30, 40 and 50 dB. Between the first and second amplifiers, and between the second third amplifiers, there are attenuations of 10 dB. Calculate the numerical values of the power ratios of each component and the overall power ratio.

SOLUTION 59

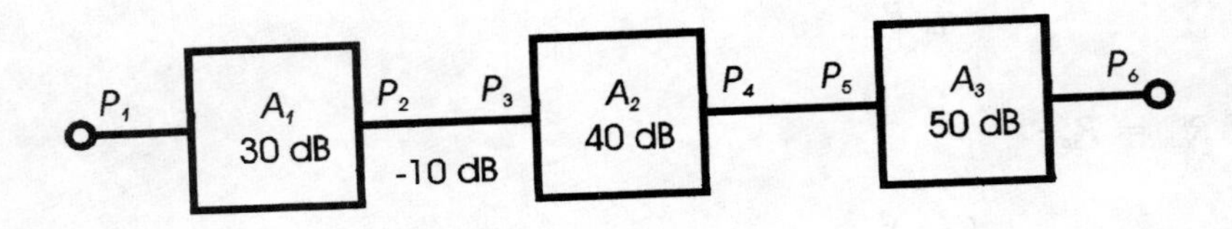

Fig. 251

$$10 \log \frac{P_2}{P_1} = 30$$

$$\frac{P_2}{P_1} = 10^3 = 1000$$

$$10 \log \frac{P_3}{P_2} = -10$$

$$\frac{P_3}{P_2} = \frac{1}{10}$$

$$10 \log \frac{P_4}{P_3} = 40$$

$$\frac{P_4}{P_3} = 10^4 = 10000$$

$$10 \log \frac{P_5}{P_4} = -10$$

$$\frac{P_5}{P_4} = \frac{1}{10}$$

$$10 \log \frac{P_6}{P_5} = 50$$

$$\frac{P_6}{P_5} = 10^5 = 100000$$

$$10 \log \frac{P_6}{P_1} = 30 - 10 + 40 - 10 + 50 = 100$$

$$\boxed{\frac{P_6}{P_1} = 10^{10}}$$

The decibelmeter reference level (dBm)

The most useful datum power reference is the 1 mW, used in dB measurements and is given by the symbol dBm, which is defined

$$\boxed{\text{dBm} = 10 \log_{10} \frac{P}{1 \text{ mW}}}$$

A power level of P W corresponds to dBm $= 10 \log_{10} \frac{P}{10^{-3}} = 10 \log_{10} 1000P$,

a power level of 100 mW corresponds to dBm $= 10 \log_{10} \frac{100 \times 10^{-3}}{1 \times 10^{-3}} = 20$,

a power level of 0.2 mW corresponds to dBm $= 10 \log_{10} \frac{0.2}{1} = -6.99$.

The dBm is widely used in telecommunications systems in conjunction with a standard impedance value of 600 Ω. 0 dB corresponds to the voltage at which 1 mW is developed across 600 Ω resistance.

$$10 \log_{10} \frac{V^2/600}{1 \text{ mW}} = 10 \log \frac{1 \text{ mW}}{1 \text{ mW}} = 0 \text{ dBm.}$$

$$\frac{V^2}{600} = 10^{-3} \Rightarrow V = \sqrt{10^{-3} \times 600} = 0.775 \text{ volts.}$$

Relationship between voltage and dBm ranges

V	dBm (600 Ω)
0	Not defined
0.1	– 17.782
0.2	– 11.761
0.3	– 8.239
0.4	– 5.740
0.5	– 3.802
0.6	– 2.218
0.7	– 0.880
0.8	– 0.208
0.9	1.303
1.0	2.218

$$10 \log_{10} \frac{V^2/600}{1 \text{ mW}} = \text{dBm}$$

If $V = 1$ volt, $10 \log 10 \frac{1^2/600}{10^{-3}} = 2.218$ dBm

If $V = 0.9$ volt, $10 \log_{10} \frac{0.9^2/600}{10^{-3}} = 1.303$ dBm

If $V = 0.8$ volt, $10 \log_{10} \frac{0.8^2/600}{10^{-3}} = 0.208$ dBm

If $V = 0.7$ volt, $10 \log_{10} \frac{0.7^2/600}{10^{-3}} = -0.880$ dBm

If $V = 0.6$ volt, $10 \log_{10} \frac{0.6^2/600}{10^{-3}} = -2.218$ dBm

If $V = 0.5$ volt, $10 \log_{10} \dfrac{0.5^2/600}{10^{-3}} = -3.802$ dBm, etc.

If $V = 0.1$ volt, $10 \log_{10} \dfrac{0.1^2/600}{10^{-3}} = -17.78$ dBm.

WORKED EXAMPLE 60

An a.c. supply of 100 V is applied between the terminals A and C. A moving iron meter is connected across AB and gives a reading of 45 V. Calculate the voltmeter reading.

SOLUTION 60

Let R_V (KΩ) be the resistance of the voltmeter, the total resistance of the circuit is

given by $\dfrac{R_V\, 2.5}{R_V + 2.5}\, 10^3 + 2.5 \times 10^3$ Ω and

$$100 = I \left(\frac{R_V\, 2.5}{R_V + 2.5} \times 10^3 + 2.5 \times 10^3 \right) \quad \ldots (1)$$

$$45 = I \frac{R_V\, 2.5}{R_V + 2.5} \times 10^3 \quad \ldots (2)$$

Dividing (1) by (2)

$$\frac{100}{45} = \left[\frac{2.5\, R_V}{R_V + 2.5} + 2.5 \right] \Big/ \frac{R_V\, 2.5}{R_V + 2.5}$$

$$= 1 + \frac{(R_V + 2.5)}{R_V}$$

$$\frac{100}{45} = 1 + 1 + \frac{2.5}{R_V}$$

$$\frac{100}{45} - 2 = \frac{2.5}{R_V} \Rightarrow R_V = 11.25 \text{ K}\Omega$$

b. **Predicts and measures the loading effects and frequency characteristics of measuring instruments.**

This section is best illustrated by the following examples.

WORKED EXAMPLE 61

An a.c. supply of 100 V is applied between the terminals A and C.

A moving-iron meter is connected across AB and is used on the 100 V range of sensitivity (2000 Ω/V). Calculate the voltage indicated on the meter.

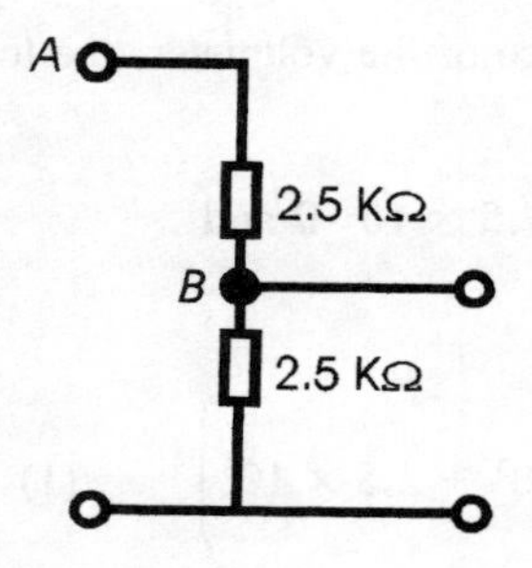

Fig. 252

SOLUTION 61

The resistance of the moving-iron meter on the 100 V range is given by

$2000 \frac{\Omega}{V} \times 100\ \text{V} = 200\ \text{K}\Omega$. This is in parallel with the 2.5 KΩ,

$$\frac{200 \times 2.5}{202.5}\ \text{K}\Omega = 2.469\ \text{K}\Omega$$

$$I = \frac{100}{(2.469 + 2.5)\,10^3} = 20.125\ \text{mA}$$

$$V = I \times 2.469 \times 10^3 = 20.125 \times 2.469 = 49.7\ \text{volts.}$$

c. **<u>Measures power in balanced and unbalanced three-phase loads using 1, 2 and 3 watt meters and uses the two-watt meter method to estimate power factor of a balanced load.</u>**

The Wattmeter is connected in a single-phase circuit.

The wattmeter has two coils, the current coil and the voltage coil. The current coil has markings M (mains) and L (load), and the voltage coil has markings V^+ and V; terminal V^+ is either connected to M or to L, in the diagram V^+ is connected to M. Observe that the current coil is connected in series with the load and the voltage coil is connected in parallel with the load.

If V^+ is connected to L, the instrument is suitable for low-voltage, high current.

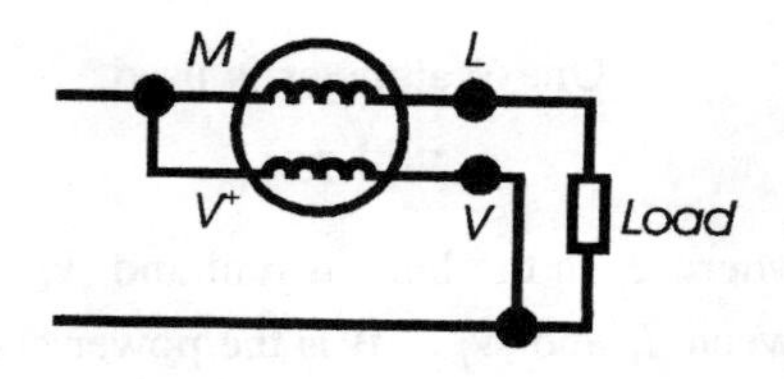

Fig. 253

$P = IV \cos \phi$

$I =$ the current through the current coil

$V =$ the voltage across the voltage coil

$\cos \phi$ = the cosine of the angle between I and V.

Measurement of power in a balanced three-phase load

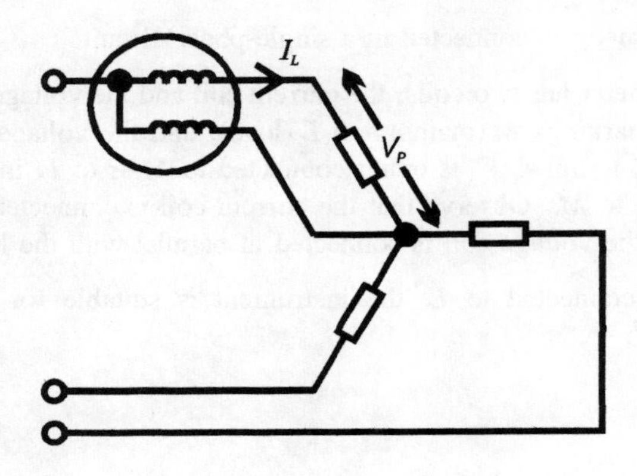

One Wattmeter is used.

Fig. 254

W reads $I_P I_L \cos\phi$ where I_L is the line current and V_P is the phase voltage and ϕ is the phase angle between I_L and V_P. W is the power consumed by one phase and hence the total power is three times for a balanced system.

The two-wattmeter method

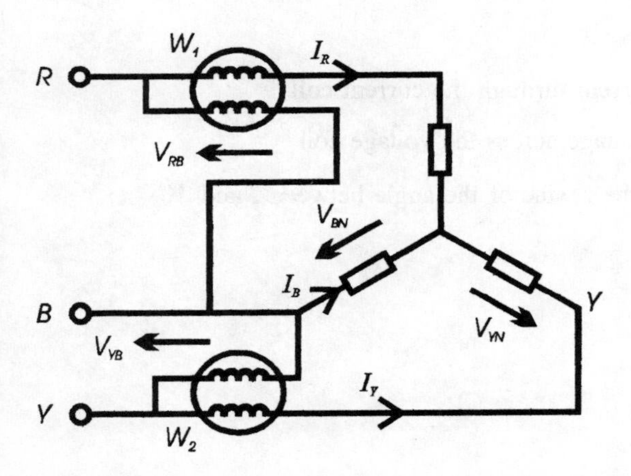

Fig. 255

This is useful for a star and delta connection, for a balanced or unbalanced system. This is the more useful method of measuring power in a 3-phase system

$$\left(v_{RN} - v_{BS}\right) i_R \quad \text{measured by } W_1$$

$$\left(v_{YN} - v_{BN}\right) i_Y \quad \text{measured by } W_2.$$

$$p_Y = i_Y v_{YN}$$

$$p_B = i_B v_{BN}$$

$$p_T = p_R + p_Y + p_B = i_R v_{RN} + i_Y v_{YN} + i_B v_{BN}$$

For a three wire system

$$i_R + i_Y + i_B = 0$$

$$i_B = -\left(i_R + i_Y\right)$$

$$p_T = i_R\left(v_{RN} - v_{BN}\right) + i_Y\left(v_{YN} - v_{BN}\right) = W_1 + W_2$$

where lower cases of p, i indicate instantaneous values.

Demonstates the effect of circuit time constants on rectangular waveforms using an oscilloscope and relates waveforms to simple concepts of integration and differentiation.

Consider a square wave of period T where the space to mark ratio is unity

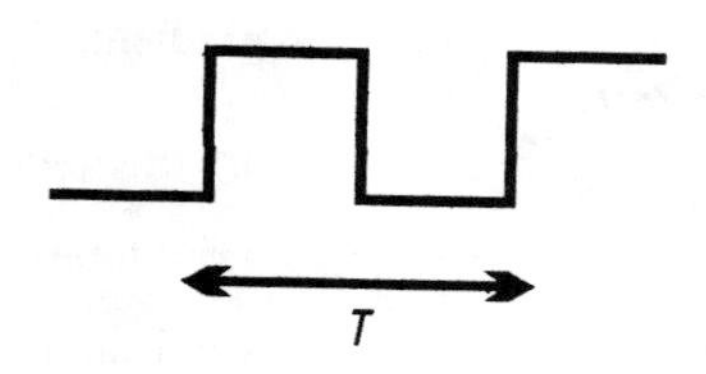

Fig. 256

The time constant of an RC network is given $\tau = RC$

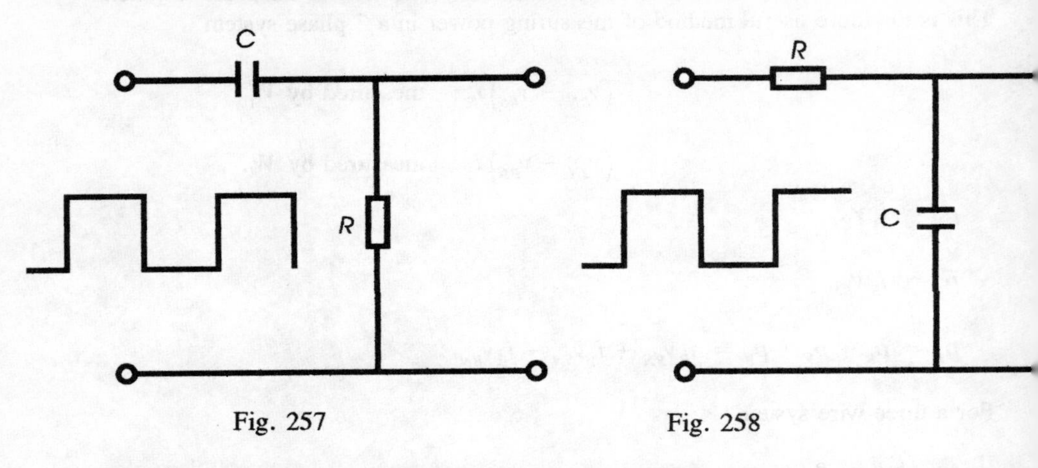

Fig. 257 Fig. 258

The output can be taken across R or across C as shown. The input is a pulse of period T.

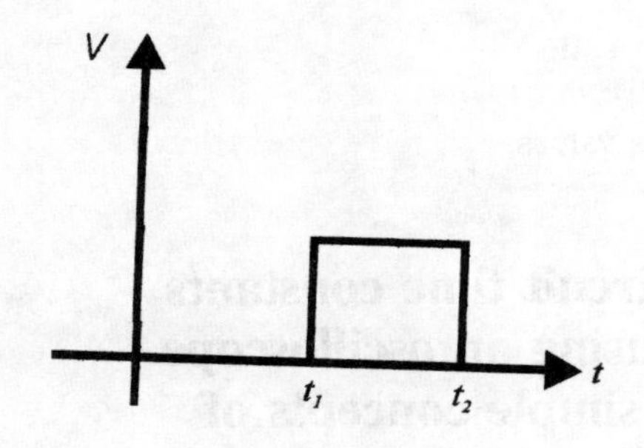

The rate of change of $V\left(\frac{dV}{dt}\right)$ is either infinite or zero.

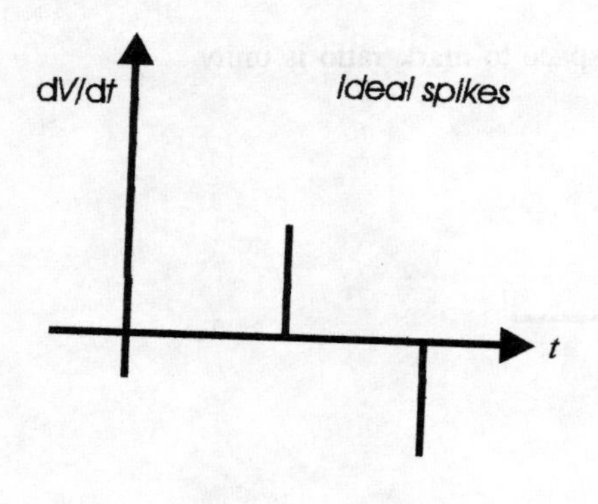

Fig. 259

At t_1, $\frac{dV}{dt} \rightarrow \infty$, on the leading and trailing edges of the pulse during the time from t_1 to t_2 the gradient, $\frac{dV}{dt} \rightarrow 0$. These concepts are theoretical, in practice there is a rise time on the leading edge, and a fall time on the trailing edge.

Passive Differentiating Circuit

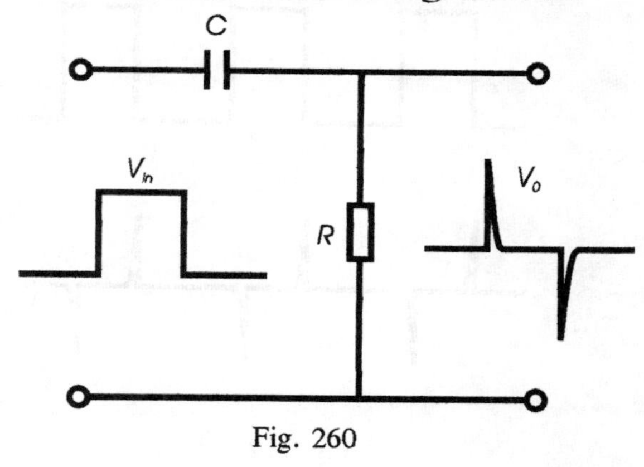

Fig. 260

If $T_1 \gg CR$, the output voltage across R is the derivative of the rectangular pulse.

$$V_O \propto \frac{d}{dt}(V_{in}).$$

Passive Integrating Circuit

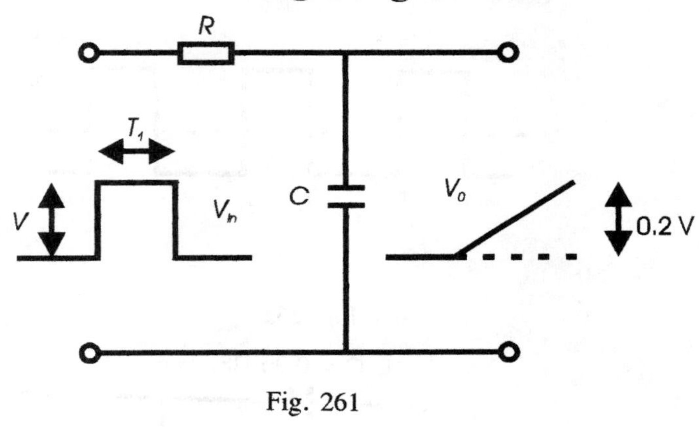

Fig. 261

If $T_1 \ll CR$, the voltage across C will rise slowly and reaches a portion of V, about 0.2 V.

$$V_O \propto \int V_{in}\, dt.$$

A pulse train such as that shown when it is applied to a differentiating circuit.

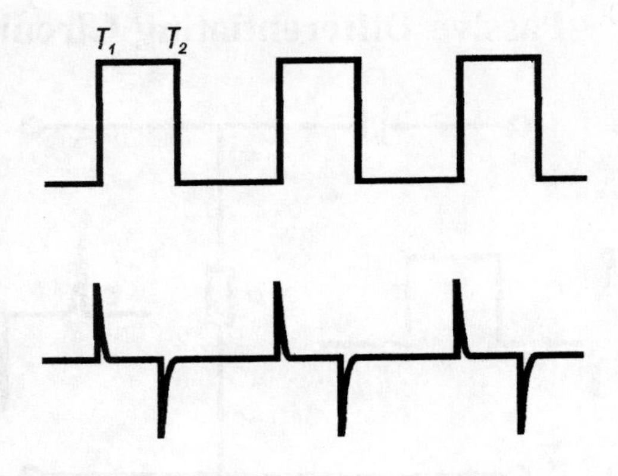

Fig. 262

A pulse train such as that shown when it is applied to an integrating circuit $\tau \gg T_1$.

$$\bar{V}_{in} = \frac{T_1}{T_1 + T_2} \cdot V_{in}$$

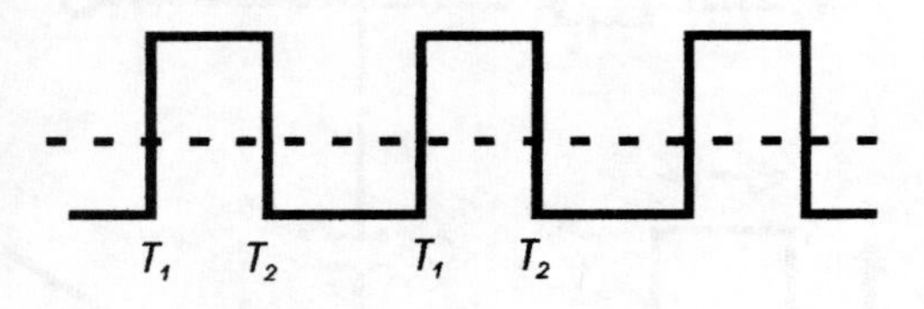

Fig. 263

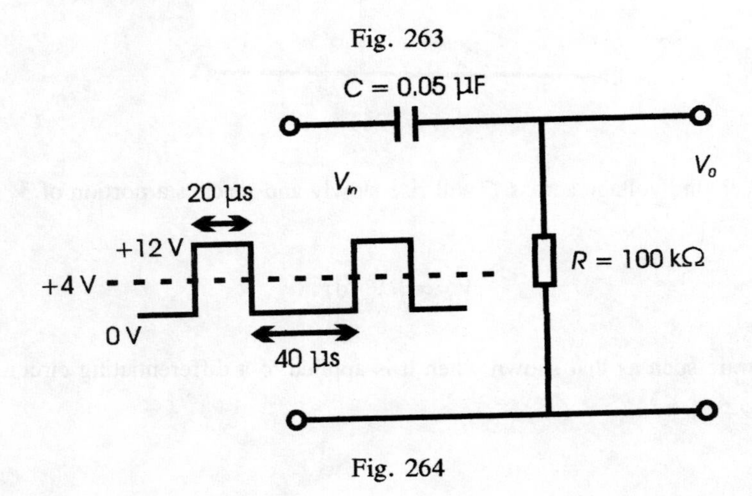

Fig. 264

$\tau = RC = 100 \times 10^3 \times 0.05 \times 10^{-6}$

$\tau = 5$ ms

$\tau =$ the time constant

τ is long compared to 40 μs.

$$V_{mean} = \frac{T_1}{T_1 + T_2} V = \frac{20 \times 10^{-6}\ V}{(20 + 40) \times 10^{-6}} = \frac{20}{60} V = \frac{1}{3} V$$

$$= \frac{1}{3} \times 12 = 4 \text{ volts.}$$

The wave varies between − 4 V and + 8 V.

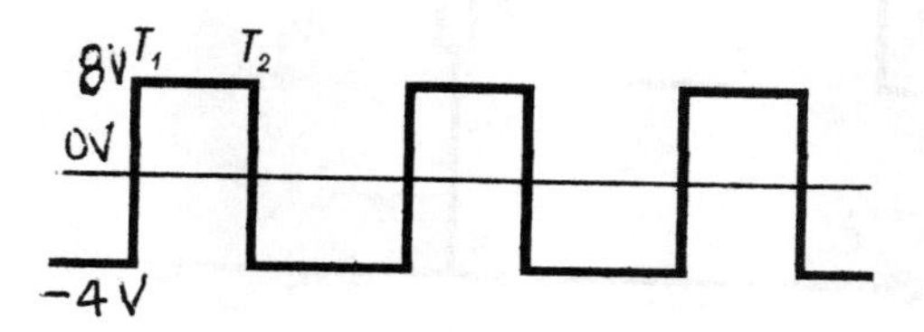

Fig. 265

Active Differentiator

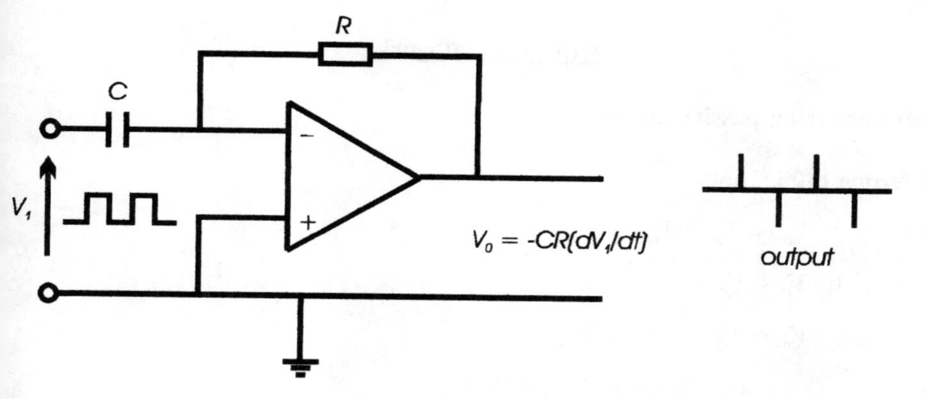

Fig. 266

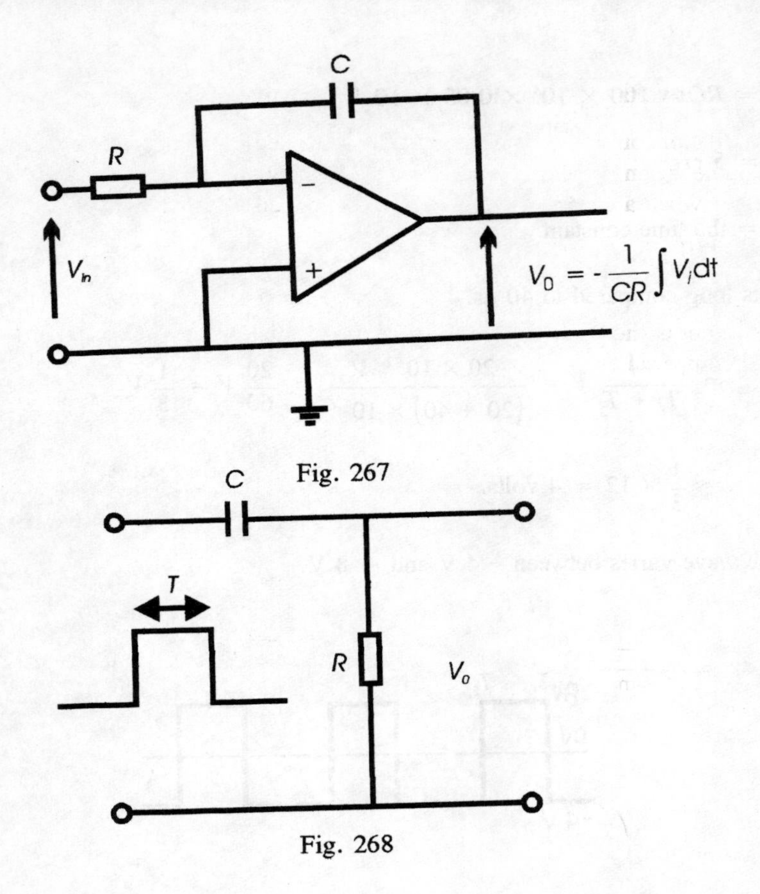

Fig. 267

Fig. 268

Sketch V_O for (i) $T > CR$

(ii) $T < CR$

RESPONSE OF AN RC NETWORK TO PULSE WAVEFORM

Experimental work

Differentiating passive circuit

Referring to Fig. 260

$$V_{in} = V_R + V_C$$

If $V_C \gg V_R$

$$V_{in} \approx V_C$$

$$i = \frac{dq}{dt} = C\frac{dV_c}{dt} \approx C\frac{d(V_{in})}{dt}$$

$$V_{out} = iR = CR\frac{d}{dt}(V_{in}).$$

The output voltage is proportional to the derivative of the input voltage.

An oscillator (on square wave output) is connected to the network input of Fig. 260. A double beam oscilloscope is used to display the input and output waveforms together when a square wave of 1 KHz is applied as input signal.

$C = 0.001\ \mu F$, $R = 1\ K\Omega$; the procedure is repeated for $R = 10\ K\Omega$, $100\ K\Omega$ and $500\ K\Omega$.

The output is shown to tend to the time differential of the input as $\tau = RC$ becomes small compared to the input signal period.

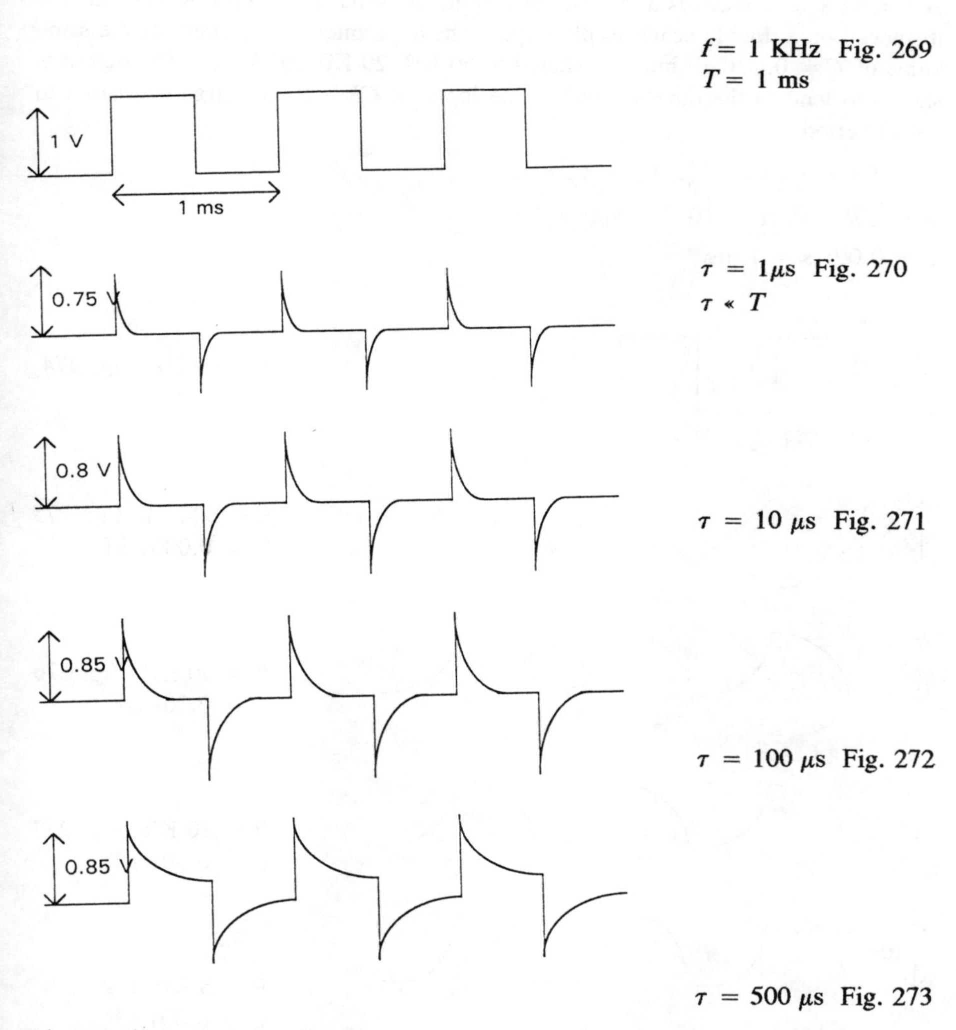

$f = 1$ KHz Fig. 269
$T = 1$ ms

$\tau = 1 \mu s$ Fig. 270
$\tau \ll T$

$\tau = 10\ \mu s$ Fig. 271

$\tau = 100\ \mu s$ Fig. 272

$\tau = 500\ \mu s$ Fig. 273

Observe the various amplitudes.

Integrating Passive Circuit

Referring to Fig. 261.

$R = 500\ \text{K}\Omega,\ C = 0.001\ \mu\text{F}$

$$V_{in} = V_R + V_C = iR + \frac{1}{C}\int i\,dt$$

$$V_{in} \approx V_R = iR$$

$$V_{out} = V_C = \frac{Q}{C} = \frac{1}{C}\int i\,dt = \frac{1}{CR}\int V_{in}\,dt$$

A 1 KHz square wave is applied at the input, the input and output waveforms are displayed on a double beam oscilloscope. The experiment is repeated for the same value of $C = 0.001\ \mu\text{F}$ but R is altered to 50 KΩ, 20 KΩ and 5 KΩ. The output is shown to tend to the time integral of the input as CR becomes large compared to signal period.

$\tau = CR \gg T\ (V_C \ll V_R,\ V_{in} \approx V_R = iR)$

$\tau = CR = 0.01 \times 10^{-6} \times 500 \times 10^3$

$T = 0.001\ \text{s} = 1\ \text{ms}.$

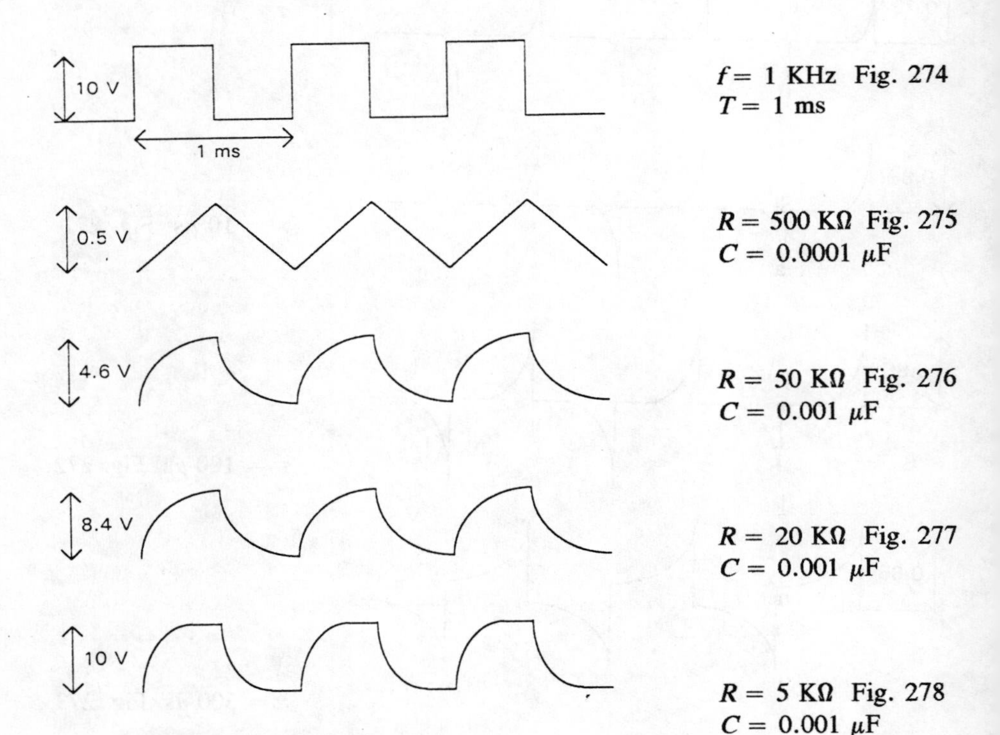

$f = 1$ KHz Fig. 274
$T = 1$ ms

$R = 500$ KΩ Fig. 275
$C = 0.0001\ \mu\text{F}$

$R = 50$ KΩ Fig. 276
$C = 0.001\ \mu\text{F}$

$R = 20$ KΩ Fig. 277
$C = 0.001\ \mu\text{F}$

$R = 5$ KΩ Fig. 278
$C = 0.001\ \mu\text{F}$

Observe the various amplitudes.

d. Uses a C.R.O. to demonstrate the presence of harmonics in various waveforms.

What are harmonics?

Harmonic frequency is one having an integral multiple of the fundamental frequency. If the fundamental frequency of an a.c. supply is f_1, then the even and odd harmonics are $2f$, $4f$, $6f$ etc. and $3f$, $5f$, $7f$ etc. correspondingly.

If f_1 is fundamental frequency, the second harmonic frequency, f_2, is $2f_1$, the fifth harmonic frequency, f_5 is $5f_1$ and so on.

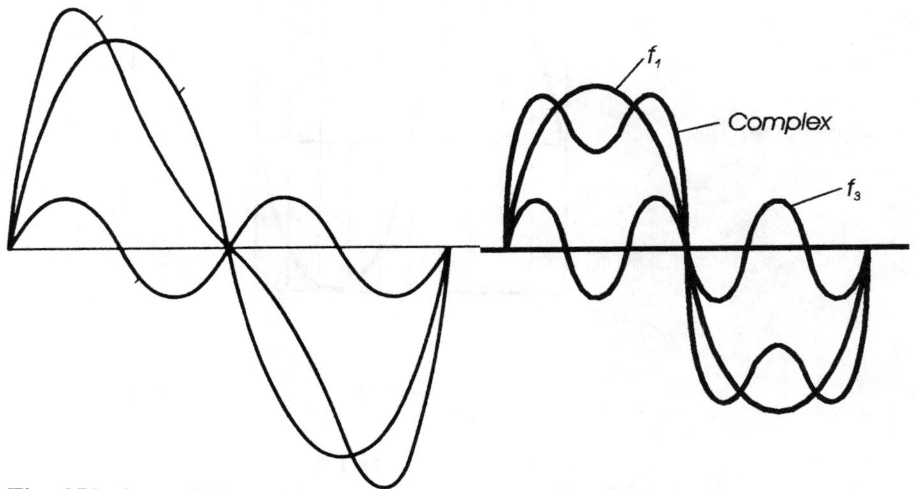

Fig. 279 Second Harmonics

Fig. 280 Third Harmonics

Observe that the second and third harmonics have much smaller amplitude. Combining the two waves f_1 and f_2 and f_1 and f_3 results into complex waveforms. A complex wave has odd or even harmonics or both.

e. Uses a C.R.O. to measure sine, square and pulse waveforms and phase differences.

<u>The Cathode-Ray Oscilloscope</u> (C.R.O.)

The C.R.O. is used to measure current, voltage, frequency, phase difference, rise-time and full-time of square waveforms, and to observe various other waveforms and their harmonics.

The C.R.O. is therefore the most versatile instrument.

To measure the period and frequency of sinusoidal, square and pulse waveforms.

The internal timebase of the oscilloscope is used for this purpose. Adjust the timebase to display one complete cycle of a sinusoidal waveform which is to be measured across the face at the tube which is about 10 cm horizontal and 10 cm vertically.

The time-base is firstly calibrated against a reliable reference frequency.

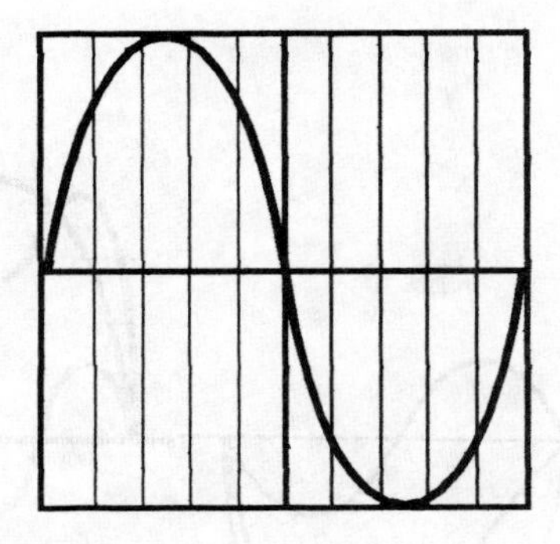

Fig. 281

$$f = \frac{1}{T} \text{ (Hz)}$$

$$T = \frac{1}{f} \text{ (s)}$$

where T is the period and f is the frequency.

WORKED EXAMPLE 62

The sensitivity of the timebase is given as (i) 100 μs/cm,

(ii) 5 ms/cm,

(iii) 50 ms/cm.

The horizontal distance of the sinewave is 10 cm and the timebase is used at (i) 100 μs/cm. Calculate the period of the waveform and hence the frequency. Repeat for (ii) and (iii).

SOLUTION 62

(i) $T = 100 \times 10^{-6}$ (s/cm) $\times 10 = 1000 \times 10^{-6} = 1$ ms.

$$f = \frac{1}{10^{-3}} = 1000 \text{ Hz.}$$

(ii) $T = 5 \times 10^{-3}$ (s/cm) $\times 10 = 50$ ms

$$f = \frac{1}{50 \times 10^{-3}} = \frac{10^3}{50} = 20 \text{ Hz}$$

(iii) $T = 50 \times 10^{-3}$ (s/cm) $\times 10 = 500 \times 10^{-3}$ s

$$f = \frac{1}{500 \times 10^{-3}} = 2 \text{ Hz.}$$

WORKED EXAMPLE 63

For the square voltage waveform displayed on an C.R.O. shown in Fig. 282, find:-

(i) its frequency and

(ii) its peak to peak voltage.

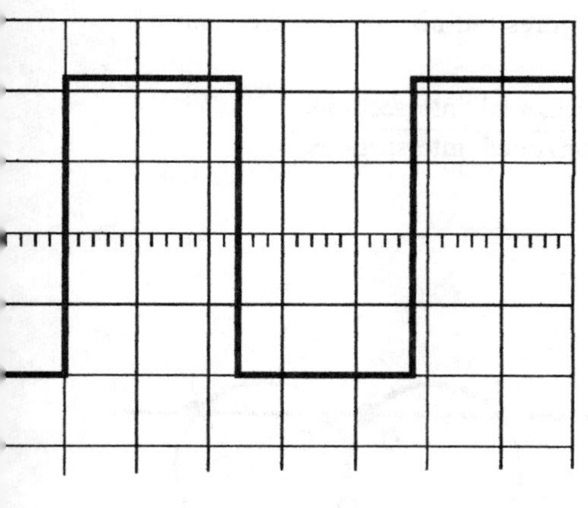

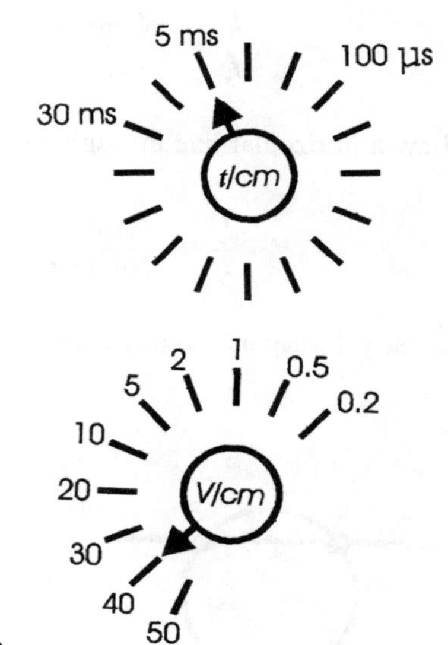

Fig. 282

SOLUTION 63

The voltage sensitivity is shown to be 40 V/cm and the time sensitivity is shown to be 5 ms/cm. The voltage is shown vertically and time horizontally.

The space-to-mark ratio is equal to 1.2 cm : 1.2 cm, the period is 2.4 cm. The peak-to-peak voltage is 2.4 cm.

(i) $f = \frac{1}{T}$ where $T = 2.4 \times 5 \times 10^{-3} = 12$ ms

$$f = \frac{1}{12 \times 10^{-3}} = 83.3 \text{ Hz.}$$

(ii) The peak-to-peak voltage = $40 \times 2.4 = 96$ volts.

LISSAJOUS'S FIGURES

Lissajous's Figures are used to measure frequencies. To obtain these figures, the time base is switched off.

The C.R.O. has two amplifiers, the *X*-amplifier and *Y*-amplifier.

A signal generator of known frequency is placed across the *X*-plates and the unknown frequency signal is placed across the *Y*-plates. The ratio of f_y to f_x is expressed as the ratio of the number of peaks under the *Y*-plates to the number of peaks under the *X*-plates.

$$\frac{f_y}{f_x} = \frac{\text{Number of peaks under the } Y\text{-plates}}{\text{Number of peaks under the } X\text{-plates}}$$

or draw a horizontal line through these figures but not through the centres.

$$\frac{f_y}{f_x} = \frac{\text{Number of horizontal intersections}}{\text{Number of horizontal intersections}}.$$

Some easy Lissajous's figures are

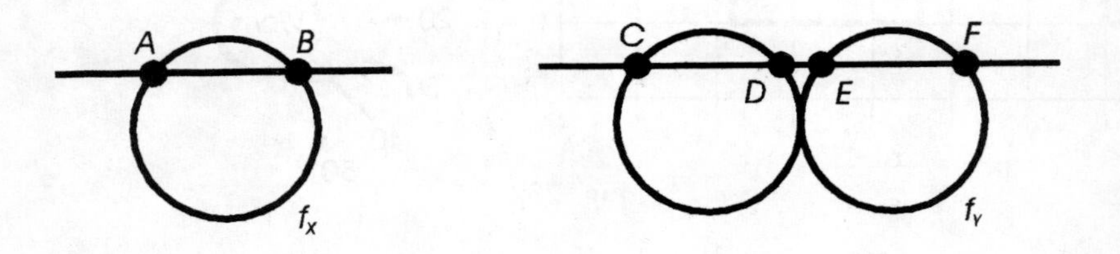

Fig. 283 Fig. 284

If a horizontal line is drawn as shown for the *X*-signal, there are two intersections *A* and *B*, and for the *Y*-signal, there are four intersections *C, D, E* and *F*, therefore

$$\frac{f_y}{f_x} = \frac{4}{2} = 2$$

$$\therefore\ f_y = 2f_x.$$

The frequency of the *Y*-signal is twice that of the *X*-signal.

Measurement of Phase angle

We would like to measure the phase angle between two sinusoidal waveforms of the same amplitude and frequency but differ by ϕ degrees or radius in phase angle.

$$v_1 = A \sin \omega t$$

$$v_2 = A \sin (\omega t \pm \phi)$$

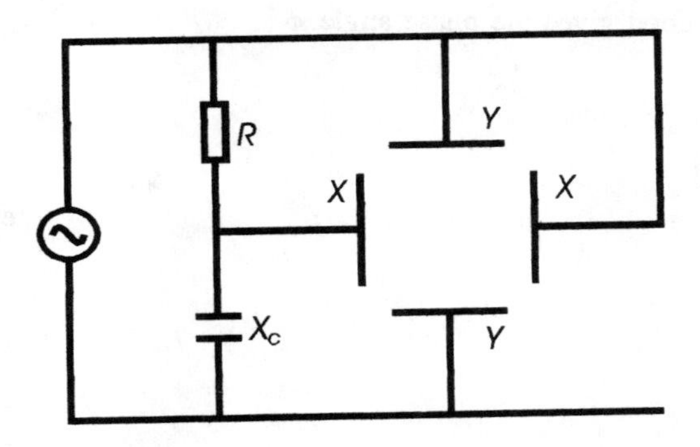

Fig. 285

The *X*-plates are connected across *R* and *Y*-plates are connected across X_C as shown in Fig. 285.

The voltage controls are adjusted to have the same gains. If v_1 is applied to the *X*-plates and v_2 is applied to the *Y*-plates.

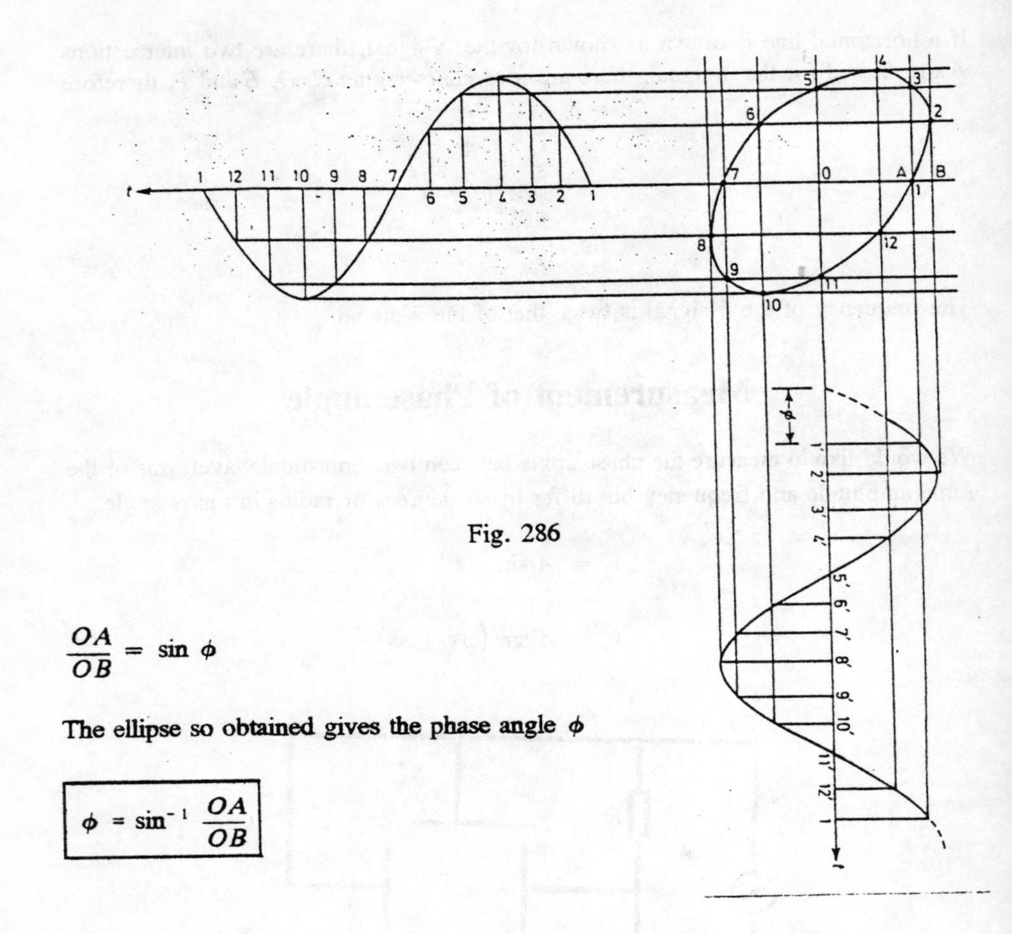

Fig. 286

$$\frac{OA}{OB} = \sin \phi$$

The ellipse so obtained gives the phase angle ϕ

$$\boxed{\phi = \sin^{-1} \frac{OA}{OB}}$$

Fig. 287

f. **Uses a commercial bridge for measuring inductance capacitance and *Q*-factor.**

THE PRINCIPLE OF WHEATSTONE BRIDGE

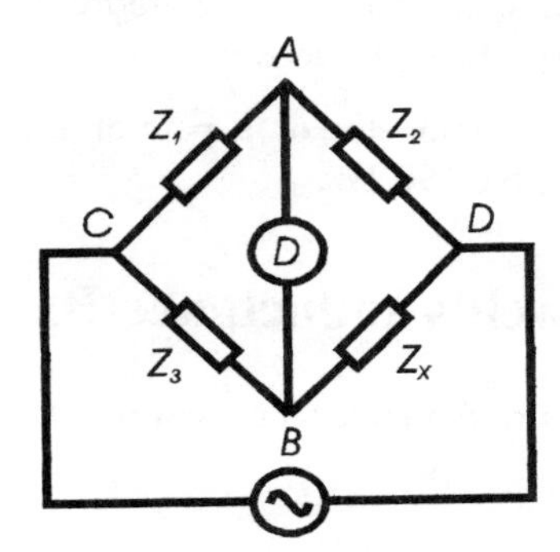

Fig. 288

D is a sensitive instrument called the Detector. If $V_D = 0$, $V_A = V_B$, *A* and *B* are at the same potential.

$V_{AC} = V_{BC}$, $\qquad V_{AD} = V_{AB}$

$I_1 Z_1 = I_2 Z_3$ ... (1) $\qquad I_1 Z_2 = I_2 Z_x$ ... (2)

Dividing (1) by (2) we have

$$\frac{I_1 Z_1}{I_1 Z_2} = \frac{I_2 Z_3}{I_2 Z_x}$$

$$\boxed{\frac{Z_1}{Z_2} = \frac{Z_3}{Z_x}}$$

If Z_1 and Z_2 are two high stability resistances of equal values then

$$Z_x = Z_3$$

$$\frac{1}{jwC_x} = \frac{1}{jwC}$$

$$\therefore\ C_x = C$$

This is the DeSauty bridge which is used for the measurement of a loss capacitor.

Therefore the d.c. mode of operation of the Wheatstone bridge examined in the Electrical and Electronic Principles II book is extended here to the a.c. mode of operation. We have measured previously the resistance of an unknown resistance using d.c. mode of operation. In the a.c. mode of operation we can measure the following qualities.

Inductance, capacitance, frequency, the *Q*-factor of components, impedance and power factor.

The measurement of these quantities requires more than one adjustment and therefore is more difficult to achieve balance in an a.c. measurement.

The detector may be an a.c. voltmeter or a pair of headphones at the sensitive frequency of the human ear, the 796 Hz.

Maxwell's Inductance Bridge

This bridge is used to measure the inductance and resistance of an inductor or coil and hence its quality factor.

P and *Q* are two precise pure resistances.

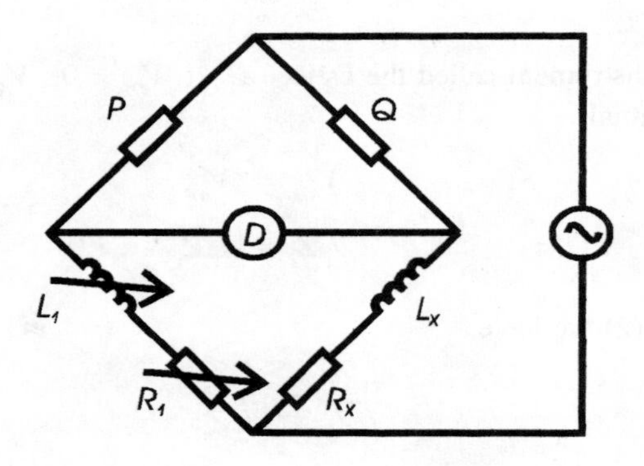

Fig. 289

At balance we have

$$P\left(R_x + j\omega L_x\right) = Q\left(R_1 + j\omega L_1\right) \quad \dots (1)$$

where $R_x + j\omega L_x$ and $R_1 + j\omega L_1$ are the complex impedances of the unknown coil and well calibrated values of R_1 and L_1.

Equating real and imaginary terms of (1), we have

$$PR_x = QR_1$$

$$PL_x = QL_1$$

hence $\boxed{L_x = \dfrac{QL_1}{P}}$ and $\boxed{R_x = \dfrac{QR_1}{P}}$

The ratio of $\dfrac{Q}{P}$ can be 1:1, 10:1, 100:1 or 1:10, 1:100.

This type of bridge is rather difficult as in practice we cannot achieve a good quality calibrated inductor. A capacitor of high Q-factor can be used instead. The arrangement of the bridge is as shown in Fig. 290.

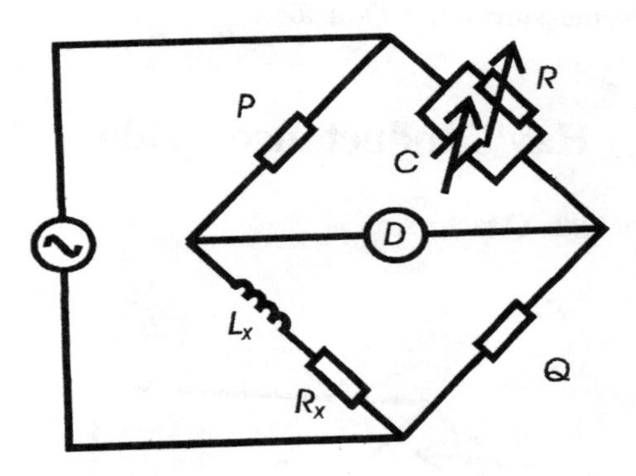

Fig. 290

P and Q are placed diagonally opposite. Observe also that the a.c. supply is placed across a diagonal and the detector across the other diagonal.

At balance $PQ = (R_x + j\omega L_x) \cdot \left(\dfrac{R \dfrac{1}{j\omega C}}{R + \dfrac{1}{j\omega C}} \right)$

$$PQ = (R_x + j\omega L_x)\left(\frac{R}{1 + j\omega CR}\right)$$

$$PQ(1 + j\omega CR) = RR_x + j\omega L_x R$$

$$PQ + j\omega CRPQ = RR_x + j\omega L_x R.$$

Equating real and imaginary terms

$$PQ = RR_x \qquad \ldots (1)$$

$$\omega CRPQ = \omega L_x R \ldots (2)$$

From (1) $\boxed{R_x = \frac{PQ}{R}}$

From (2) $\boxed{L_x = CPQ}$

Both sets of equations are independent of the frequency.

The former relies on the ratio arm and the latter on the product arm.

This bridge is used to measure small Q-factors.

Hay's Inductance Bridge

This bridge measures high Q-factors.

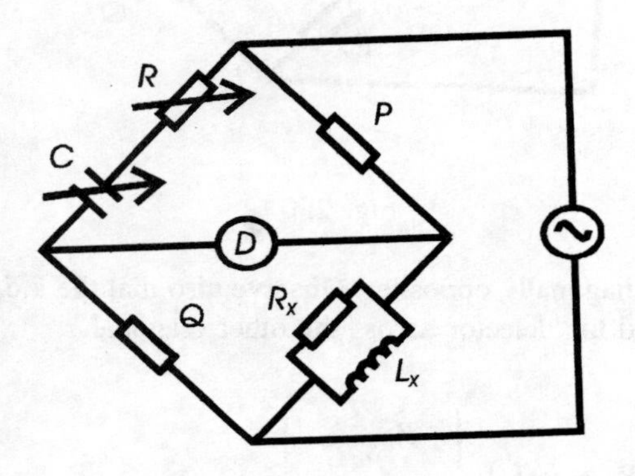

Fig. 291

$$\text{At balance } PQ = \left(R + \frac{1}{j\omega C}\right) \cdot \left(\frac{R_x j\omega L_x}{R_x + j\omega L_x}\right)$$

$$PQj\omega C(R_x + j\omega L_x) = (Rj\omega C + 1)(R_x j\omega L_x)$$

$$PQj\omega CR_x - \omega^2 L_x CPQ = Rj\omega CR_x j\omega L_x + R_x j\omega L_x$$

Equating real and imaginary terms

$$PQ\omega CR_x = R_x \omega L_x$$

$$\omega^2 L_x CPQ = RR_x \omega^2 CL_x$$

$$\boxed{L_x = PQC}$$

$$\boxed{R_x = \frac{PQ}{R}}$$

WORKED EXAMPLE 64

The product arm in the Hay's Bridge is given as 100000 and the frequency of the a.c. supply is 796 Hz at balance and $R = 5000\ \Omega$ and $C = 5\ \mu F$.

Calculate:- (a) The unknown resistance of the coil.

(b) The unknown inductance of the coil.

(c) The Q-factor at 796 Hz.

SOLUTION 64

(a) $R_x = \frac{PQ}{R} = \frac{100000}{5000} = 20\ \Omega$

(b) $L_x = PQC = 100000 \times 5 \times 10^{-6} = 0.5$ H

(c) $Q = \frac{\omega L_x}{R_x} = \frac{2\pi f 0.5}{20} = \frac{2\pi 796 \times 0.5}{20} = 125.$

WORKED EXAMPLE 65

Distinguish between Maxwell's Bridge and Hay's Bridge in measuring the self-inductance and resistance of a coil. State the formulae for the two bridges and comment on their results.

SOLUTION 65

For the Maxwell's Bridge

$$\boxed{L_x = \frac{Q}{P} L_1} \qquad \boxed{R_x = \frac{Q}{P} R_1} \qquad \ldots (1)$$

For the Hay's Bridge

$$\boxed{L_x = CPQ} \qquad \boxed{R_x = \frac{PQ}{R}} \qquad \ldots (2)$$

From equations (1), L_x and R_x depend on the ratio-arm of the resistances and the values of L_1 and R_1, in practice the self inductance L_1 is not pure it contains an effective resistance which is not taken into account, the Q-factor of the best of the coils are in the order of 10, in this case we assume infinite value.

From equations (2), L_x and R_x depend on the product arm of the resistances and the values of C and R which are of high quality, in practice the Q-factor of C is very high and therefore the measured values are more reliable

g. Calculates simple errors due to the insertion of instruments.

Error in Instruments.

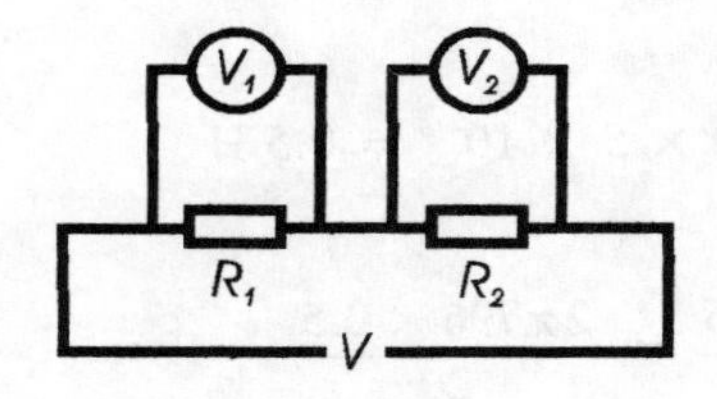

Fig. 292

WORKED EXAMPLE 66

If $V_1 = 15\text{ V} \pm 1\%$, $V_2 = 50\text{ V} \pm 3\%$. Determine the limits of the voltage across both resistors.

SOLUTION 66

$$V = V_1 + V_2$$

$$= (15 \pm 1\%) + (50 \pm 3\%)$$

$$V_{max} = 15 + \frac{1}{100} \times 15 + 50 + \frac{3}{100} \times 50$$

$$= 15.15 + 51.50$$

$$= 66.55 \text{ volts}$$

$$V_{min} = \left(15 - \frac{1}{100} \times 15\right) + \left(50 - \frac{3}{100} \times 50\right)$$

$$= 15 - 0.15 + 50 - 1.5 = 65 - 1.65$$

$$= 63.35 \text{ volts.}$$

Q-factor measurement

The *Q*-meter

It is required to measure the *Q*-factor of a coil.

A coil is connected in series with a good or high quality-factor capacitor *C*.

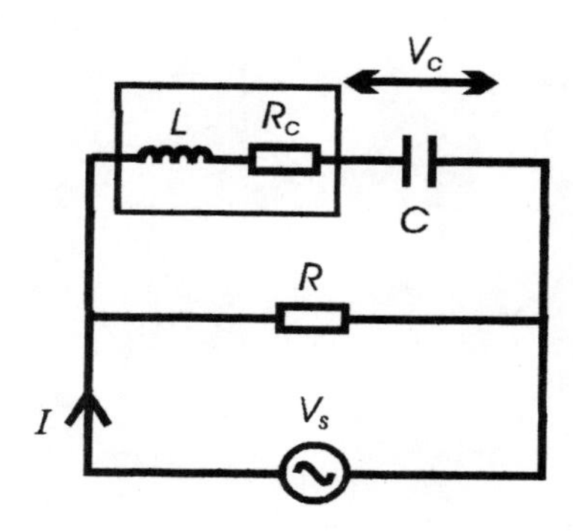

Fig. 293

The frequency of the oscillator is adjusted so that resonance occurs.

V_s and V_c are measured where V_s is the a.c. supply voltage.

V_c should be maximum and equals to QV_s

$$\therefore \ Q = \frac{V_c}{V_s} \ \dots (1)$$

f_o is noted and L is the self inductance of the coil

$$Q = \frac{2\pi f_o L}{R_s} \ \dots (2)$$

R_s is calculated from (2).

This is the Q-factor of the coil at resonance.

If C is replaced by a variable capacitor then the Q-factor of the coil at arms frequency can be calculated.

f is adjusted to the frequency required and C is adjusted to give resonance at that frequency.

$$f_o = \frac{1}{2\pi\sqrt{LC}} \ \dots (3)$$

and L is calculated from (3)

$$L = \frac{1}{4\pi^2 C f_o^2}.$$

This circuit is called the Q-meter. The current is maintained constant throughout.

EXERCISES 7

1. A sinusoidal waveform is displayed on the C.R.O. and has a peak-to-peak 7.5 cm. There are three complete cycles which are spanned to 9 cm horizontally.

   If the Y-control is set to 5 V/cm and the time base-control at 1 μs/cm determine the peak-to-peak voltage of the waveform, the period and the frequency.

2. Explain the significance of the Lissajous's figures and how you would use the C.R.O. to measure an unknown frequency.

3. (i) (ii) (iii) (iv)

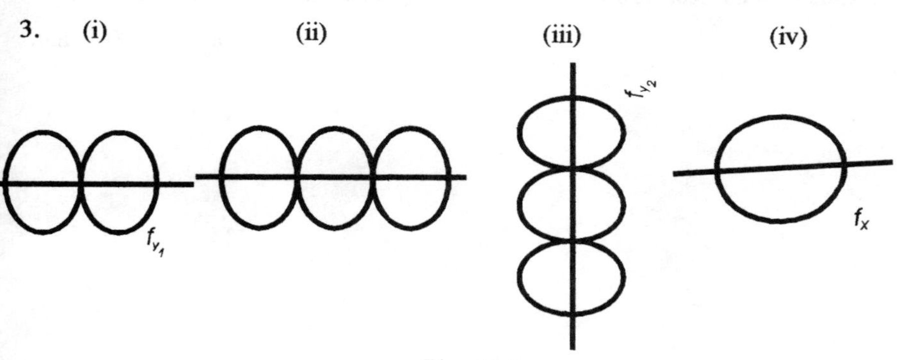

Fig. 294

From the figures above, determine the frequencies.

4.

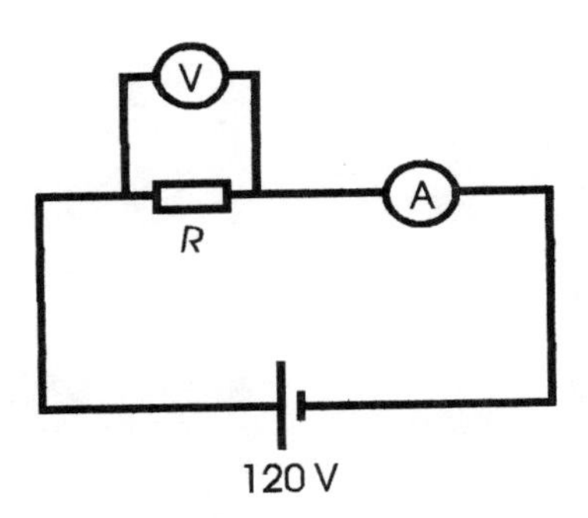

Fig. 295

It is required to measure the resistance R by an ammeter and voltmeter method. The ammeter reading is 10 mA and the p.d. across R is 40 volts, the resistance of the voltmeter is 50 KΩ. Calculate the exact value of the resistance.

5. Hay's Bridge is used to measure the inductance and its effective resistance of a coil. If $P = 10\ \Omega$ and $Q = 100\ \Omega$, $C = 50\ \mu F$, $R = 100\ \Omega$. Calculate L_x and R_x of the coil.

6. The tolerances of the resistors of P and Q in the Maxwell's bridge is $\pm$ 1% and R has a tolerance of $\pm$ 1%. Calculate the minimum and maximum errors in evaluating L_x and R_x.

7. An AVO when is used for a.c. measurements of voltages, the sensitivity is 200 Ω/V and when is used for d.c. measurements of voltages, the sensitivity is 20000 Ω/V.

   Calculate the resistance of the instrument when (a) measures 10 V d.c.

   (b) measures 300 V a.c.

ELECTRICAL AND ELECTRONIC PRINCIPLES III

SUMMARY

SERIES RESONANCE

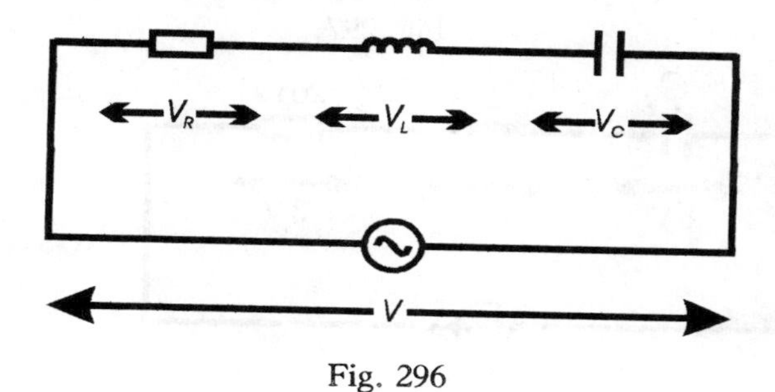

Fig. 296

$Z = \sqrt{R^2 + (X_L - X_C)^2}$

$\cos\theta = \frac{R}{Z}$, $\cos\theta$ is power factor

Power absorbed by coil $P = I^2R$, $P = IV\cos\theta$

$$f_o = \frac{1}{2\pi\sqrt{LC}}$$

At resonant frequency the impedance is a pure resistance and is a minimum for the circuit.

$Q = \frac{\omega_o L}{R}$ Q = voltage magnification

$Q = \frac{1}{R}\sqrt{\frac{L}{C}}$

Apparant Power $(S) = VI$

Reactive Power $(Q) = VI\sin\phi$

True Power $(P) = VI\cos\phi$.

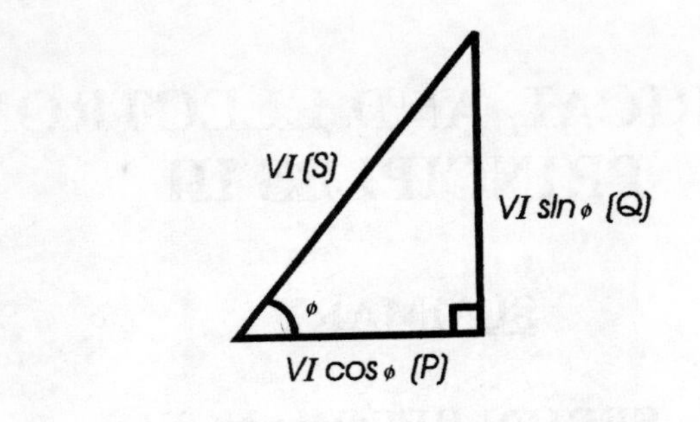

Fig. 297

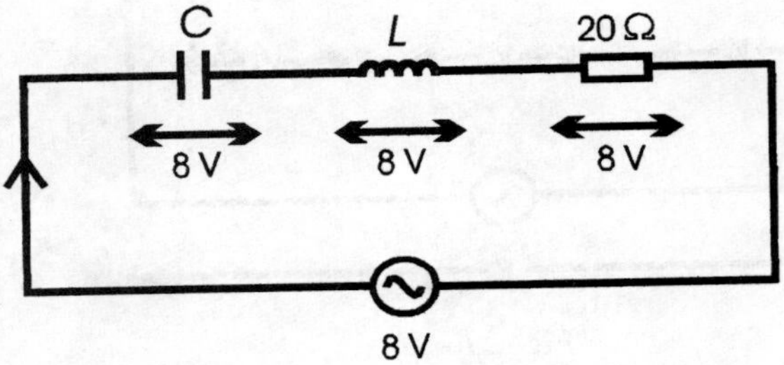

Fig. 298

INDUCTANCE AND CAPACITANCE IN PARALLEL

The current taken by the inductance is

$I_L = \dfrac{V}{X_L}$ lagging by 90°

$I_C = \dfrac{V}{X_C}$ lagging by 90°

$$f_o = \frac{1}{2\pi\sqrt{LC}}$$

$$I \text{ magnification} = \theta = \frac{\omega_o L}{R}.$$

The Q-factor is a measure of voltage magnification in a series circuit and of current magnification in a parallel circuit.

In a resonant circuit the impedance is known as a dynamic impedance.

$$Z_d = \frac{L}{CR}$$

TRANSIENTS

Charging a capacitor.

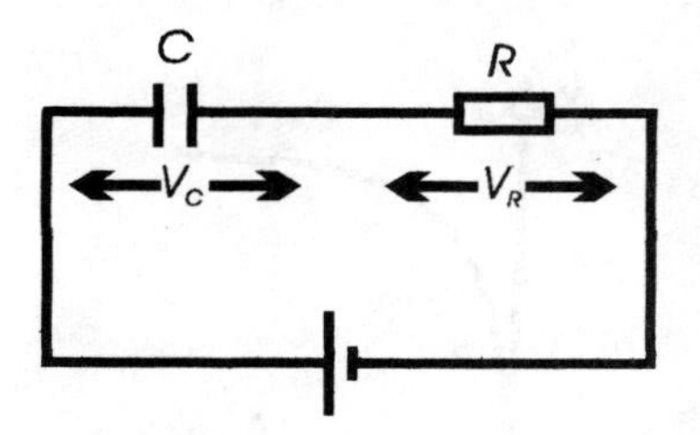

Fig. 299

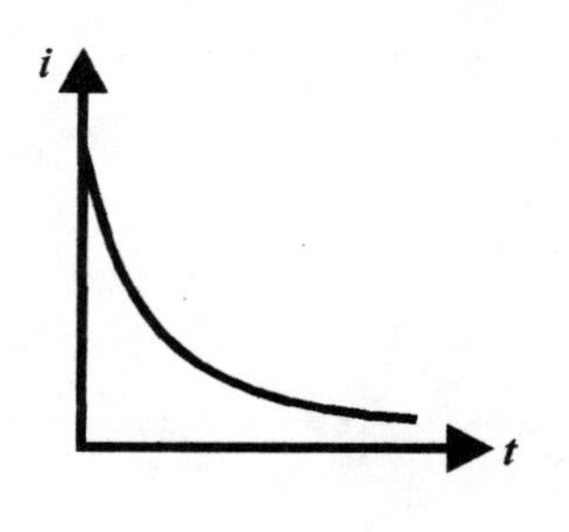

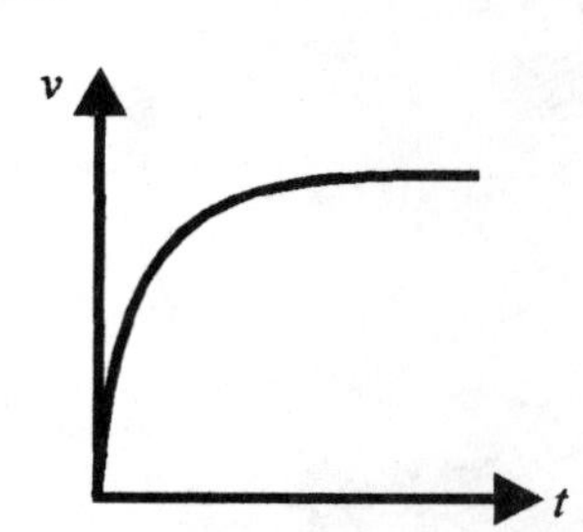

Fig. 300

$$v_C = V\left(1 - e^{-t/\tau}\right)$$

$$i = Ie^{-t/\tau}$$

$$q = Q\left(1 - e^{-t/\tau}\right)$$

$$v_R = Ve^{-t/\tau}$$

Time constant $(\tau) = RC$

$$\left(\frac{\mathrm{d}i}{\mathrm{d}t}\right)_{t=0} = -\frac{I}{\tau}$$

$$\left(\frac{\mathrm{d}i}{\mathrm{d}v}\right)_{t=0} = \frac{V}{\tau}$$

Energy stored $= \frac{1}{2}cv^2$

$$Q = CV$$

DISCHARGING A CAPACITOR

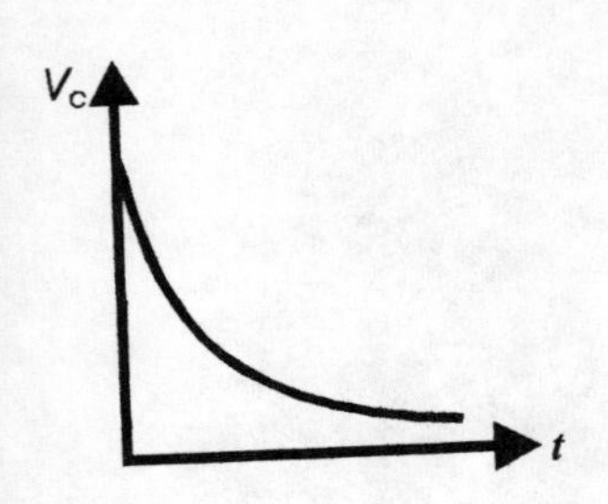

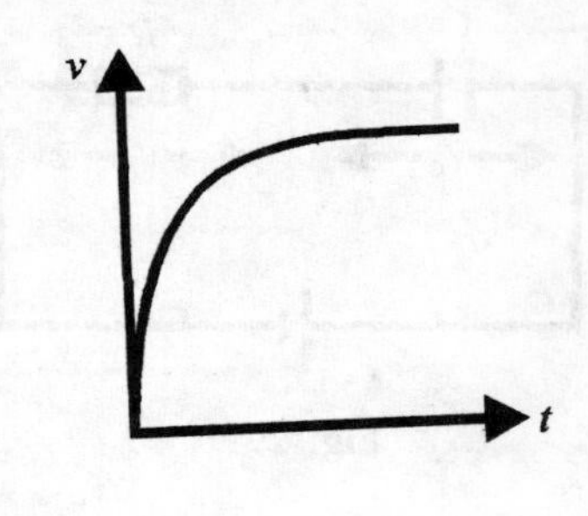

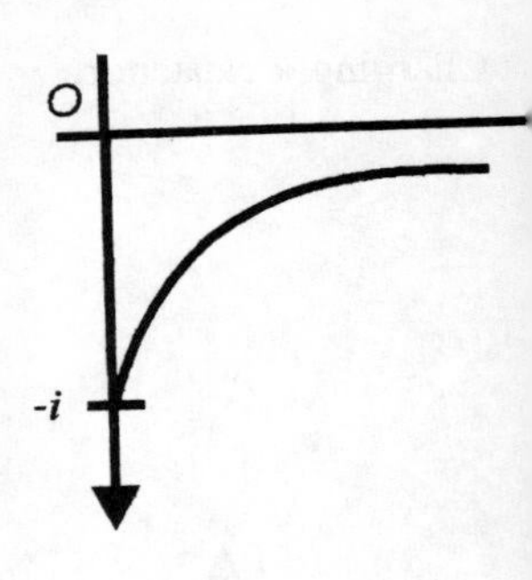

Fig. 301

$$v_C = Ve^{-t/\tau}$$

$$i = -Ie^{-t/\tau}$$

$$v_R = -Ve^{-t/\tau}$$

$$q = Qe^{-t/\tau}$$

$$W = \frac{1}{2}cv^2$$

$$Q = cv$$

LR CIRCUIT

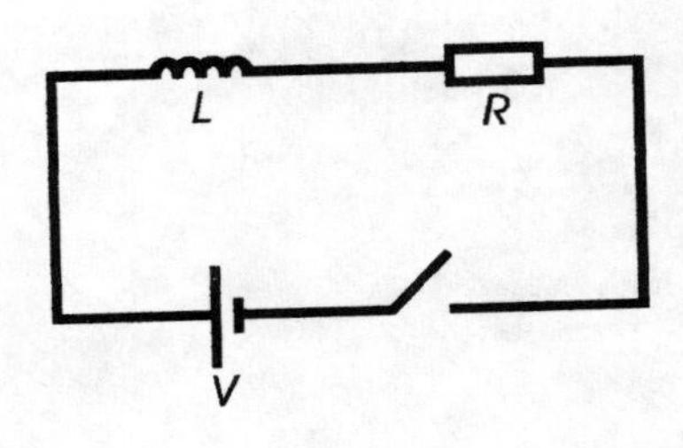

$i = 0$ initially

$i = I$ finally.

Fig. 302

$$i = I\left(1 - e^{-t/\tau}\right) \qquad \tau = \frac{L}{R}$$

Growth of current

$$v_L = Ve^{-t/\tau} \qquad v_R = V\left(1 - e^{-t/\tau}\right)$$

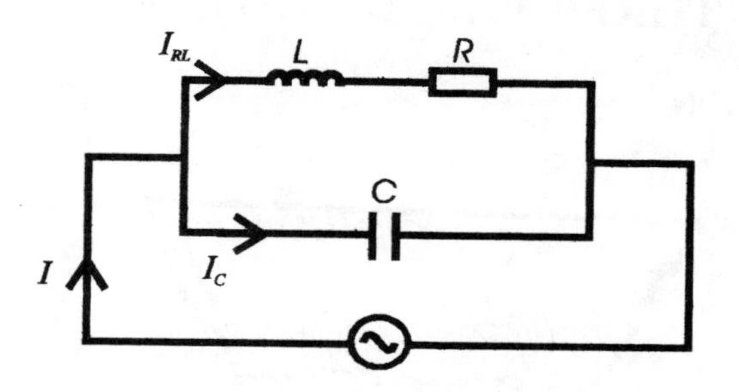

Fig. 303

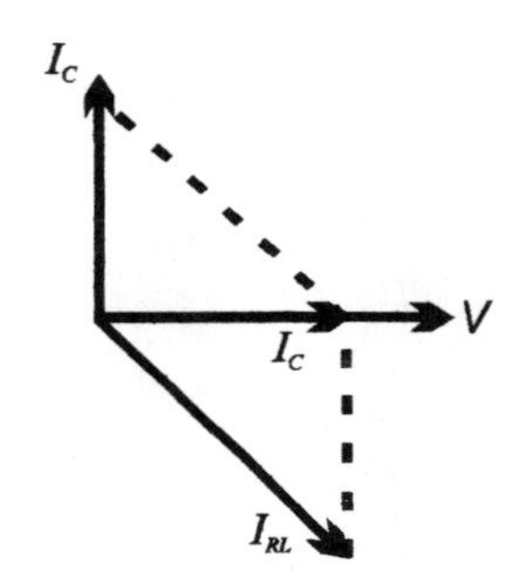

$$\tan\theta = \frac{v_L}{v_R} = \frac{\omega_o L I_{2L}}{R I_{2L}} = Q$$

$$Q = \frac{\omega_o L}{R} = \tan\theta$$

Fig. 304

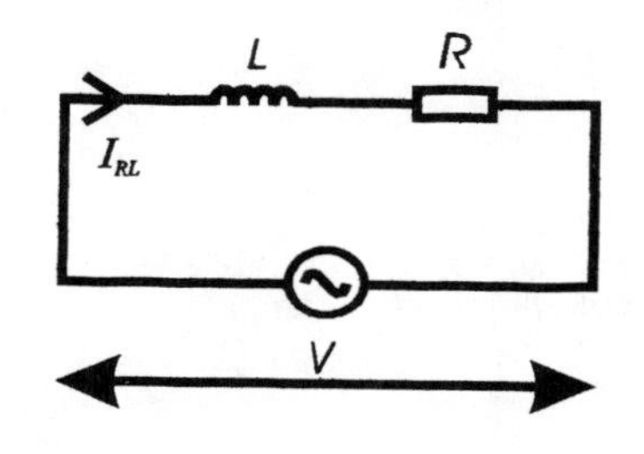

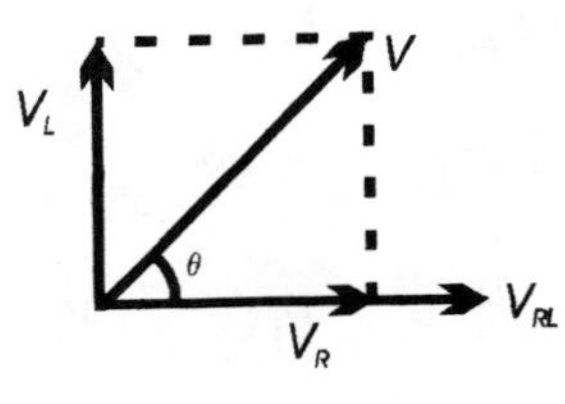

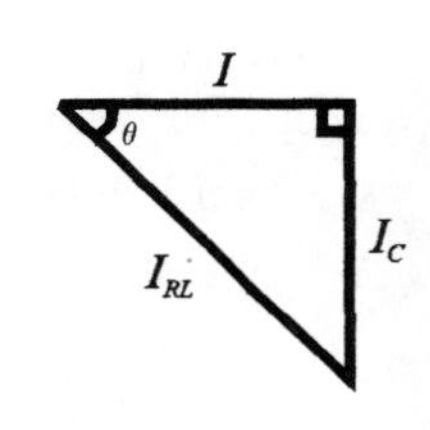

Fig. 305

$$f_o = \frac{1}{2\pi}\sqrt{\frac{1}{LC} - \frac{R^2}{L^2}} \qquad 2d = \frac{L}{CR}$$

$$\tan\theta = \frac{I_c}{I}$$

$$I_c = \tan\theta I \qquad I_c = QI$$

THREE PHASE SYSTEMS

STAR CONNECTION

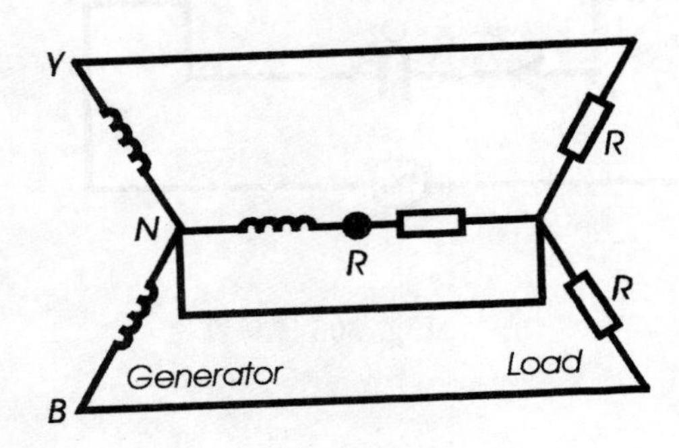

Fig. 306

$I_L = I_P$

$v_L = \sqrt{3}\, V_P$

General formula for power in an a.c. circuit is

$$P = VI \cos \phi$$

Power Per Phase $= P_P = V_P I_P \cos \phi$

Total Power $= P_T = 3 V_P I_P \cos \phi$

$$P_T = \sqrt{3}\, V_L I_L \cos \phi$$

Current in Red line $= \dfrac{V_P}{R}$

Impedance Per Phase $= Z = \sqrt{R^2 + x^2}$

Phase Current $= I_P = \dfrac{V_P}{Z_P}$

Power factor $= \cos \theta = \dfrac{R}{Z_P}$

Apparent Power $S = \sqrt{3}\ V_L I_L$

$$\text{Power factor} = \cos\phi = \frac{\text{Power (P)}}{\text{Volt-ampere (S)}}$$

Reactive Power $Q = VI \sin\phi$

$$Z \text{ per phase} = Z_P = \frac{V_P}{I_P}$$

$$P \text{ per phase} = P_P = \frac{P}{3}. \quad \text{Resistance Per Phase} = R = \frac{P_P}{I_p^2}$$

Reactance Per Phase

$$X_P = \sqrt{Z_P^2 - R_P^2}\text{^R}$$

DELTA CONNECTION

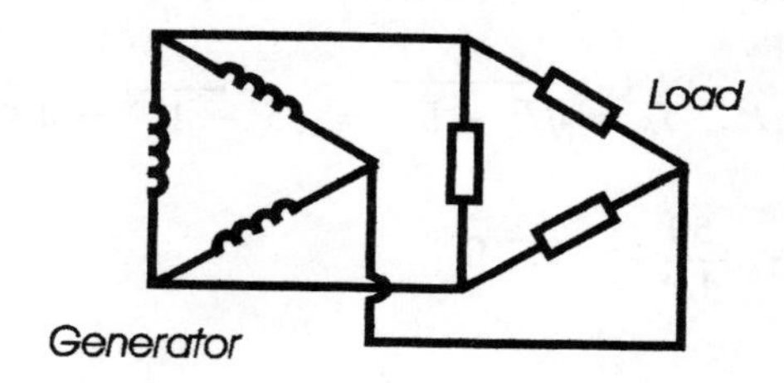

Fig. 307

$$v_L = v_P$$

$$I_L = \sqrt{3}\ I_P$$

Total active power input to motor.

$P = \sqrt{3}\ V_L I_L$

ELECTRICAL AND ELECTRONIC PRINCIPLES III

PART II

SOLUTIONS 1

1. See text

$$\frac{1}{Z} = \sqrt{\frac{1}{X_C^2} + \frac{1}{R^2}} = \sqrt{\frac{R^2 + X_C^2}{X_C^2 R^2}}$$

$$Z = \frac{RX_C}{\sqrt{R^2 + X_C^2}}$$

2. (a) $X_C = \frac{V}{I_C} = \frac{50}{1} = 50\ \Omega,\ X_C = \frac{1}{2\pi fC} = \frac{1}{2\pi 50 C} = 50$

$$C = \frac{1}{5000\pi} = 63.7\ \mu\text{F}, \quad \boxed{C = 63.7\ \mu\text{F}}$$

(b) $I_R^2 + I_C^2 = I^2,\ I_R = \sqrt{I^2 - I_C^2} = \sqrt{5^2 - 1^2} = 4.9$ A

(c) $R = \frac{V}{4.9} = \frac{50}{4.9} = 10.2\ \Omega$

(d) $IV\cos\theta = P = 5 \times 50 \times \frac{I_R}{I}$

$$= 5 \times 50 \times \frac{4.9}{5.0} = 245\ \text{W}$$

$$\cos\theta = \frac{245}{250} = 0.98.$$

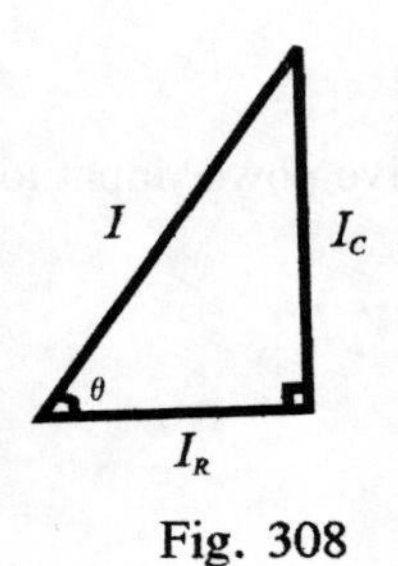

Fig. 308

3.

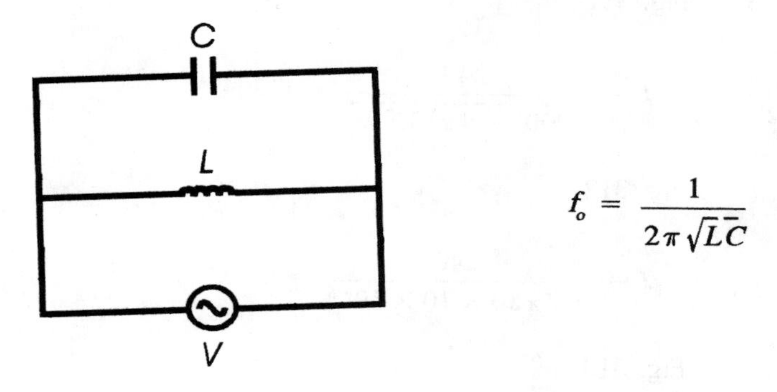

$$f_o = \frac{1}{2\pi\sqrt{LC}}$$

Fig. 309

$$4\pi^2 LC = \frac{1}{f_o^2} \qquad \text{If } C = 200 \text{ pF}, \; L = \frac{1}{4\pi^2 C f_o^2}$$

$$= \frac{1}{4\pi^2 200 \times 10^{-12}\, 10^{12}} = 12.7 \text{ mH}$$

4. See pages 13 and 14.

The resonant frequency, $f_o = \frac{1}{2\pi}\sqrt{\frac{1}{LC} - \frac{R^2}{L^2}}$.

The impedance $R_d = \frac{L}{CR}$ is resistive at resonance.

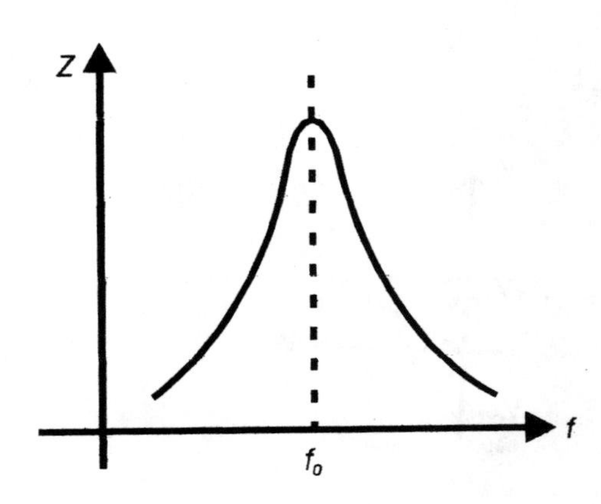

Fig. 310

$$Z = \frac{V}{I} = R_d = \text{maximum since } I \text{ is minimum.}$$

5. Fig. 311

$$I = \frac{240}{\sqrt{200^2 + 4\pi^2 50^2 2^2}} = 0.364 \text{ A}$$

Fig. 312

$$I = \frac{240}{1/2\pi 50 \times 10 \times 10^{-6}} = 0.754 \text{ A}$$

Fig. 313

$$I = \frac{240}{\sqrt{200^2 + (X_L - X_C)^2}} = \frac{240}{\sqrt{200^2 + (628.3 - 318.3)^2}} = 0.65 \text{ A}$$

where $X_L = 2\pi fL = 2\pi 50 \times 2 = 200\pi = 628.3\ \Omega$

$$X_C = \frac{1}{2\pi fC} = \frac{1}{2\pi 50 \times 10 \times 10^{-6}} = 318.3\ \Omega$$

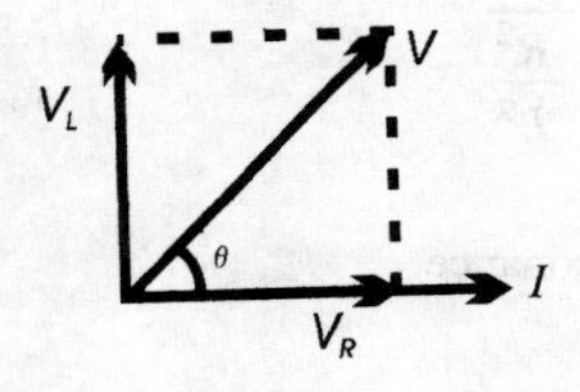

Fig. 311

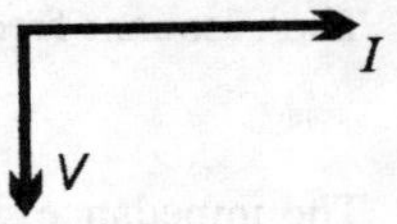

Fig. 312

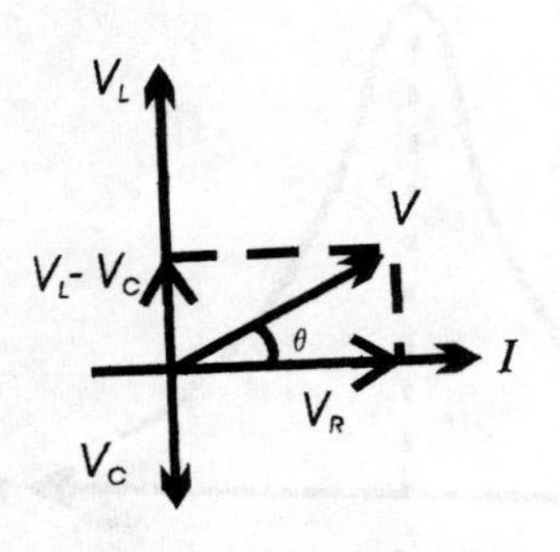

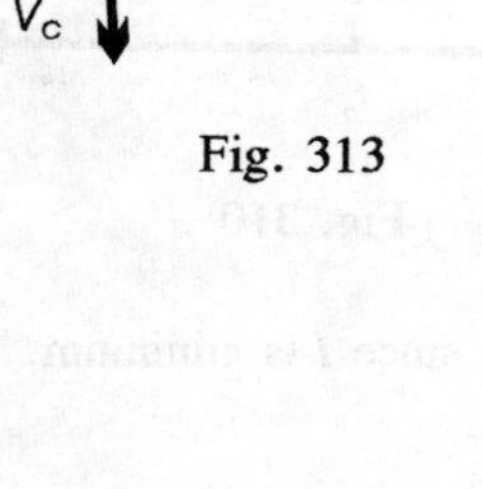

Fig. 313

For Fig. 313

$$X_L = X_C \qquad 2\pi f_o L = \frac{1}{2\pi f_o C}$$

$$f_o = \frac{1}{2\pi\sqrt{LC}} = \frac{1}{2\pi\sqrt{2 \times 10 \times 10^{-6}}} = 35.6 \text{ Hz}$$

(i) $I = \dfrac{240}{200} = 1.2 \text{ A}$

(ii) $V_R = 1.2 \times 200 = 240$ volts

(iii) $I^2R = 1.2^2 \times 200 = 1.44 \times 200 = 288 \text{ W}$

(iv) $\theta = 0°$

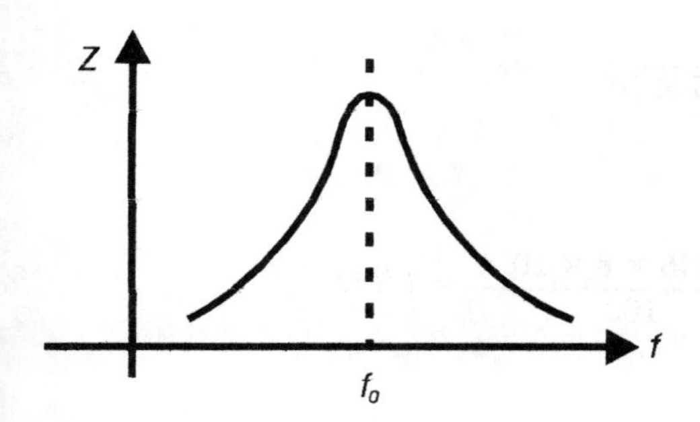

Fig. 314

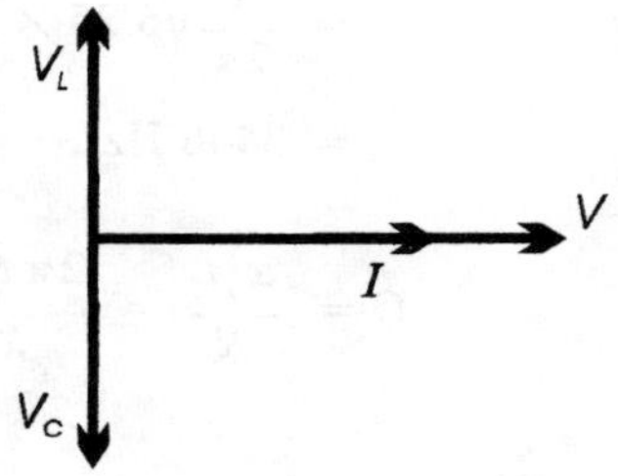

Fig. 315

6.

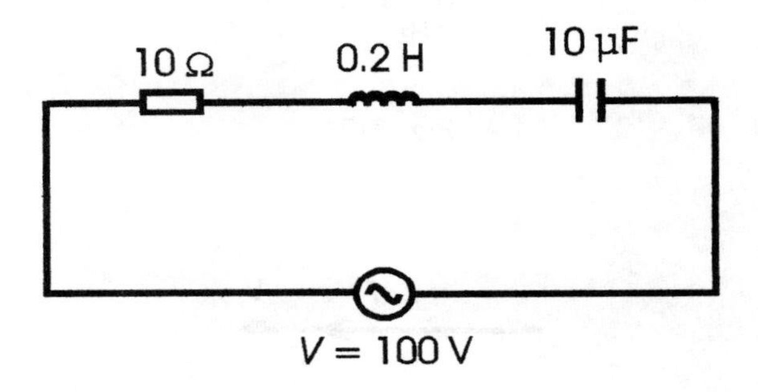

Fig. 316

Maximum current occurs at resonance, that is, when $X_L = X_C$.

$$I_o = \text{maximum current} = \frac{100}{10} = 10 \text{ A.}$$

$$Q = \frac{\omega_o L}{R} = \frac{2\pi f_o L}{R} = \frac{1}{2\pi f_o CR} = \tan\theta = 5 \Rightarrow \frac{1}{2\pi f_o C} = 5 \times 10 = 50$$

$$V_C = I_o X_c = 10 \times 50 = 500 \text{ volts.}$$

7. See text.

$$f_o = \frac{1}{2\pi}\sqrt{\frac{1}{LC} - \frac{R^2}{L^2}}$$

$$= \frac{1}{2\pi}\sqrt{\frac{1}{8 \times 10^{-3} \times 2 \times 10^{-7}} - \frac{100^2}{64 \times 10^{-6}}}$$

$$= \frac{1}{2\pi}\sqrt{6.25 \times 10^8 - 1.5625 \times 10^8}$$

$$= 3446 \text{ Hz.}$$

$$Q = \frac{\omega_o L}{R} = \frac{2\pi f_o L}{R} = \frac{2\pi\, 3446 \times 8 \times 10^{-3}}{100} = 1.732$$

8.

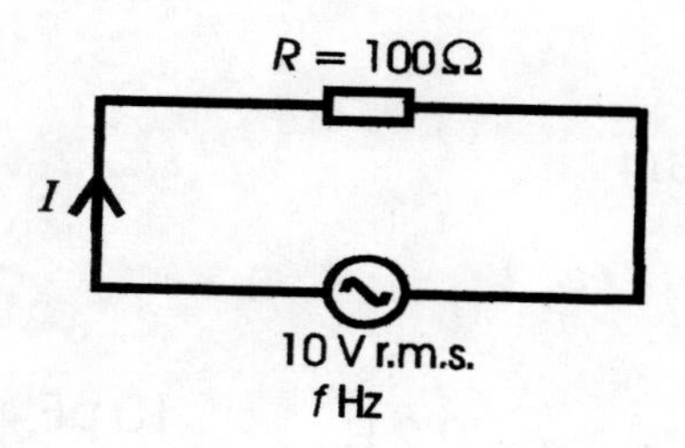

Fig. 317

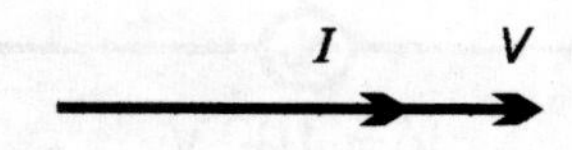

Fig. 318

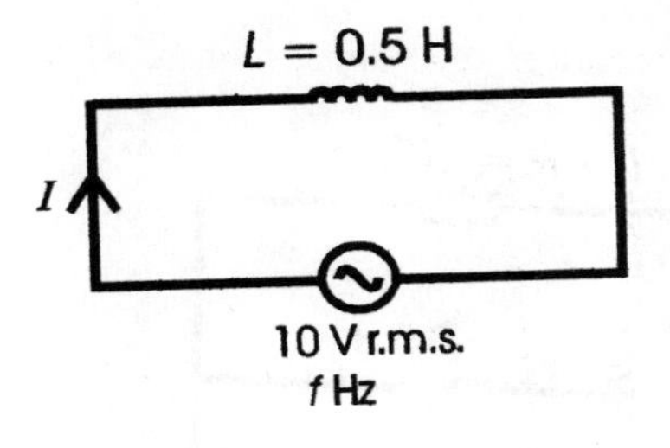

Fig. 319

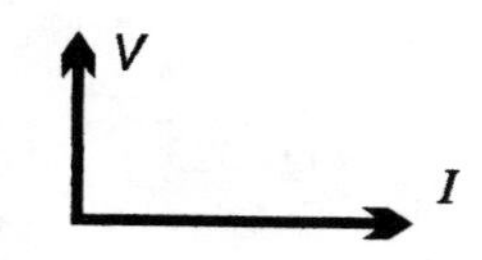

Fig. 320

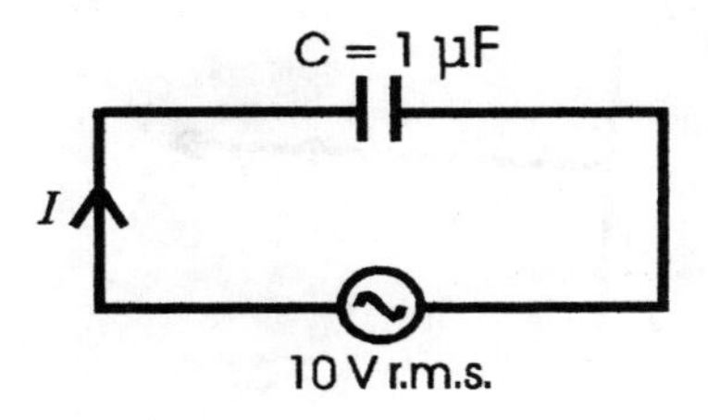

Fig. 321

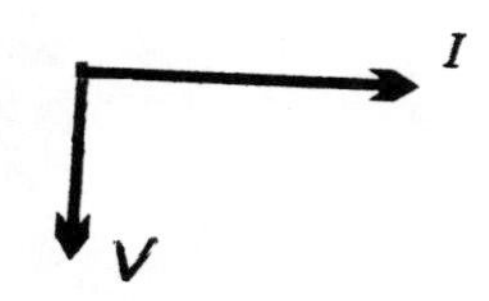

Fig. 322

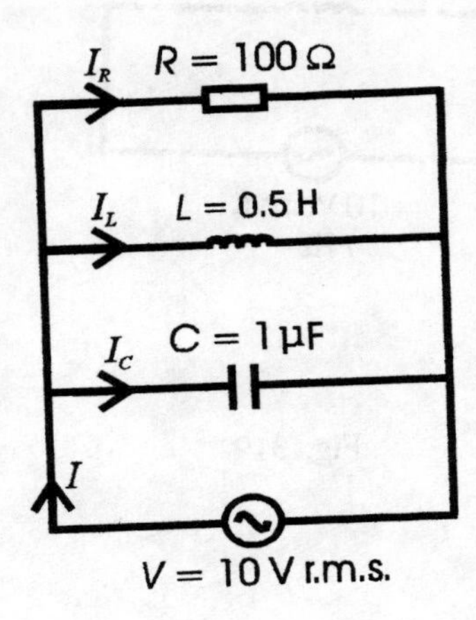

Fig. 323

The sinusoidal voltage 10 V r.m.s. is common

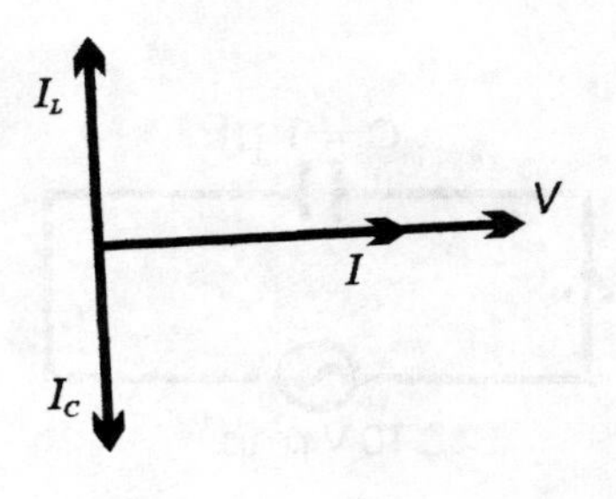

Fig. 324

At resonance

$$I_L = I_C$$

$$\frac{V}{X_L} = \frac{V}{X_C}$$

$$X_L = X_C$$

$$2\pi f_o L = \frac{1}{2\pi f_o C}$$

$$\boxed{f_o = \frac{1}{2\pi\sqrt{LC}}}$$

$$f_o = \frac{1}{2\pi\sqrt{0.5 \times 1 \times 10^{-6}}} = 225 \text{ Hz}$$

$$I_L = \frac{10}{2\pi f_o L} = \frac{10}{2\pi\, 225 \times 0.5} = 14.2 \text{ mA}$$

$$I_C = \frac{10}{X_C} = I_L = 14.2 \text{ mA}$$

$I = \dfrac{10}{100} = 0.1$ or 100 mA $= I_R$ the supply current.

At resonance $Z = R = 100\ \Omega$

$$Q = \frac{2\pi f_o L}{R} = \frac{2\pi\, 225 \times 0.5}{100} = 7.07.$$

9.

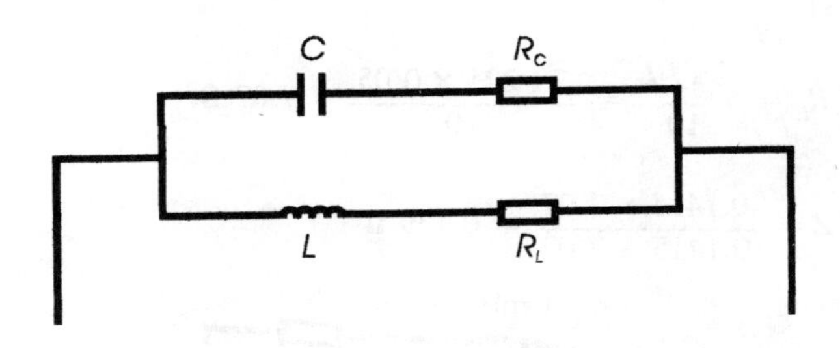

Fig. 325

$$Q_C = 500 = \frac{1}{\omega_o C R_C} \qquad Q_L = 10 = \frac{\omega_o L}{R_L}$$

$$Q = \frac{500 \times 10}{500 + 10} = 9.8$$

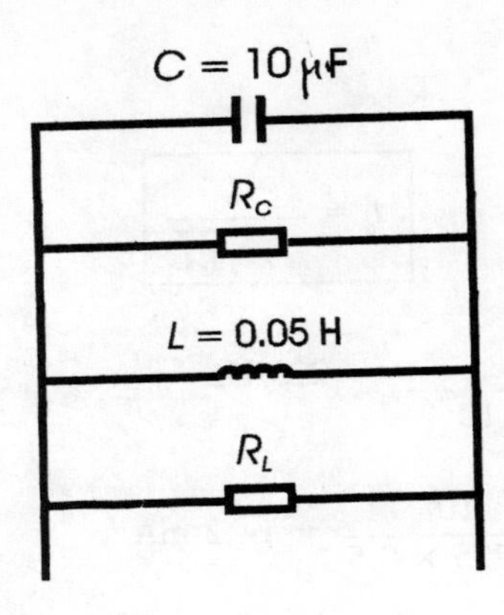

Fig. 326

$$\omega_o L = \frac{1}{\omega_o C}$$

(a) $$f_o = \frac{1}{2\pi\sqrt{LC}} = \frac{1}{2\pi\sqrt{0.05 \times 10 \times 10^{-6}}} = 225 \text{ Hz}$$

(b) $$Z = R = \frac{R_C R_L}{R_C + R_L}$$

$$R_C = \frac{1}{2\pi f_o C 500} = \frac{1}{2\pi 225 \times 10 \times 10^{-6} \times 500} = 0.1415 \ \Omega.$$

$$R_L = \frac{2\pi f_o L}{10} = \frac{2\pi 225 \times 0.05}{10} = 7.07 \ \Omega.$$

$$Z = \frac{0.1415 \times 7.07}{0.1415 + 7.07} = 0.139 \ \Omega$$

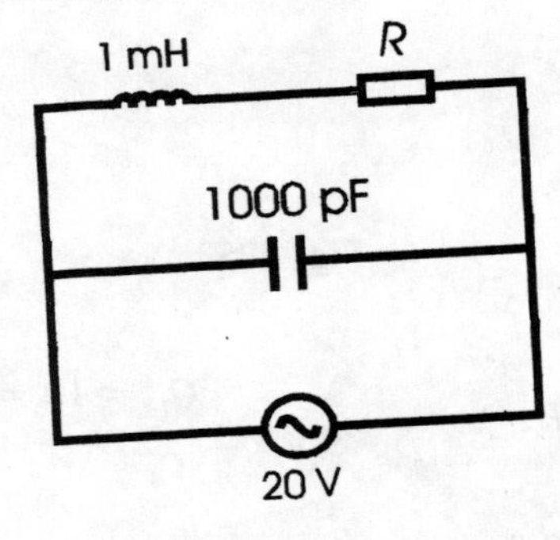

Fig. 327

10.

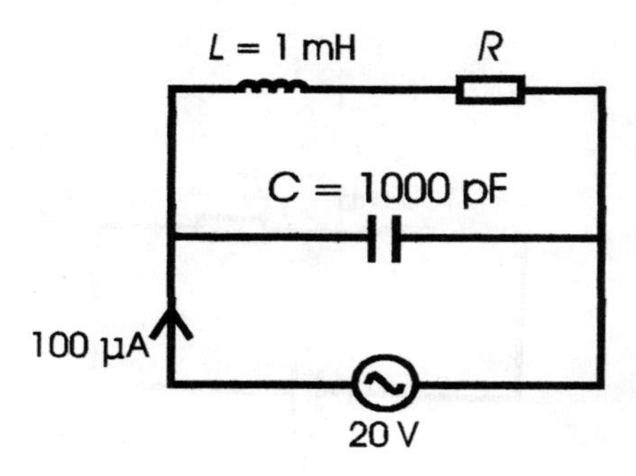

Fig. 328

$$R_d = \frac{L}{CR} = \frac{10^{-3}}{1000 \times 10^{-12} R} = 200 \times 10^3$$

$$R = \frac{10^{-3}}{1000 \times 10^{-12} \times 200 \times 10^3} = 5\ \Omega$$

(a) $$f_o = \frac{1}{2\pi}\sqrt{\frac{1}{LC} - \frac{R^2}{L^2}}$$

$$= \frac{1}{2\pi}\sqrt{\frac{1}{10^{-3} \times 1000 \times 10^{-12}} - \frac{25}{10^{-6}}}$$

$$= \frac{1}{2\pi}\sqrt{10^{12} - 25 \times 10^6} \approx 159 \text{ KHz}$$

(b) $$I_C = \frac{20}{X_C} = \frac{20}{1000} = 0.02 \text{ A}$$

$$X_C = \frac{1}{2\pi \times 159 \times 10^3 \times 1000 \times 10^{-12}} = 1000\ \Omega$$

(c) $$Q = \frac{\omega_o L}{R} = \frac{2\pi f_o L}{R} = \frac{2\pi \times 159 \times 10^3 \times 10^{-3}}{5}$$

= 200 which practically is very high.

11.

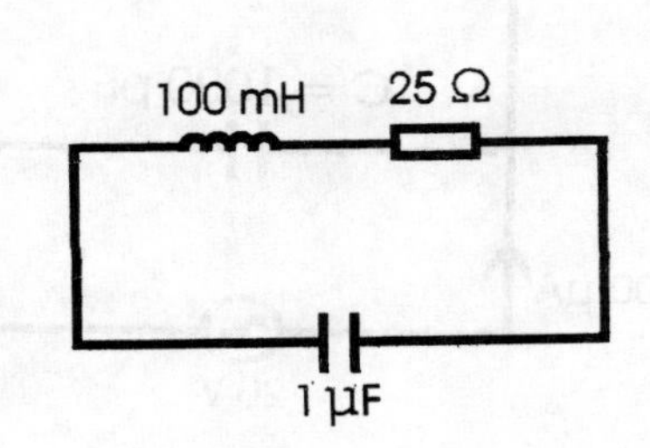

Fig. 329

(i) $$f_o = \frac{1}{2\pi}\sqrt{\frac{1}{LC} - \frac{R^2}{L^2}}$$

$$= \frac{1}{2\pi}\sqrt{\frac{1}{100 \times 10^{-3} \times 10^{-6}} - \frac{25^2}{0.01}}$$

$$= \frac{1}{2\pi}\sqrt{10^7 - 62500} = 502 \text{ Hz}$$

(ii) $$Z = \frac{L}{CR} = \frac{100 \times 10^{-3}}{1 \times 10^{-6} \times 25} = 4 \text{ K}\Omega$$

(iii) $$Q = \frac{\omega_o L}{R} = \frac{2\pi f_o L}{R} = \frac{2\pi \times 502 \times 100 \times 10^{-3}}{25} = 12.6.$$

12. $$X_L = 2\pi fL = 2\pi\, 10^3 \times 150 \times 10^{-3} = 942.5 \ \Omega$$

$$X_C = \frac{1}{2\pi fC} = \frac{1}{2\pi \times 10^3 \times 200 \times 10^{-12}} = 795775 \ \Omega.$$

Using complex numbers

$$\frac{jX_L(-jX_C)}{jX_L - jX_C} = \frac{X_L X_C}{j(X_L - X_C)} = \frac{942.5 \times 795775}{-j794833} \times \frac{j}{j} = j944 \ \Omega$$

$j944\ \Omega$, the reactance is inductive.

$$I_L = \frac{240}{942.5} = 0.2546419$$

$$I_C = \frac{240}{795775} = 3.0159279 \times 10^{-4}$$

$I_L - I_C = 0.2543403$, the total current

the total reactance $= X_L = \dfrac{240}{0.2543403} = 944\ \Omega.$

13. $$Z = \frac{10 \times 10}{\sqrt{10^2 + 10^2}} = \frac{100}{\sqrt{2} \times 10} = \frac{10}{\sqrt{2}} = 7.07\ \Omega$$

Using complex numbers

$$Z = \frac{10(-j10)}{10 - j10} = \frac{-j100}{10(1 - j)} = \frac{-j10}{1 - j} \times \frac{1 + j}{1 + j} = -5j(1 + j)$$

$$= -5j - 5j^2 = 5 - 5j$$

$$|Z| = \sqrt{25 + 25} = 7.07\ \Omega.$$

14. $$X_L = 2\pi fL = 2\pi 1000 \times 20 \times 10^{-3} = 40\pi = 126\ \Omega$$

$$X_C = \frac{1}{2\pi fC} = \frac{1}{2\pi 10^3 \times 2.5 \times 10^{-6}} = 63.7\ \Omega.$$

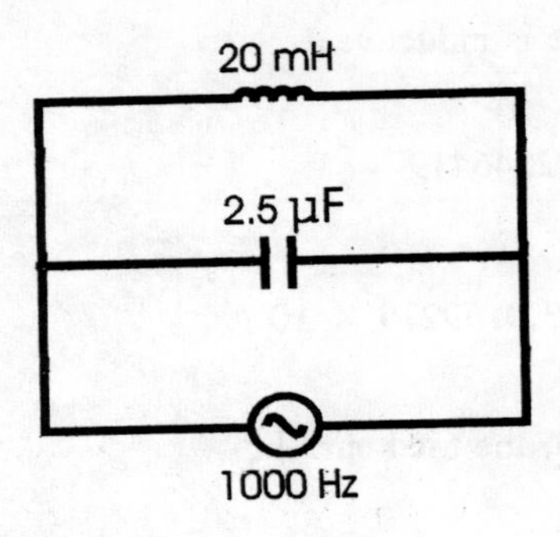

Fig. 330

$$\frac{j126 \times (-j63.7)}{j126 - j63.7} = \frac{8026.2}{j62.3} = -j129.$$

The circuit is capacitive.

15. $R_d = \frac{L}{CR} = \frac{5 \times 10^{-3}}{15 \times 10^{-12} \times 5} = \frac{10^9}{15} = 66.7 \text{ M}\Omega.$

16. $R_d = \frac{L}{CR} = \frac{5 \times 10^{-3}}{15 \times 10^{-12} \times 5} = 66.7 \text{ M}\Omega.$

$$I = \frac{100}{R_d} = \frac{100}{66.7 \times 10^6} = 14.9 \ \mu\text{A}.$$

17.

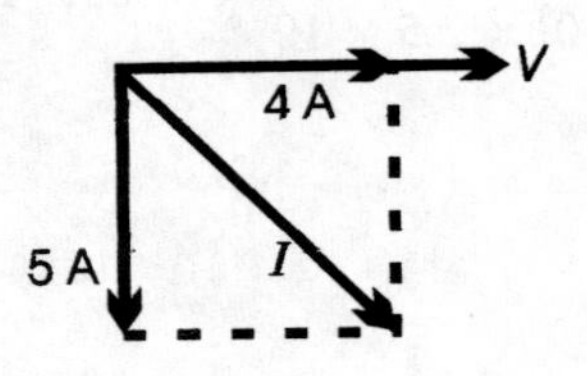

Fig. 331

$$I = \sqrt{4^2 + 5^2} = 6.4 \text{ A}.$$

18. $Z = \dfrac{X_C R}{\sqrt{R^2 + X_C^2}} = \dfrac{10 \times 10}{\sqrt{10^2 + 10^2}} = \dfrac{10}{\sqrt{2}} = 7.07\ \Omega.$

19.

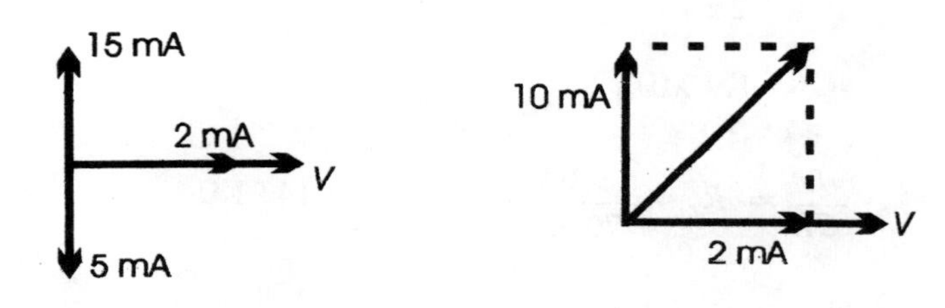

Fig. 332 Fig. 333

$$I = \sqrt{(2 \times 10^{-3})^2 + (10 \times 10^{-3})^2} = 10.2 \text{ mA}.$$

20. $X_L = 2\pi 1000 \times 40 \times 10^{-3} = 251.3\ \Omega$

$$X_C = \frac{1}{2\pi 1000 \times 2 \times 10^{-6}} = 79.6\ \Omega$$

$$I = \frac{200}{\sqrt{100^2 + (251.3 - 79.6)^2}} = 1 \text{ A}.$$

21. (i) Phase angle = 0°.

(ii) $Q = \dfrac{\omega_o L}{R} = 120 = \dfrac{X_L}{R}$ $\qquad V = I_O R = 40 \times 10^{-3}$

$$V_C = I_O X_C = \frac{40 \times 10^{-3}}{R} \cdot X_L$$

$$V_C = \frac{40 \times 10^{-3}}{R} 120R$$

$$= 40 \times 10^{-3} \times 120$$

$$= 4.8 \text{ volts}.$$

22. (i) $f_o = \frac{1}{2\pi}\sqrt{\frac{1}{100\times10^{-6}\times70\times10^{-12}} - \frac{10^2}{(100\times10^{-6})^2}}$

$= \frac{1}{2\pi}\sqrt{1.429\times10^{14} - 1\times10^{10}}$

$\approx$ 1.9 MHz

(ii) $\frac{L}{CR} = R_d = \frac{100\times10^{-6}}{70\times10^{-12}\times10} = 143\ \text{K}\Omega$

(iii) $\frac{I_C}{I} = 180,\ I_C = 180\times25\times10^{-6} = 4.5\ \text{mA}$

(iv) $I = \frac{V}{R_d} = \frac{100}{143\times10^3} = 0.699\ \text{mA}.$

23. See text.

24. See text.

$$Q = \tan\theta = \frac{\omega_o L}{R}$$

$$\omega_o = \sqrt{\frac{1}{LC} - \frac{R^2}{L^2}} = \sqrt{\frac{1}{8\times10^{-3}\times2\times10^{-7}} - \frac{100^2}{(8\times10^{-3})^2}}$$

$$= \sqrt{6.25\times10^8 - 1.5625\times10^8} = 21650.635$$

$$Q = \frac{21650.635\times8\times10^{-3}}{100} = 1.732.$$

25. $I_o = \frac{100}{10} = 10\ \text{A}$

$V_C = I_O X_C = 10\times50 = 500$ volts

where $\tan\theta = 5 = \frac{\omega_o L}{R} = \frac{1}{R\omega_o C}$

$$X_c = \frac{1}{\omega_o C} = 5R = 5\times10 = 50$$

26. (a) $f_o = \dfrac{1}{2\pi}\sqrt{\dfrac{1}{LC} - \dfrac{R^2}{L^2}}$

$$= \frac{1}{2\pi}\sqrt{\frac{1}{50 \times 10^{-6} \times 80 \times 10^{-12}} - \frac{5^2}{(50 \times 10^{-6})^2}}$$

$$= \frac{1}{2\pi}\sqrt{2.5 \times 10^{14} - 1.0 \times 10^{10}}$$

$$\approx 2.52 \text{ MHz}$$

$$\boxed{f_o = 2.52 \text{ MHz}}$$

(b) $R_d = \dfrac{L}{CR} = \dfrac{50 \times 10^{-6}}{80 \times 10^{-12} \times 5} = \boxed{125 \text{ K}\Omega}$

(c) $I = \dfrac{50 \text{ mV}}{125 \times 10^3} = \boxed{0.4 \times 10^{-6} \text{ A}}$

$$Q = \frac{I_C}{I} = \frac{63.3 \times 10^{-6}}{0.4 \times 10^{-6}} = \boxed{158}$$

(d) $I_C = \dfrac{V}{X_C} = \dfrac{50 \times 10^{-3}}{790} = \boxed{63.3 \times 10^{-6} \text{ A}}$

where $X_C = \dfrac{1}{2\pi f_o C} = \dfrac{1}{2\pi \times 2.52 \times 10^6 \times 80 \times 10^{-12}}$

$$\boxed{\approx 790\ \Omega}$$

(e) $P = \dfrac{V^2}{R_d} = \dfrac{\left(50 \times 10^{-3}\right)^2}{125 \times 10^3} = 20$ nW or

$P = VI = 50 \times 10^{-3} \times 0.4 \times 10^{-6} = 20$ nW

27. (i) $R_d = \dfrac{L}{CR} = \dfrac{80 \times 10^{-3}}{0.2 \times 10^{-6} \times 100} = 4000\ \Omega$

(ii) $I = \dfrac{80}{R_d} = \dfrac{80}{4000} = 0.02$ A or 20 mA

(iii) $f_o = \dfrac{1}{2\pi}\sqrt{\dfrac{1}{LC} - \dfrac{R^2}{L^2}}$

$$= \frac{1}{2\pi}\sqrt{\frac{1}{80 \times 10^{-3} \times 0.2 \times 10^{-6}} - \frac{10^4}{80^2 \times 10^{-6}}} = 1242.4 \text{ Hz}$$

(iv) $X_C = \dfrac{1}{2\pi f_o C} = \dfrac{1}{2\pi \times 1242.4 \times 0.2 \times 10^{-6}} = 640.5\ \Omega$

(v) $I_C = \dfrac{V}{X_C} = \dfrac{80}{640.5} = 125$ mA

(vi) $I_L = \sqrt{I_C^2 + I^2} = \sqrt{(0.125)^2 + (0.02)^2} = 127$ mA

alternative $I_L = \dfrac{80}{\sqrt{100^2 + (2\pi\, 1242.4 \times 80 \times 10^{-3})^2}} = 127$ mA

(vii) $Q = \dfrac{I_C}{I} = \dfrac{125 \times 10^{-3}}{20 \times 10^{-3}} = 6.25$

(viii)

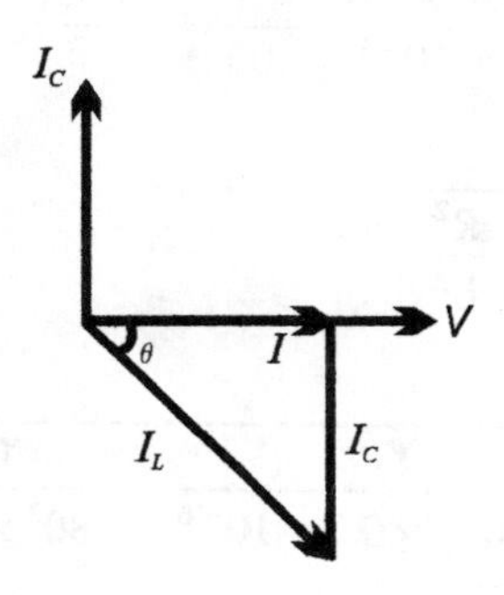

Fig. 334

$$\cos\theta = \frac{I}{I_L} = \frac{20 \text{ mA}}{127 \text{ mA}}$$

$$\theta = 80.9^\circ.$$

28. (i) $$f_o = \frac{1}{2\pi}\sqrt{\frac{1}{50\times10^{-6}\times80\times10^{-12}} - \left(\frac{5}{50\times10^{-6}}\right)^2}$$

$$= \frac{1}{2\pi}\sqrt{\frac{10^{15}}{4} - \frac{25}{2500\times10^{-12}}}$$

$$= \frac{1}{2\pi}\sqrt{2.5\times10^{14} - 10^9} \approx \frac{1.58\times10^7}{2\pi}$$

$$= 2.52 \text{ MHz}$$

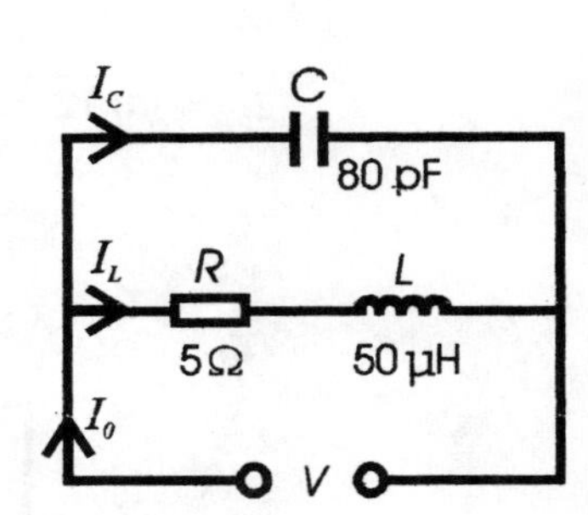

(ii) $$Q = \frac{I_C}{I_O}$$

(iii) $$\frac{L}{CR} = Z_O = \frac{50\times10^{-6}}{80\times10^{-12}\times5} = 125 \text{ K}\Omega$$

$$I_o = \frac{V}{Z_o} = \frac{50\times10^{-3}}{125\times10^3} = 0.4\ \mu\text{A}.$$

(iv) $$\frac{V^2}{Z_O} = \frac{50^2\times10^{-6}}{125\times10^3} = 20 \text{ nW}$$

29. (i) $R_d = \frac{L}{CR} = \frac{80 \times 10^{-3}}{0.2 \times 10^{-6} \times 100} = \frac{8000}{2} = 4000\ \Omega$

(ii) $f_o = \frac{1}{2\pi}\sqrt{\frac{1}{LC} - \frac{R^2}{L^2}}$

$$= \frac{1}{2\pi}\sqrt{\frac{1}{80 \times 10^{-3} \times 0.2 \times 10^{-6}} - \frac{10^4}{80^2 \times 10^{-6}}}$$

$$= \frac{1}{2\pi}\sqrt{62500000 - 1562500}$$

$$= 1242.4 \text{ Hz}$$

(iii) $I = \frac{80}{R_d} = \frac{80}{4000} = \frac{1}{50} = 0.02 \text{ A}$

(iv) $I_C = \frac{80}{X_C} = \frac{80}{640.5} = 0.125 \text{ A}$

$$X_C = \frac{10^6}{2\pi\, 1242.4 \times 0.2} = 640.5\ \Omega$$

(v) $I_L^2 = I_C^2 + I^2$

$$I_L = \sqrt{0.125^2 + 0.02^2} = 0.127 \text{ A}$$

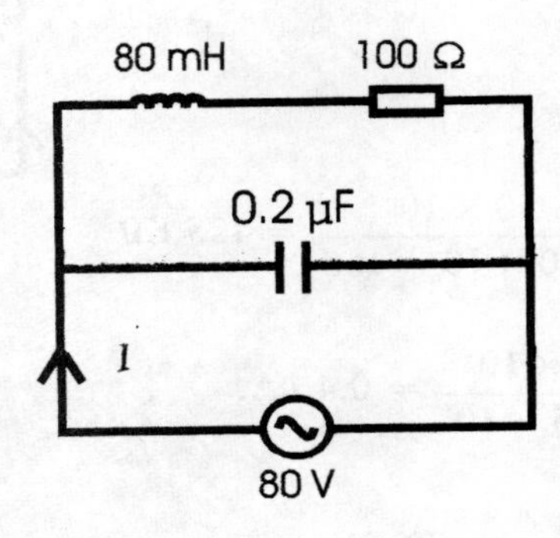

SOLUTIONS 2

1. (a) Replacing the constant current generator of 10 A, we have.

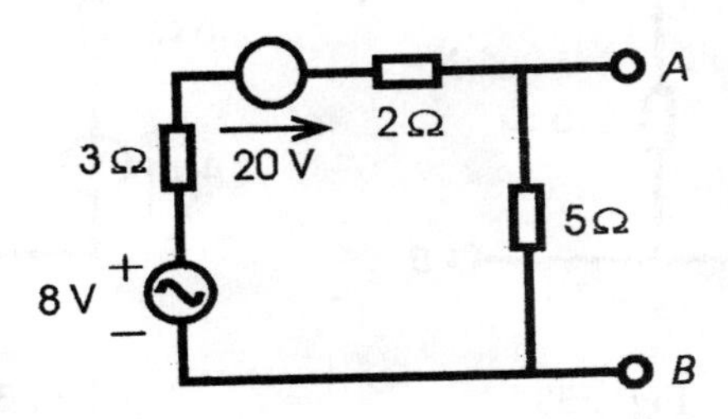

Fig. 335

Adding the e.m.f.s and the resistors, we have.

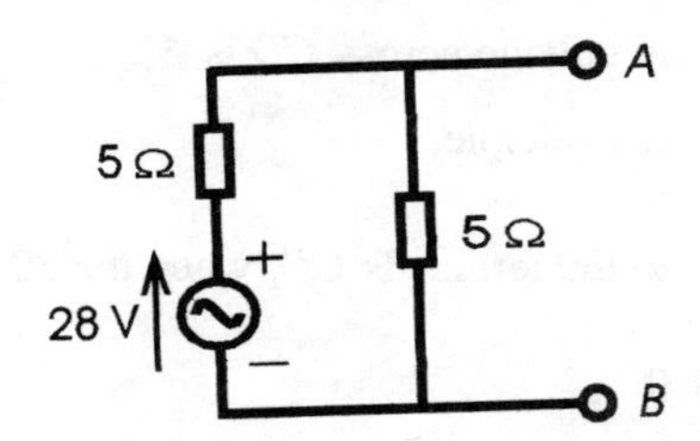

Fig. 336

Replacing the constant voltage generator of 28 V, we have.

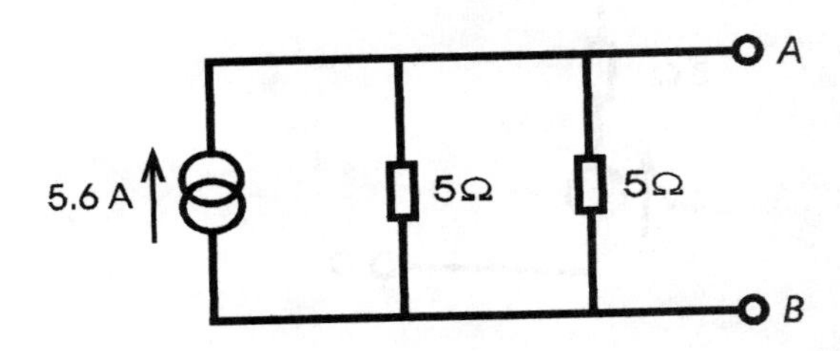

Fig. 337

The equivalent circuit is now.

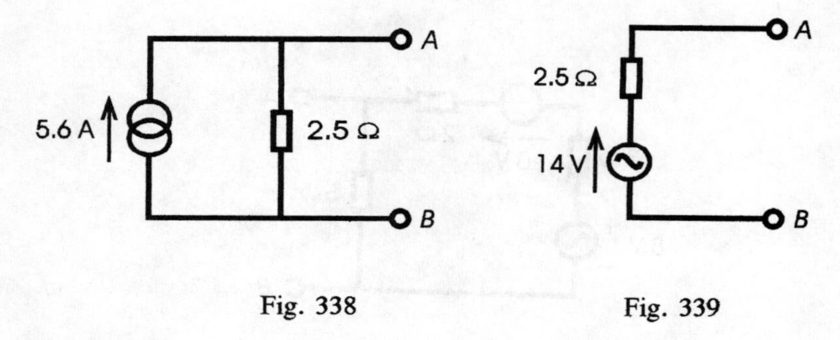

Fig. 338　　　　Fig. 339

(b) For maximum power, the resistive load will be 2.5 Ω and the maximum power will be $P = \dfrac{E^2}{4r} = \dfrac{14^2}{10} = 19.6$ W.

2. Referring to Fig. 145, the voltage across CD is $V_{CD} = \dfrac{40}{50} \times 30 = 24$ volts, using the potential divider principle.

The resistance looking into the terminals CD, when the 30 V is suppressed, is

$$\frac{40 \times 10}{40 + 10} = \frac{400}{50} = 8\ \Omega$$

Fig. 145 is now replaced by the open circuit voltage $V_{CD} = 24$ V and the equivalent Thévenin resistance of 8 Ω.

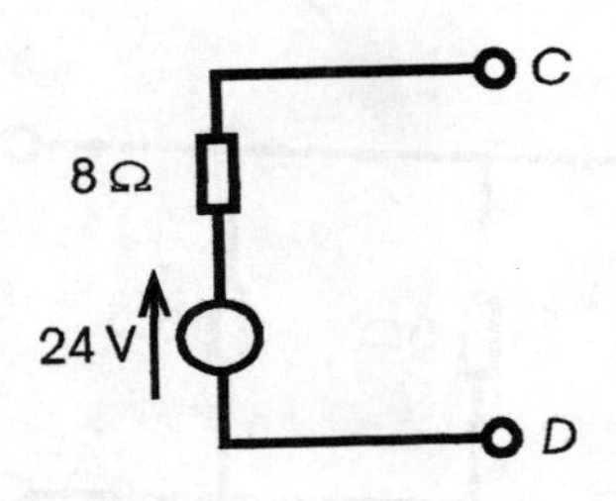

Fig. 340

(b)

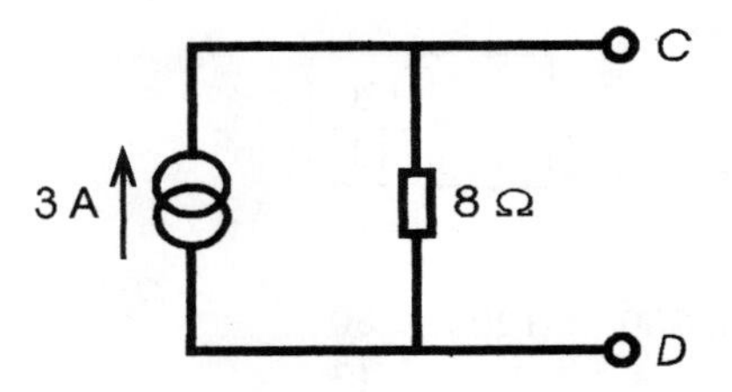

Fig. 341

(c) The load resistance to be connected between C and D for maximum power transfer is 8 Ω.

(d)

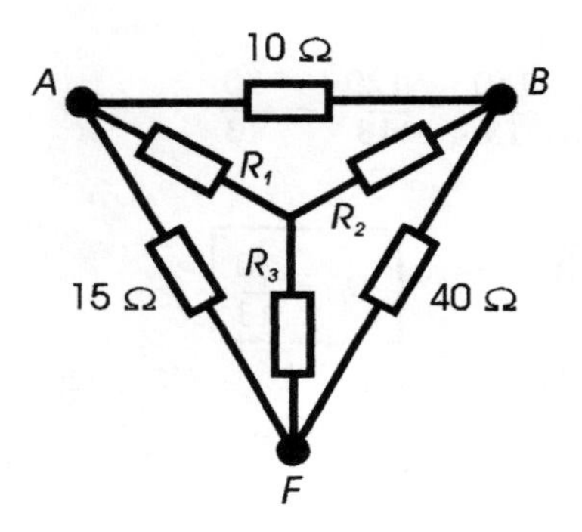

Fig. 342

$$R_{AF} = \frac{15 \times 50}{65} = \frac{150}{13} = R_1 + R_3 \quad \ldots (1)$$

$$R_{AB} = \frac{10 \times 55}{65} = \frac{110}{13} = R_1 + R_2 \quad \ldots (2)$$

$$R_{BF} = \frac{40 \times 25}{65} = \frac{200}{13} = R_2 + R_3 \quad \ldots (3)$$

(1) − (2) $\qquad R_3 - R_2 = \dfrac{40}{13} \quad \ldots (4)$

(3) $\qquad R_2 + R_3 = \dfrac{200}{13} \quad \ldots (5)$

(4) + (5) $\quad 2R_3 = \dfrac{240}{13}$

$$\boxed{R_3 = \frac{120}{13}}$$

From (3) $\quad R_2 = \dfrac{200}{13} - \dfrac{120}{13} = \dfrac{80}{13}$

$$\boxed{R_2 = \frac{80}{13}}$$

$$R_1 + R_3 = \frac{150}{13}$$

$$R_1 = \frac{150}{13} - \frac{120}{13} = \frac{30}{13}$$

$$\boxed{R_1 = \frac{30}{13}}$$

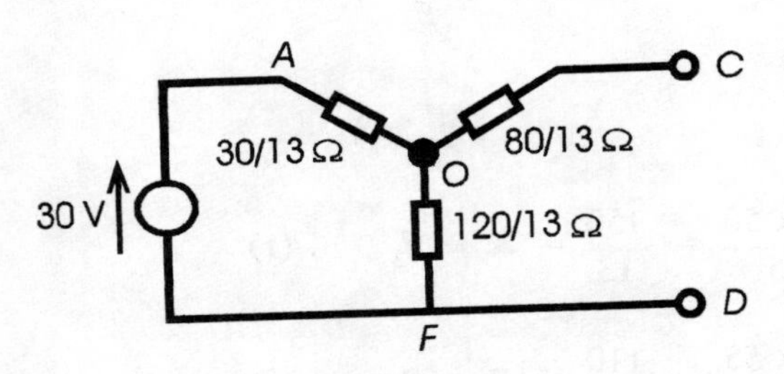

Fig. 343

(e)

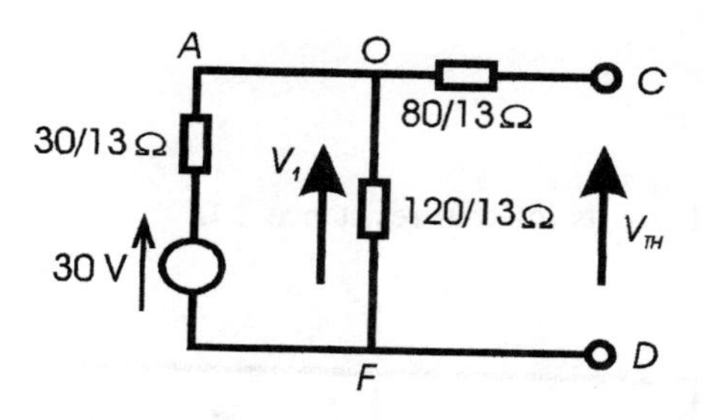

Fig. 344

$$V_{TH} = V_1 = \frac{\frac{120}{13}}{\frac{30}{13} + \frac{120}{13}} \times 30 = \frac{120}{150} \times 30 = \frac{4}{5} \times 30 = 24 \text{ volts}$$

$$R_{TH} = \frac{80}{13} + \frac{\frac{30}{13} \times \frac{120}{13}}{\frac{30}{13} + \frac{120}{13}} = \frac{80}{13} + \frac{30 \times 120}{150 \times 13} = \frac{80}{13} + \frac{24}{13} = \frac{104}{13} = 8\ \Omega$$

3. Replace the 24 V e.m.f. by its internal resistance 4 Ω.

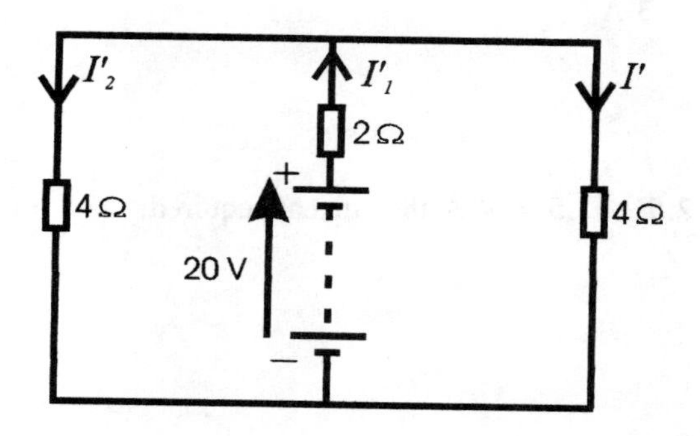

Fig. 345

$$I_1' = \frac{20}{4} = 5 \text{ A}$$

$$I_2' = \frac{4}{8} \times 5 = 2.5 \text{ A}$$

$$I' = 2.5 \text{ A}$$

Replace the 20 V e.m.f. by its internal resistance 2 Ω.

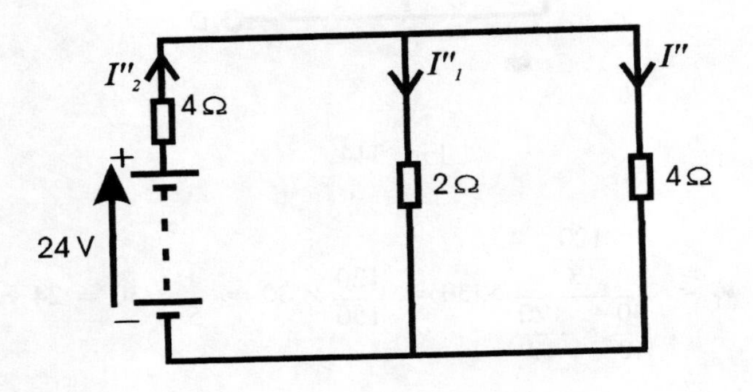

Fig. 346

$$I_2'' = \frac{24}{4 + \frac{8}{6}} = \frac{24 \times 6}{32} = 4.5 \text{ A}$$

$$I_1'' = \frac{4}{6} \times 4.5 = 3 \text{ A}$$

$$I'' = 1.5 \text{ A}$$

$I = I' + I'' = 2.5 + 1.5 = 4$ A the current required.

4. Remove the 5 Ω resistor from Fig. 147

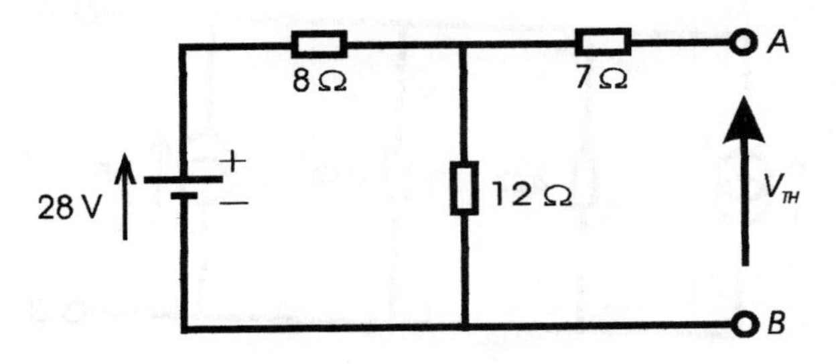

Fig. 347

Measure the open circuit voltage, V_{TH}.

$$V_{TH} = \frac{12}{20} \times 28 = 16.8 \text{ volts},$$

using the potential divider principle. Find the resistance, looking into the terminals AB by suppressing the e.m.f. voltage.

$$R_{TH} = 7 + \frac{8 \times 12}{8 + 12} = 11.8 \ \Omega$$

The Thévenin equivalent circuit is

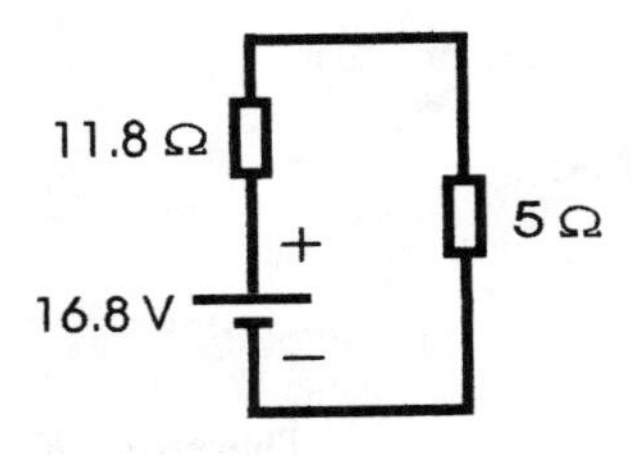

Fig. 348

The current through the 5 Ω resistor is $\frac{16.8}{16.8} = 1$ A.

5. Replace the constant voltage generator of Fig. 148 by a constant current generator.

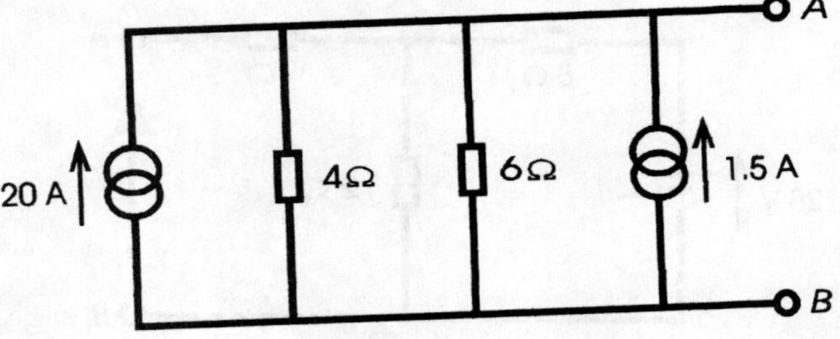

Fig. 349

This is replaced by

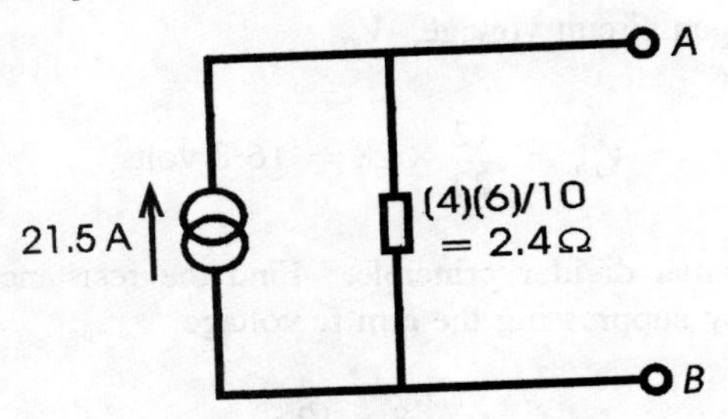

Fig. 350

Therefore $I_N = 21.5$ A, $R_{TH} = 2.4\ \Omega$.

The constant voltage generator is

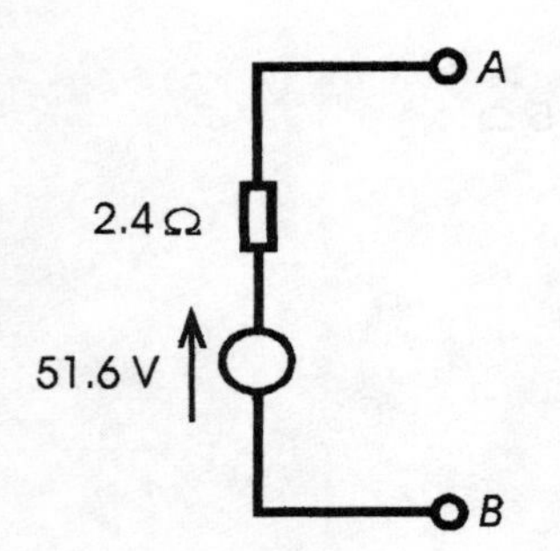

Therefore $R_{TH} = 2.4\ \Omega$, $V_{TH} = 51.6$ V.

Fig. 351

6. Remove the 4 Ω detector of Fig. 151 and measure the open circuit voltage V_{AB}.

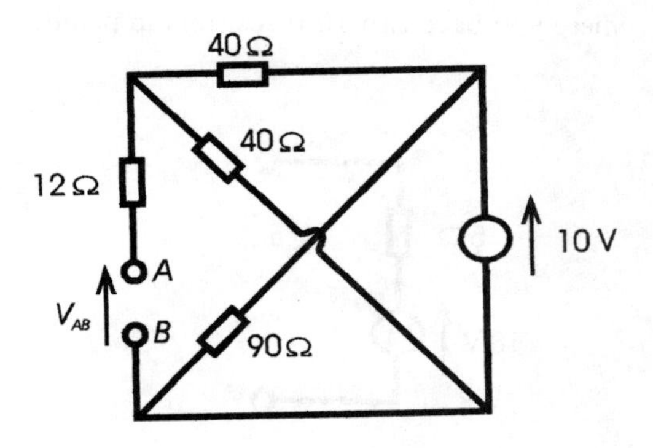

Fig. 352

V_{AB} is the voltage across the 40 Ω resistor and is half of the 10 V.

$V_{AB} = 5$ V.

Suppressing the 10 V e.m.f. which short circuits the 90 Ω resistor.

$$R_{TH} = 12 + \frac{40 \times 40}{80} = 32\ \Omega.$$

The equivalent Thévenin circuit is

$$I = \frac{5}{36} = 0.139\ \text{A}.$$

Fig. 353

7. From Fig. 152, $V_{AB} = \frac{10}{20} \times 20 = 10$ volts the open circuit voltage.

Suppressing the 20 V e.m.f.

$R_{AB} = 5\ \Omega$, where you have two $10\ \Omega$ resistors in parallel.

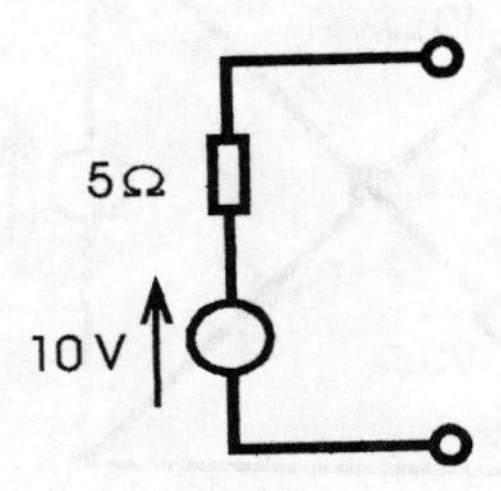

Fig. 354

From Fig. 153, $V_{AB} = \frac{2}{2+2} \times 10 = 5$ volts

$R_{TH} = 1\ \Omega$.

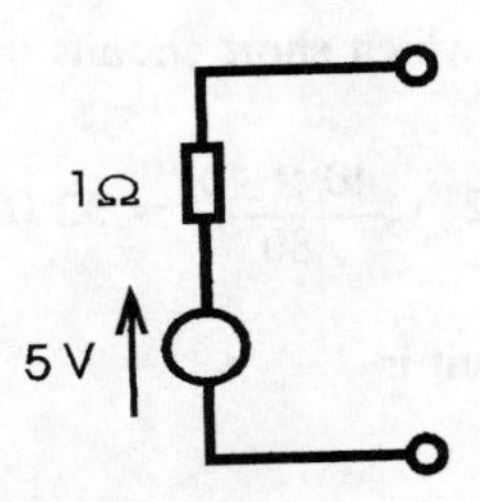

Fig. 355

From Fig. 154, the circulating current is $\frac{25 - 16}{5 + 4} = \frac{9}{9} = 1$ A.

$V_{AB} = 16 + 4 \times 1 = 20$ V.

Suppressing the e.m.f.s

$$R_{TH} = R_{AB} = \frac{4 \times 5}{4 + 5} = \frac{20}{9} = 2.22\ \Omega.$$

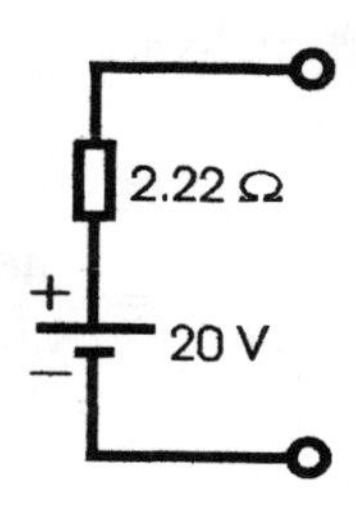

Fig. 356

8. Remove the 5 Ω resistor from Fig. 155

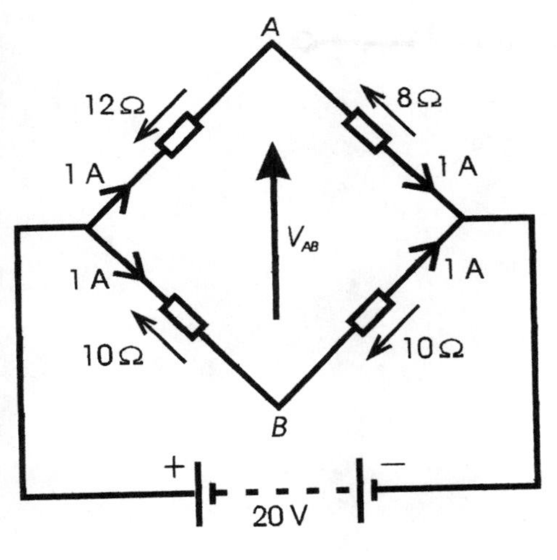

Fig. 357

$V_{AB} = 10 - 12 = -2$ V

$V_{AB} = 8 - 10 = -2$ V

Therefore the voltage V_{AB} is reversed.

Suppressing the 20 V e.m.f.

$$R_{TH} = \frac{12 \times 8}{12 + 8} + \frac{10 \times 10}{10 + 10}$$

$$= 4.8 + 5 = 9.8\ \Omega.$$

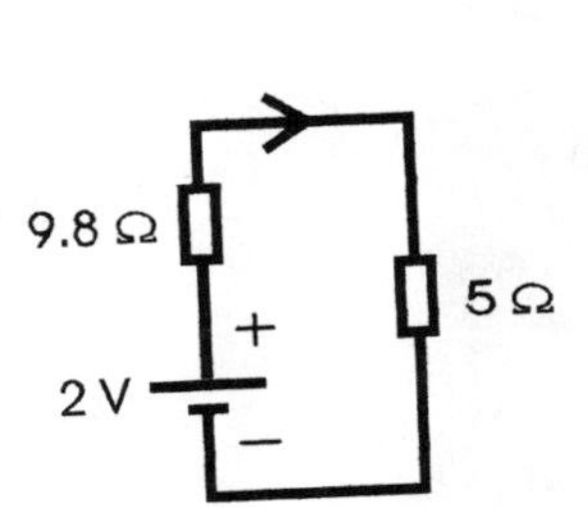

Fig. 358

$$I = \frac{2}{14.8} = 0.135 \text{ A}$$

is the current through the 5 Ω resistor.

9. From Fig. 156, suppressing the e.m.f.s

$$R_{AB} = \frac{\left(\frac{2 \times 6}{2 + 6} + 10\right)4}{\frac{2 \times 6}{2 + 6} + 10 + 4} = \frac{46}{15.5} = 2.97\ \Omega.$$

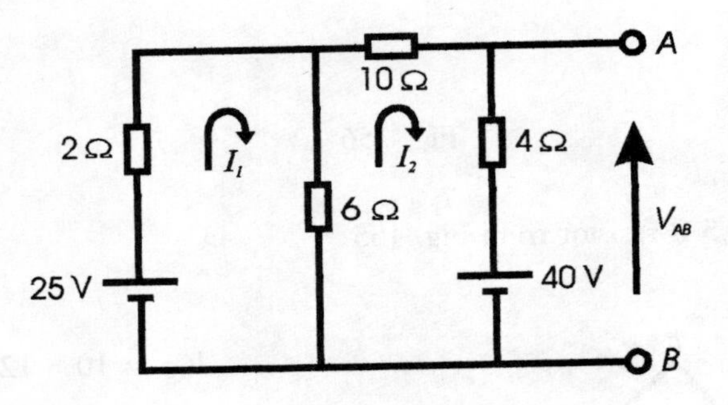

Fig. 359

To find I_1 and I_2

$$8I_1 - 6I_2 = 25$$

$$-6I_1 + 20I_2 = -40$$

$$\frac{I_1}{\Delta_1} = -\frac{I_2}{\Delta_2} = \frac{1}{\Delta}$$

$$\Delta_1 = \begin{vmatrix} -6 & -25 \\ 20 & 40 \end{vmatrix} = -240 + 500 = 260$$

$$\Delta_2 = \begin{vmatrix} 8 & -25 \\ -6 & 40 \end{vmatrix} = 320 - 150 = 170$$

$$\Delta = \begin{vmatrix} 8 & -6 \\ -6 & 20 \end{vmatrix} = 120 - 36 = 84$$

$$I_2 = -\frac{\Delta_2}{\Delta} = -\frac{170}{84} = -2.02\ \text{A}$$

$V_{AB} = 40 - 4 \times 2.02 = 31.9$ volts.

The equivalent Thévenin circuit is

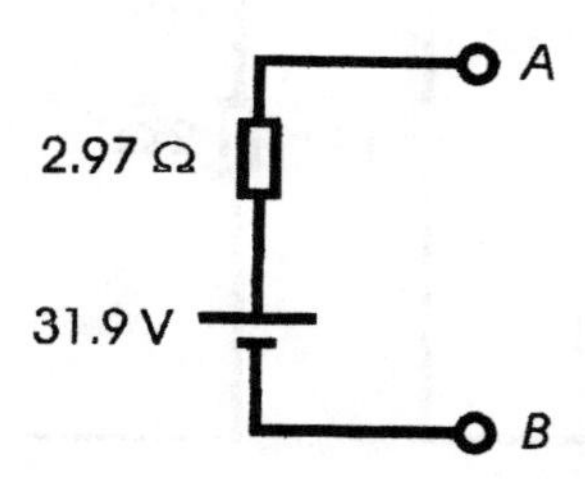

Fig. 360

The equivalent Norton circuit is

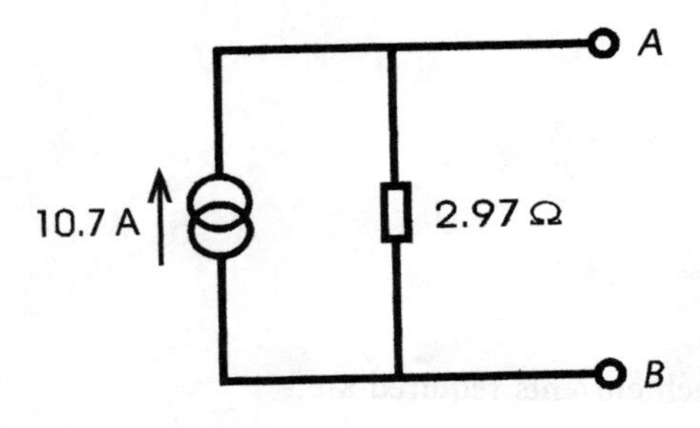

Fig. 361

10. Due to the 10 V supply, the branch currents are:-

$$I_1' = \frac{10}{2 + \frac{15 \times 1}{16}} = 3.4 \text{ A}, \quad I' = \frac{1}{16} \times 3.4 = 0.21 \text{ A},$$

$$I_2' = \frac{15}{16} \times 3.4 = 3.19 \text{ A}$$

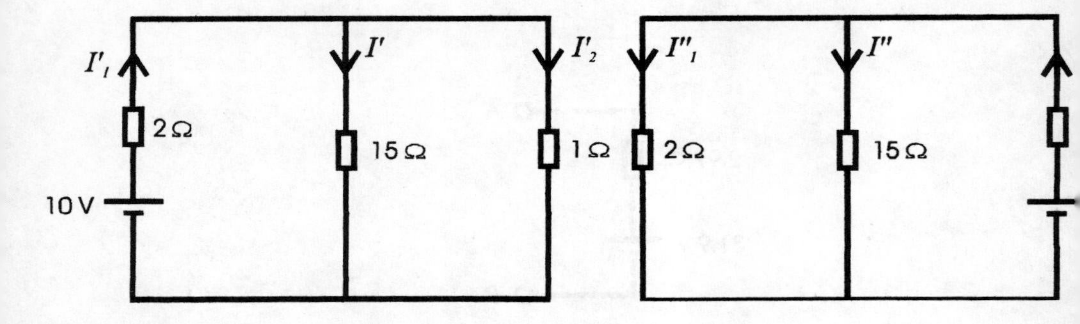

Fig. 362

Due to the 7 V supply, the branch currents are:-

$$I_2'' = \frac{7}{1 + \dfrac{2 \times 15}{17}} = 2.53 \text{ A}$$

$$I'' = \frac{2}{17} \times 2.53 = 0.30 \text{ A}$$

$$I_1'' = \frac{15}{17} \times 2.53 = 2.23 \text{ A.}$$

Therefore, the branch currents required are:-

$$I_1 = I_1' - I_1'' = 3.4 - 2.23 = 1.17 \text{ A}$$

$$I_2 = -I_2' + I_2'' = -3.19 + 2.53 = -0.66 \text{ A}$$

$$I_1 + I_2 = 1.17 - 0.66 = 0.51 \text{ A.}$$

11. From Fig. 158, suppressing the two e.m.f.s

$$R_{AB} = \frac{\left(\dfrac{1 \times 15}{16} + 5\right)2}{\dfrac{1 \times 15}{16} + 5 + 2} = \frac{11.875}{7.9375} = 1.5\ \Omega.$$

Determine the currents in the circuit

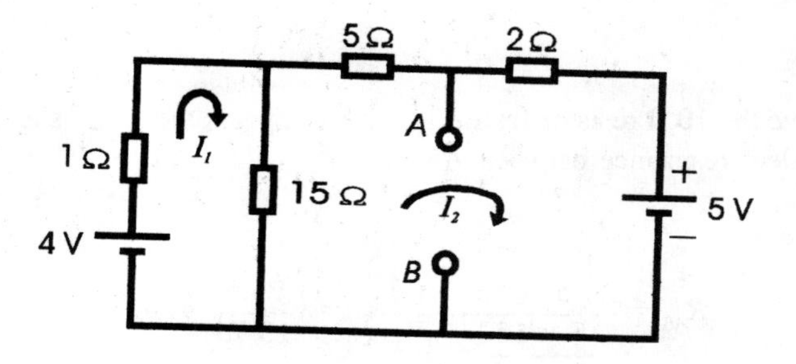

Fig. 363

$$16I_1 - 15I_2 = 4$$
$$-15I_1 + 22I_2 = -5$$

$$\frac{I_1}{\Delta_1} = -\frac{I_2}{\Delta_2} = \frac{1}{\Delta}$$

$$I_1 = \frac{\Delta_1}{\Delta} = \frac{\begin{vmatrix} -15 & -4 \\ 22 & 5 \end{vmatrix}}{\begin{vmatrix} 16 & -15 \\ -15 & 22 \end{vmatrix}} = \frac{13}{127} = 0.102 \text{ A}$$

$$I_2 = -\frac{\Delta_2}{\Delta} = -\frac{\begin{vmatrix} 16 & -4 \\ -15 & 5 \end{vmatrix}}{127} = -\frac{20}{127} = -0.157 \text{ A}$$

$V_{AB} = 5 - 0.157 \times 2 = 4.69$ volts.

The Thévenin equivalent circuit

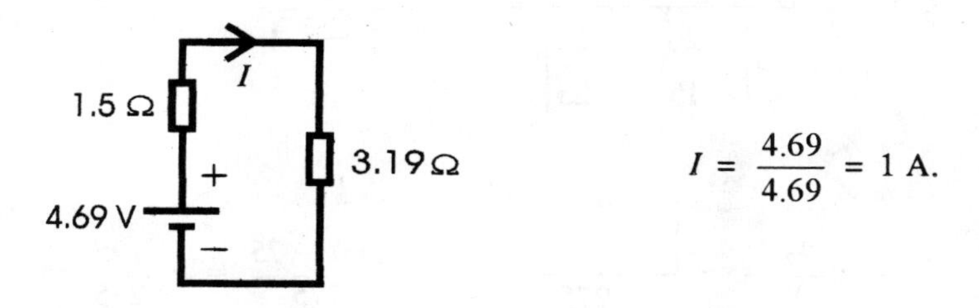

$$I = \frac{4.69}{4.69} = 1 \text{ A.}$$

Fig. 364

12. Remove the 10 Ω resistor from Fig. 159, suppress the e.m.f.s and calculate the equivalent resistance between *AB*.

$$R_{AB} = \frac{\left(\frac{5 \times 15}{5 + 15} + 8\right)2}{\frac{5 \times 15}{5 + 15} + 8 + 2} = \frac{23.5}{13.75} = 1.71\ \Omega$$

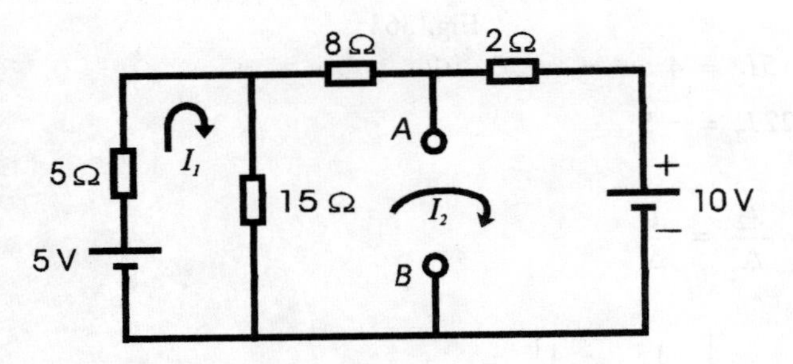

Fig. 365

Determine the currents

$$20I_1 - 15I_2 = 5$$

$$-15I_1 + 25I_2 = -10$$

$$\frac{I_1}{\Delta_1} = -\frac{I_2}{\Delta_2} = \frac{1}{\Delta}$$

$$I_1 = \frac{\Delta_1}{\Delta} = \frac{\begin{vmatrix} -15 & -5 \\ 25 & 10 \end{vmatrix}}{\begin{vmatrix} 20 & -15 \\ -15 & 25 \end{vmatrix}} = \frac{-150 + 125}{500 - 225} = -\frac{25}{275} = -\frac{1}{11}$$

$$I_2 = -\frac{\Delta_2}{\Delta} = -\frac{\begin{vmatrix} 20 & -5 \\ -15 & 10 \end{vmatrix}}{275} = -\frac{200 - 75}{275} = -\frac{125}{275}$$

$$I_2 = -\frac{5}{11}$$

$$V_{AB} = 10 - \frac{10}{11} = \frac{110 - 10}{11} = \frac{100}{11} \text{ volts.}$$

The equivalent Thévenin circuit is

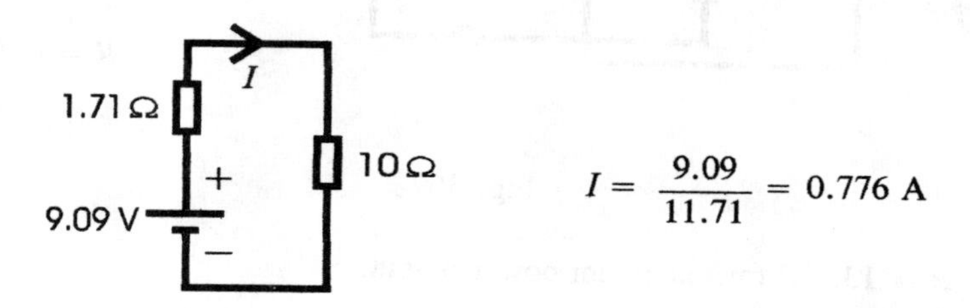

$$I = \frac{9.09}{11.71} = 0.776 \text{ A}$$

Fig. 366

The current through the 10 Ω resistor is 0.776 A.

13.

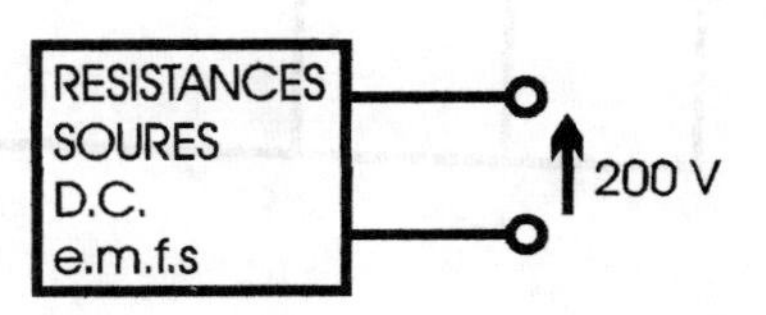

Fig. 367

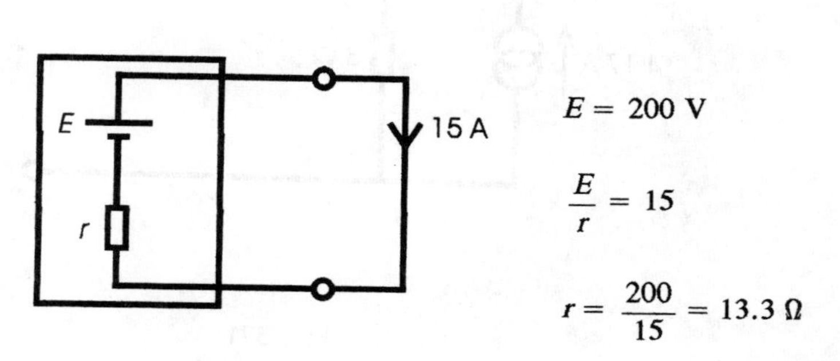

$E = 200$ V

$$\frac{E}{r} = 15$$

$$r = \frac{200}{15} = 13.3 \ \Omega$$

Fig. 368

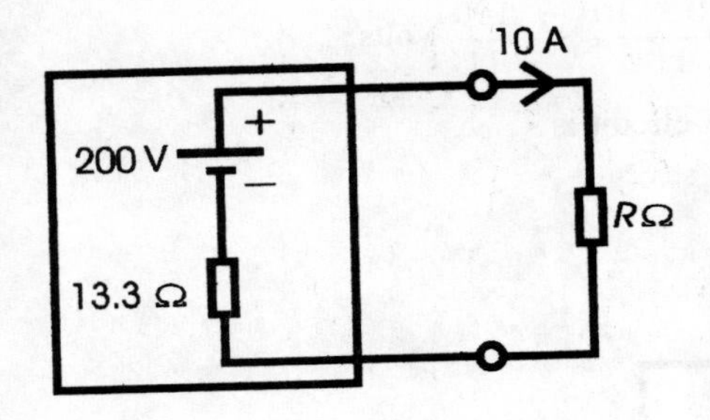

$$10 = \frac{200}{13.3 + R}$$

$$13.3 + R = \frac{200}{10}$$

$$R = 6.7\ \Omega$$

Fig. 369

$R = 13.3\ \Omega$ **for maximum power output.**

14. Fig. 160 becomes

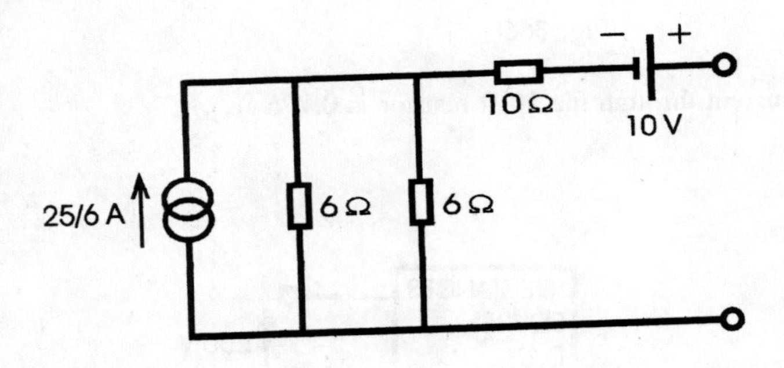

Fig. 370

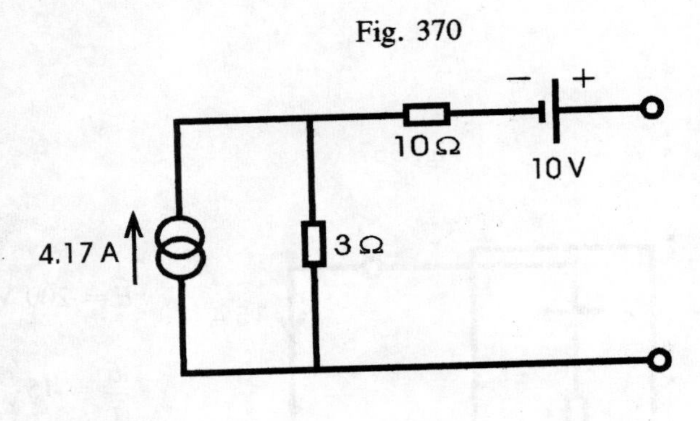

Fig. 371

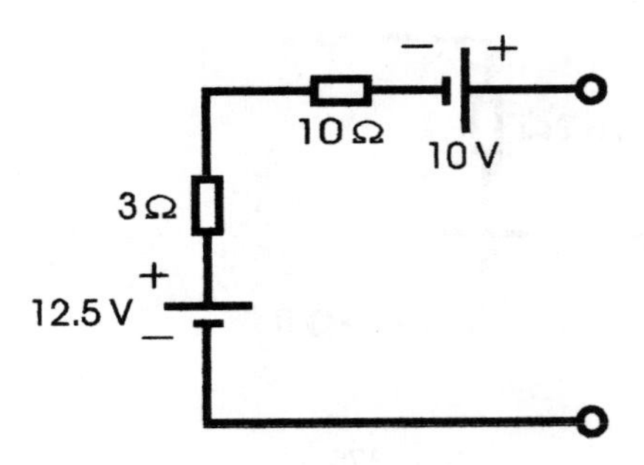

Fig. 372

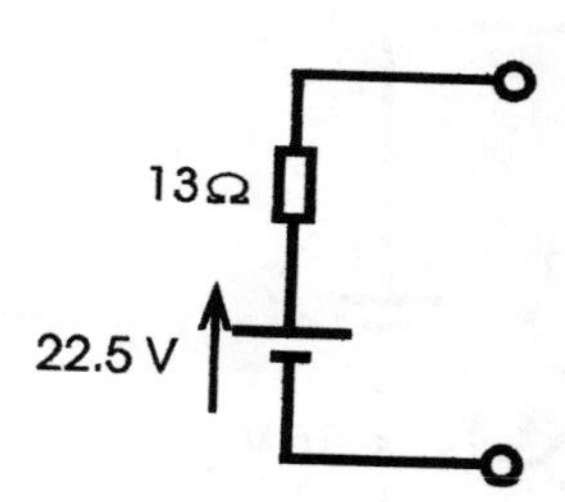

Fig. 373

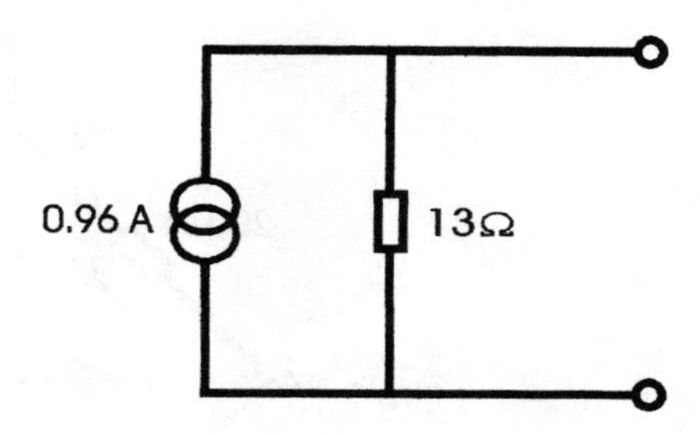

Fig. 374

15. In Fig. 161, $I_1 = \dfrac{30}{150} = 0.2$ A

$$I_2 = \frac{30}{350} = 0.086 \text{ A}$$

$$V_{AB} = 50I_2 - 50I_1$$

$$= 50 \times \frac{3}{35} - 50 \times \frac{3}{15} = 4.29 - 10 = -5.71 \text{ volts.}$$

Suppressing the 30 V e.m.f.

$$R_{AB} = \frac{50 \times 100}{150} + \frac{50 \times 300}{350} = 33.3 + 42.9 = 76.2\ \Omega$$

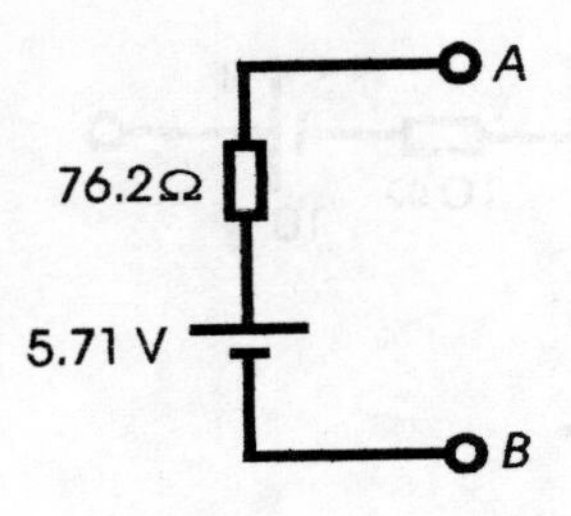

Fig. 375

16. Remove the ammeter of Fig. 162.

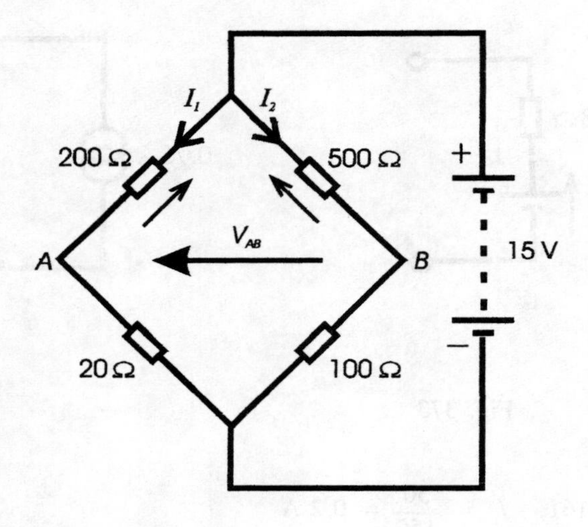

Fig. 376

$$I_1 = \frac{15}{220} = 0.068 \text{ A}$$

$$I_2 = \frac{15}{600} = 0.025 \text{ A}$$

$$V_{AB} = 500 I_2 - 200 I_1 = 0.068 \times 500 - 200 \times 0.025 = 29 \text{ V}.$$

Suppressing the 1.5 V e.m.f.

$$R_{AB} = \frac{200 \times 20}{220} + \frac{500 \times 100}{600} = 18.18 + 83.33 = 102 \ \Omega.$$

The Thévenin's equivalent circuit

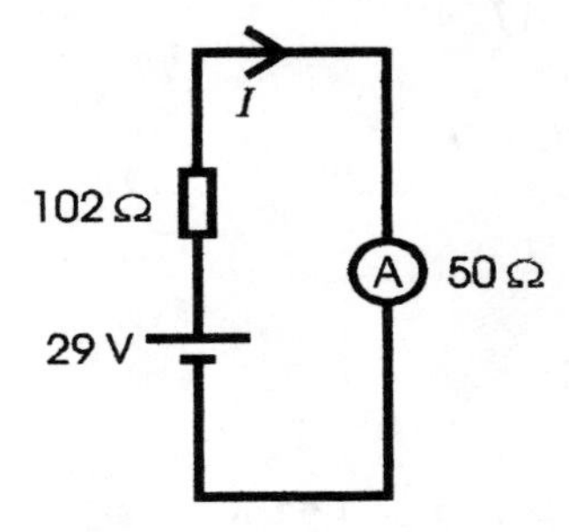

Fig. 377

$$I = \frac{29}{152} = 0.19 \text{ A the current through the ammeter.}$$

17. (i)

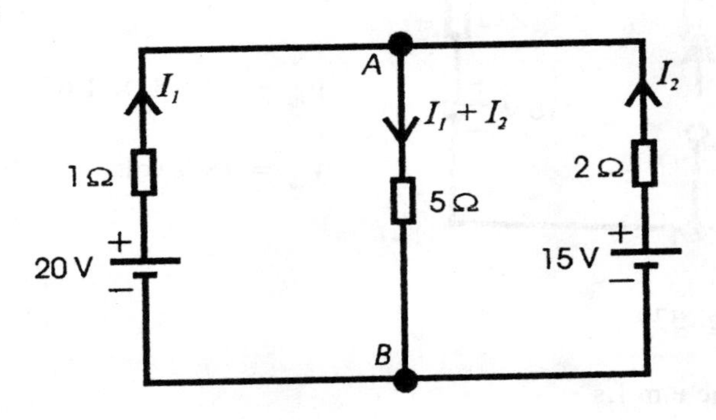

Fig. 378

From Fig. 163,

$$20 = I_1 + 5I_1 + 5I_2$$

$$15 = 2I_2 + 5I_1 + 5I_2$$

$$6I_1 + 5I_2 = 20 \ldots (\times 5)$$

$$5I_1 + 7I_2 = 15 \ldots (\times -6)$$

$$30I_1 + 25I_2 = 100$$

$$-30I_1 - 42I_2 = -90$$

$$-17I_2 = 10$$

$$I_2 = -\frac{10}{17} = -0.59 \text{ A}$$

$$6I_1 + 5(-0.59) = 20$$

$$6I_1 = 20 + 2.95$$

$$I_1 = \frac{22.95}{6} = 3.825$$

$$I = I_1 + I_2 = 3.825 - 0.59 = 3.23 \text{ A}$$

(ii) Remove the 5 Ω resistor

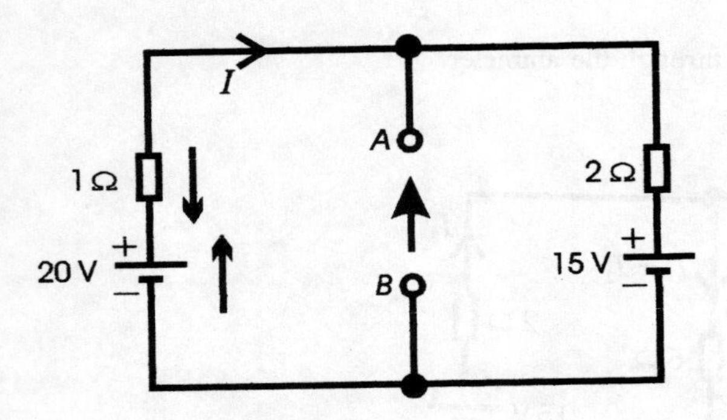

$$I = \frac{20-15}{3} = \frac{5}{3} = 1.67 \text{ A}$$

$$V_{AB} = 20 - 1 \times 1.67$$

$$V_{AB} = 18.3 \text{ volts}$$

Fig. 379

Suppressing the e.m.f.s

$$R_{AB} = \frac{1 \times 2}{3} = 0.67 \ \Omega$$

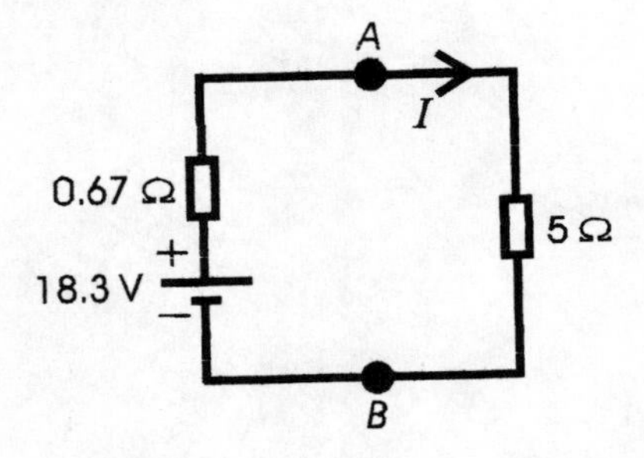

$$I = \frac{18.3}{5.67} = 3.23 \text{ A}$$

Fig. 380

(iii) Remove the 5 Ω resistor from Fig. 163. Short circuit AB, let I_N be the short circuit current.

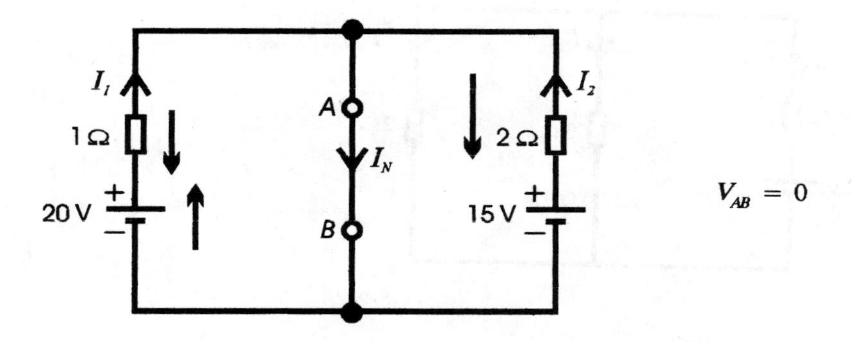

$V_{AB} = 0$

Fig. 381

$$20 - I_1 = 0 \quad \Rightarrow I_1 = 20 \text{ A}$$

$$15 - 2I_2 = 0 \quad \Rightarrow I_2 = 7.5 \text{ A}$$

$$I_N = I_1 + I_2 = 27.5 \text{ A}.$$

Suppress the e.m.f.s we have

$$R_{AB} = \frac{1 \times 2}{3} = \frac{2}{3} \ \Omega.$$

The Norton's equivalent circuit is

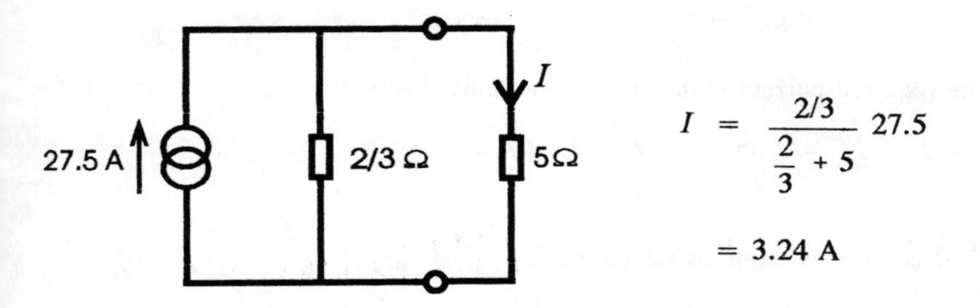

$$I = \frac{2/3}{\frac{2}{3} + 5} 27.5$$

$$= 3.24 \text{ A}$$

Fig. 382

(iv)

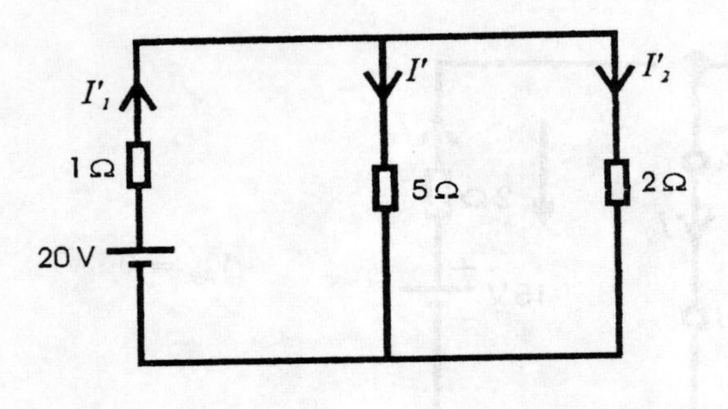

Fig. 383

Suppress the 15 V e.m.f.

$$I_1' = \frac{20}{1 + \frac{5 \times 2}{7}} = 8.24 \text{ A}$$

$$I' = \frac{2}{7} \times 8.24 = 2.35 \text{ A}$$

$$I_2' = \frac{5}{7} \times 8.24 = 5.89 \text{ A}$$

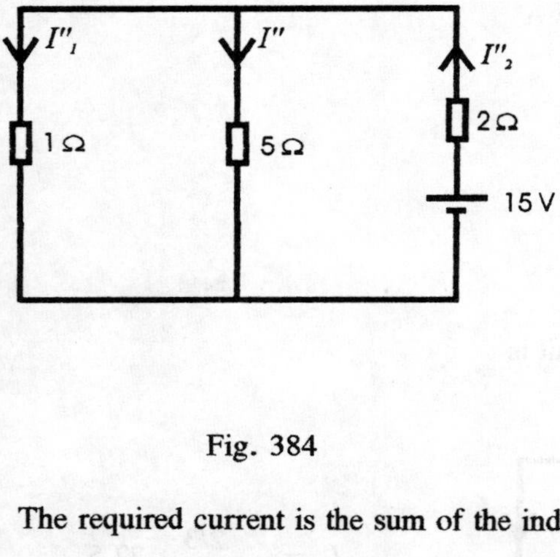

Suppress the 20 V e.m.f.

$$I_2'' = \frac{15}{2 + \frac{1 \times 5}{6}} = 5.29 \text{ A}$$

$$I'' = \frac{1}{6} \times 5.29 = 0.88 \text{ A}$$

$$I_1'' = \frac{5}{6} \times 5.29 = 4.41 \text{ A}$$

Fig. 384

The required current is the sum of the individual contributions of the e.m.f.s

$I = I' + I'' = 2.35 + 0.88 = 3.23$ A.

18. $\frac{2}{3}$ Ω is the new value of resistance

$$P_{max} = \frac{E^2}{4r} = \frac{18.3^2}{4 \times 0.67} = 126 \text{ W.}$$

19. (i) Fig. 164 is replaced by constant voltage generators

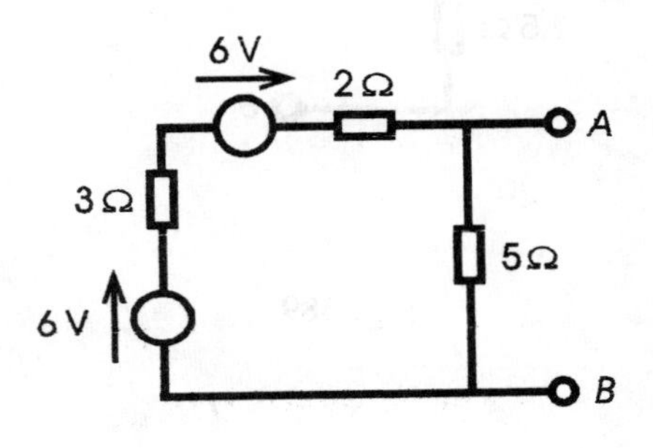

Fig. 385

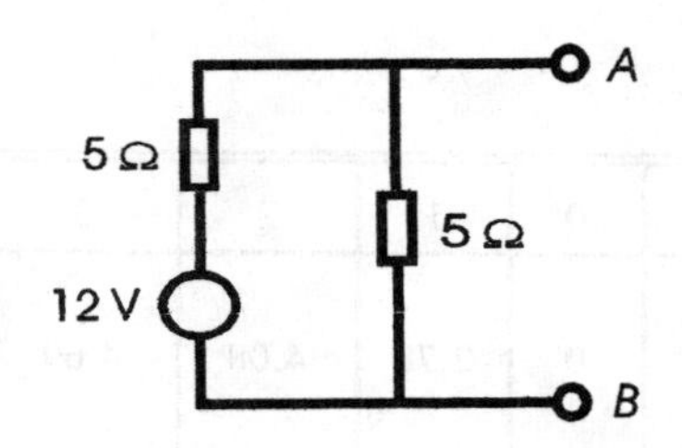

Fig. 386

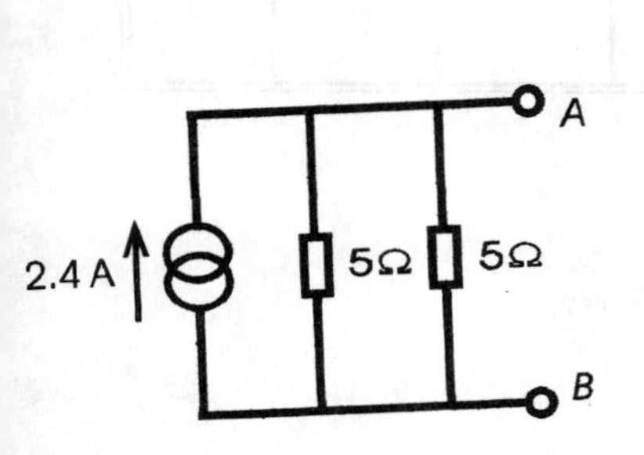

Fig. 387

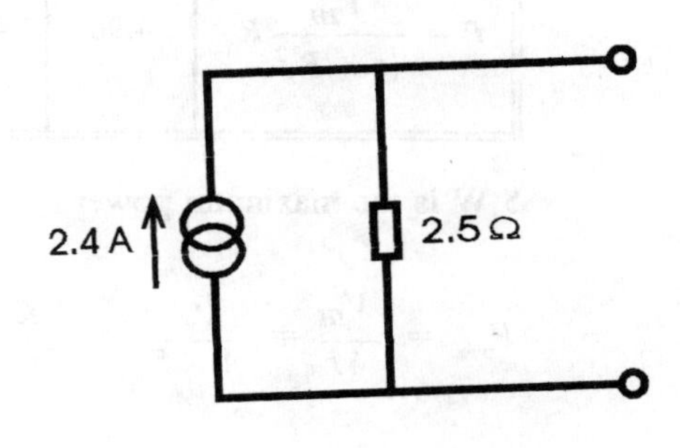

Fig. 388

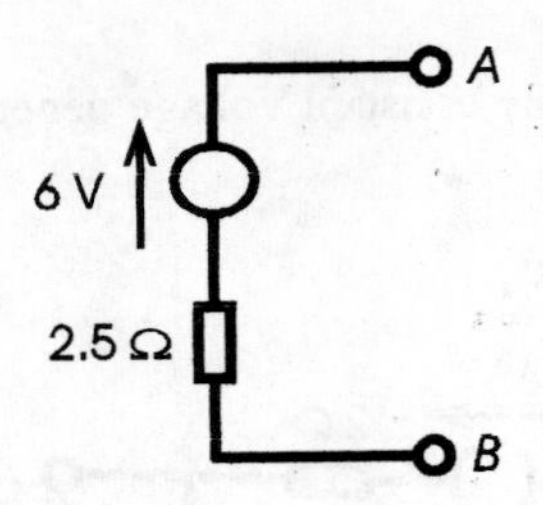

Fig. 389

(ii) $R = 2.5\ \Omega$

$$P_{max} = \frac{E^2}{4r} = \frac{6^2}{4 \times 2.5} = \frac{36}{10} = 3.6\ \text{W}.$$

20. (i) See text

(ii)

$R\ (\Omega)$	0	1	2	3	4	5
$P = \dfrac{V_{TH}^2}{(r + R)^2} R$	0	2.78	4.08	4.69	4.94	5

$R\ (\Omega)$	6	7	8	9	10
$P = \dfrac{V_{TH}^2}{(r + R)^2} R$	4.96	4.86	4.73	4.59	4.44

5 W is the maximum power

$$P_{max} = \frac{V_{TH}^2}{4r} = \frac{10^2}{4 \times 5} = \frac{100}{20} = 5\ \text{W}.$$

21. (i) From Fig. 167

The circulating current

$$I = \frac{24 - 20}{10} = 0.4\ \text{A}$$

$$V_{TH} = 24 - 0.4 \times 6 = 24 - 2.4 = 21.6 \text{ V}.$$

(ii)

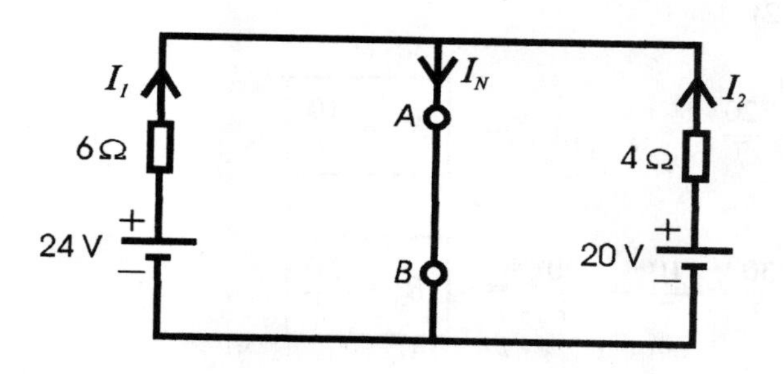

Fig. 390

$$V_{AB} = 0$$

$$24 - 6I_1 = 0$$

$$I_1 = 4 \text{ A}$$

$$20 - 4I_2 = 0$$

$$I_2 = 5 \text{ A}$$

$$I_N = I_1 + I_2 = 9 \text{ A}$$

$$R_{TH} = \frac{V_{TH}}{I_N} = \frac{21.6}{9} = 2.4 \ \Omega.$$

22. From Fig. 168 and Fig. 169

$$R_{AC} = \frac{10 \times 25}{35} = \frac{50}{7} = R_A + R_C \quad \ldots (1)$$

$$R_{AB} = \frac{5 \times 30}{35} = \frac{30}{7} = R_A + R_B \quad \ldots (2)$$

$$R_{BD} = \frac{20 \times 15}{35} = \frac{60}{7} = R_B + R_C \quad \ldots (3)$$

(1) - (3)

$$R_A - R_B = \frac{50}{7} - \frac{60}{7} = -\frac{10}{7} \quad \ldots (4)$$

$$R_A + R_B = \frac{30}{7} \quad \ldots \quad (2)$$

(4) + (2)

$$2R_A = \frac{20}{7} \qquad \Rightarrow \qquad \boxed{R_A = \frac{10}{7}}$$

$$R_B = \frac{30}{7} - \frac{10}{7} = \frac{20}{7} \qquad \Rightarrow \qquad \boxed{R_B = \frac{20}{7}}$$

$$R_C + R_B = \frac{60}{7}$$

$$R_C = \frac{60}{7} - \frac{20}{7} = \frac{40}{7} \qquad \Rightarrow \qquad \boxed{R_C = \frac{40}{7}}$$

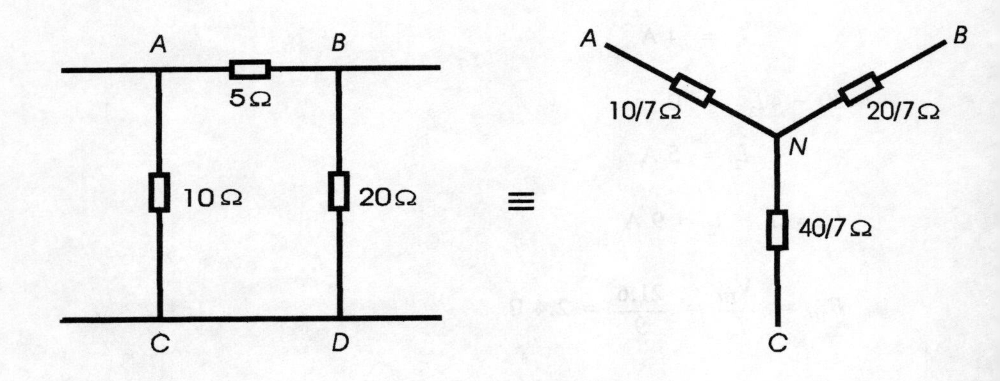

Fig. 391

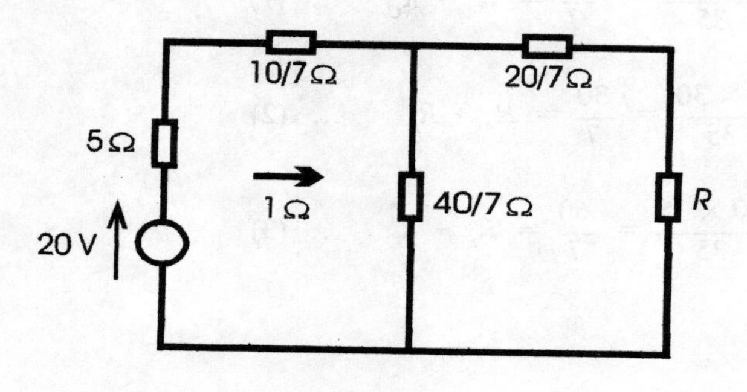

Fig. 392

$$\frac{\left(R + \frac{20}{7}\right)\frac{40}{7}}{R + \frac{20}{7} + \frac{40}{7}} + \frac{10}{7} = 5$$

$$\frac{(7R + 20)\,40}{(7R + 20 + 40)\,7} + \frac{10}{7} = 5$$

$$\frac{(7R + 20)\,40}{(7R + 60)\,7} = \frac{25}{7}$$

$$(7R + 20)\,40 = 25(7R + 60)$$

$$280R + 800 = 175R + 1500$$

$$105R = 700$$

$$R = \frac{700}{105}$$

$$R = 6.67\ \Omega$$

$$P_{max} = \frac{E^2}{4r} = \frac{20^2}{4 \times 5} = 20\text{ W}$$

23. Remove the 10 Ω resistor in Fig. 171

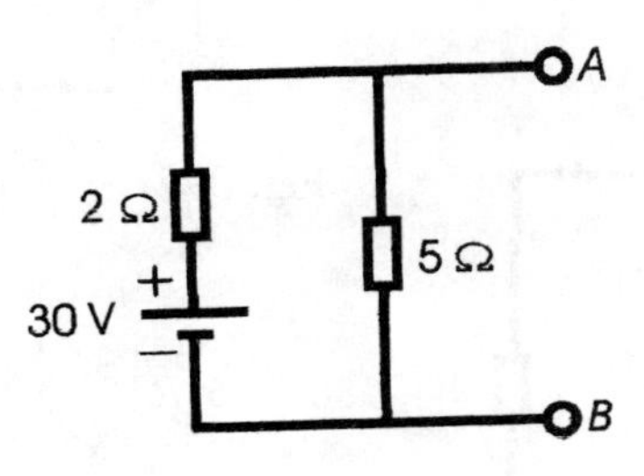

Fig. 393

Suppress the 30 V e.m.f.

$$R_{AB} = \frac{2 \times 5}{7} = \frac{10}{7}\ \Omega$$

$$V_{AB} = \frac{5}{7} \times 30 = \frac{150}{7}\text{ volts}$$

using the potential divider principle.

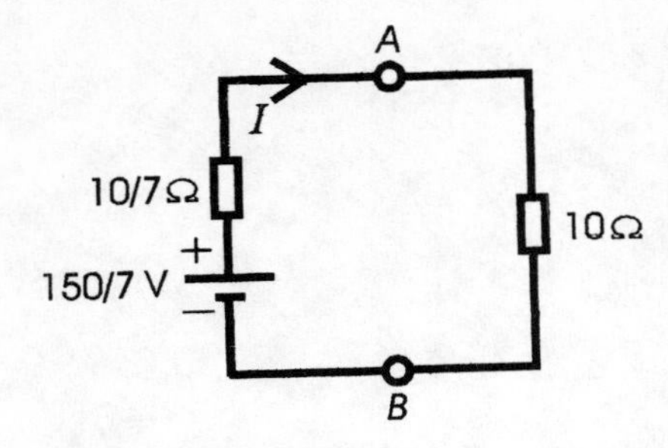

$$I = \frac{150/7}{10 + \frac{10}{7}} = \frac{150}{80}$$

$$I = \frac{15}{8} = 1.875 \text{ A}$$

Fig. 394

Using the current divider principle.

$$\text{The total current, } I = \frac{30}{2 + \frac{5 \times 10}{15}} = \frac{450}{80} = \frac{45}{8}$$

$$\text{The current through the 10 Ω resistor is } \frac{5}{15} \times \frac{45}{8} = \frac{15}{8} = 1.875 \text{ A.}$$

24.

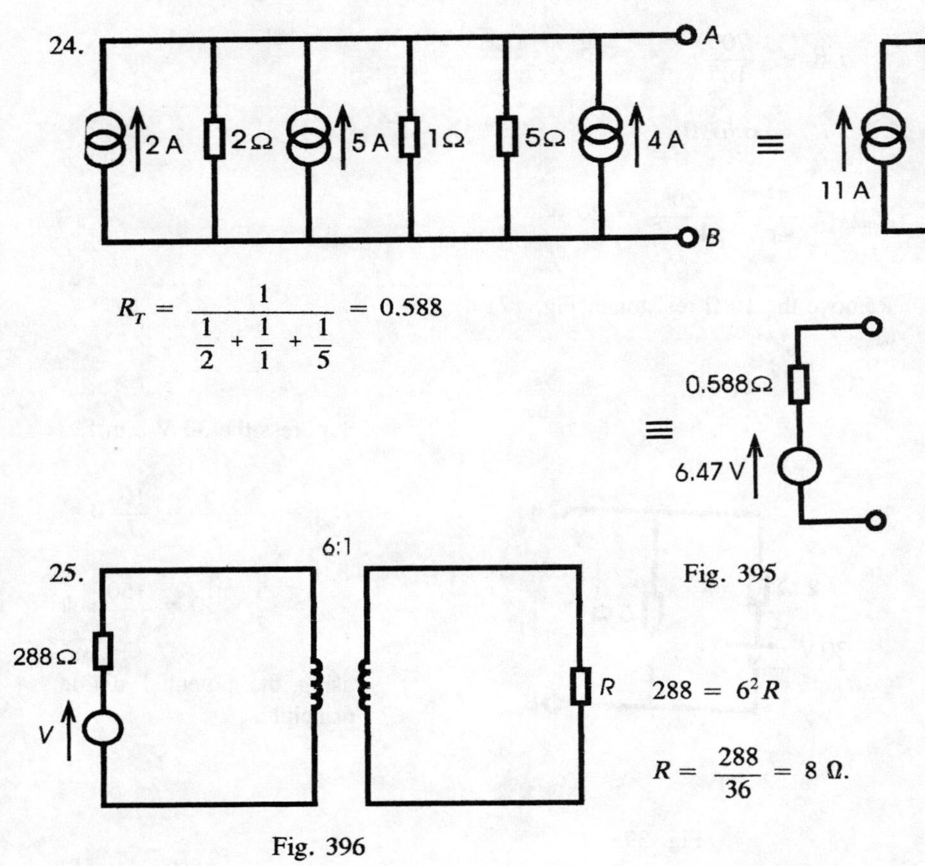

$$R_T = \frac{1}{\frac{1}{2} + \frac{1}{1} + \frac{1}{5}} = 0.588$$

Fig. 395

25.

$$288 = 6^2 R$$

$$R = \frac{288}{36} = 8 \text{ Ω.}$$

Fig. 396

SOLUTIONS 3

1.

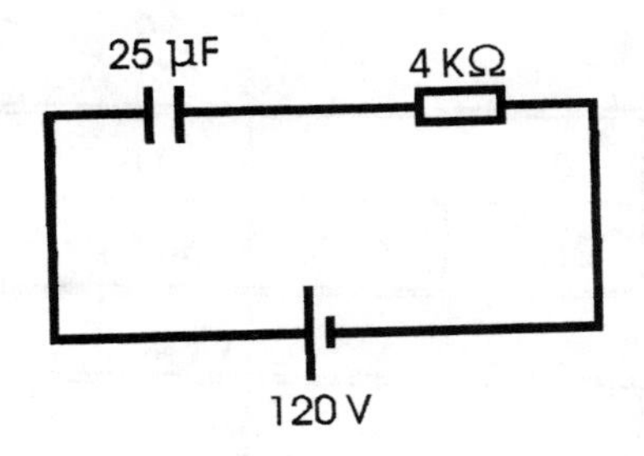

Fig. 397

$i = Ie^{-t/\tau}$ $\tau = CR = 25 \times 10^{-6} \times 4 \times 10^3 = 0.1$ s

$v_C = V\left(1 - e^{-\frac{t}{\tau}}\right)$ $\tau = CR$

t	0	0.01	0.05	0.1	0.15
i (mA)	30	27	18.2	11	6.69
V	0	11.41	47.21	75.85	93.23

t	0.2	0.3	0.4	0.5	0.8
i (mA)	4.06	1.49	0.55	0.2	0.01
V	103.76	114	117.8	119.2	120

2.

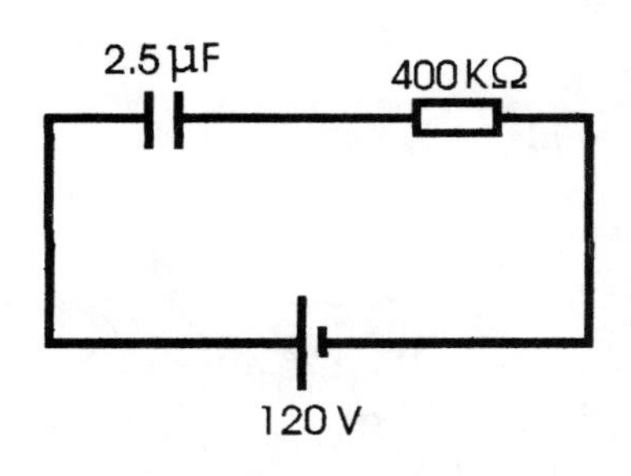

Fig. 398

$\tau = CR = 2.5 \times 10^{-6} \times 400 \times 10^{3} = 1$ s.

$v_C = V\left(1 - e^{-t/\tau}\right)$

t	0	0.2	0.4	0.6	0.8	1
v	0	21.75	39.56	54.14	66.08	75.85

t	1.2	1.4	1.6	1.8	2	2.3
v	83.85	90.4	95.77	100.16	103.75	107.96

3. (a)

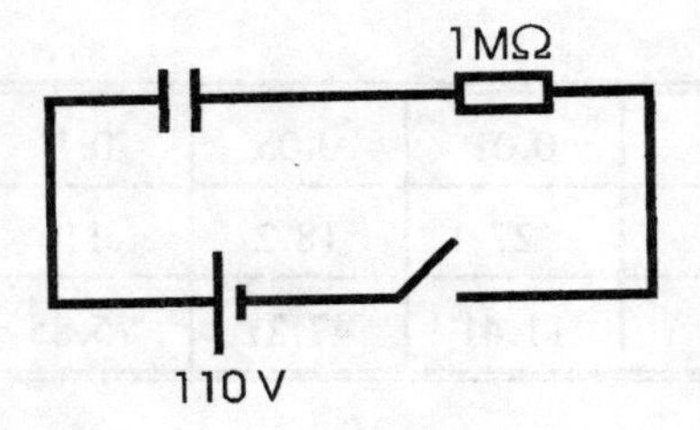

Fig. 399

$$\left(\frac{dv_C}{dt}\right)_{t=0} = \frac{V}{\tau} = 27.5 \text{ v/s}$$

$$\frac{100}{27.5} = \tau \quad \approx \tau = 4 \text{ s}$$

$$\tau = CR \quad \Rightarrow \quad C = \frac{\tau}{R}$$

$$C = \frac{4}{10^{+6}} = 4\ \mu\text{F}$$

(b) $\tau = \dfrac{L}{R}$

$$\left(\frac{di}{dt}\right)_{t=0} = \frac{I}{\tau} = 1.76 \text{ A/s}$$

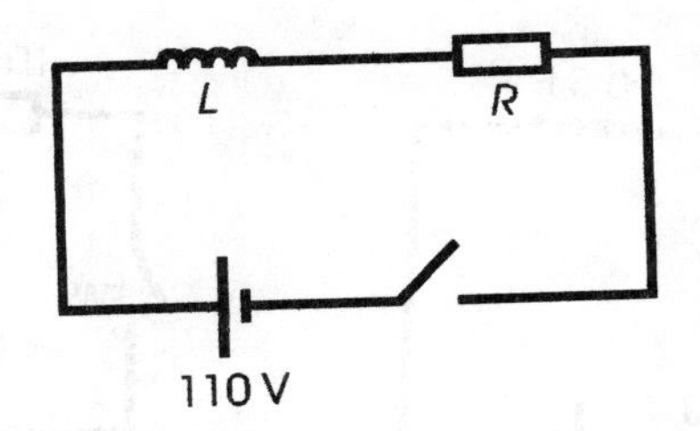

Fig. 400

$$\tau = \frac{L}{R}$$

$$I = 1.76 \times 3.125 = 5.5 \text{ A}$$

$$I = \frac{V}{R} = \frac{100}{R} = 5.5 \text{ A}$$

$$R = \frac{100}{5.5} = 20 \ \Omega$$

$$L = \tau R = 3.125 \times 20 = 62.5 \text{ H}$$

4.

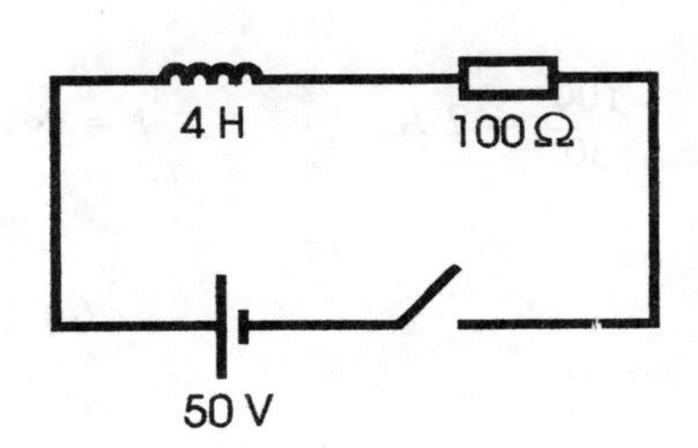

Fig. 401

(i) $\tau = \frac{L}{R} = \frac{4}{100} = 0.04 \text{ s}$

(ii) $i = I\left(1 - e^{-t/\tau}\right)$ after 10 ms

$$I = \frac{V}{R} = \frac{50}{100} = 0.5 \text{ A}$$

$$i = 0.5\left(1 - e^{-10 \times 10^{-3}/0.04}\right) = 0.11 \text{ A}$$

5. (a)

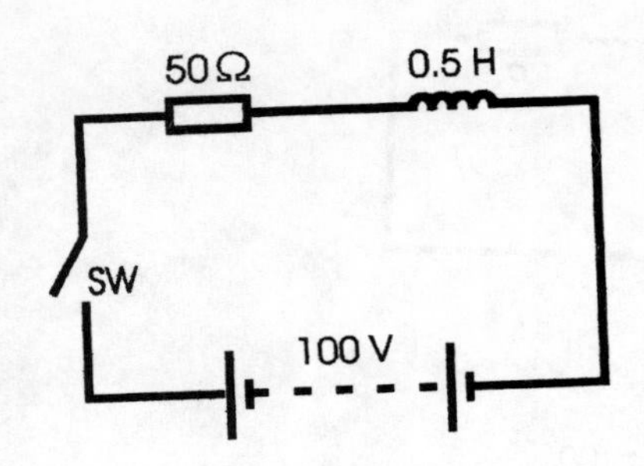

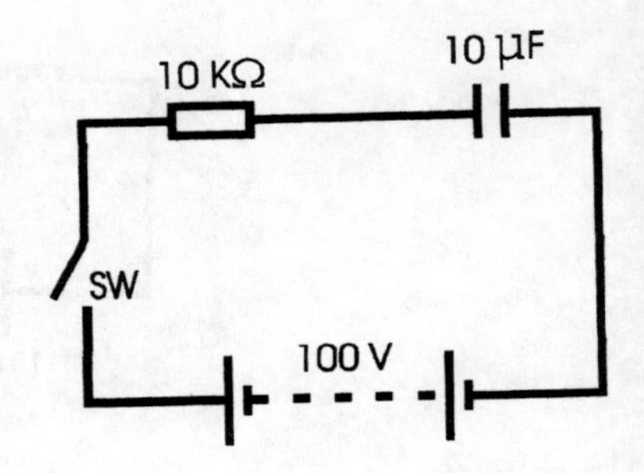

Fig. 402

(i) The current grows
at $t = 0_+$, $i = 0$

The current grows across C
at $t = 0_+$, $v = 0$

$$I = \frac{V}{R} = \frac{100}{10 \times 10^3}$$

$$= 0.01 \text{ A}$$

(ii) $\tau = \dfrac{L}{R} = \dfrac{0.5}{50} = 10$ ms

$$\tau = RC$$

$$= 10000 \times 10 \times 10^{-6}$$

$$= 100 \text{ ms}$$

(iii) at $t = \infty$, $I = \dfrac{100}{50} = 2$ A

$t = \infty$, $i = 0$

(b)

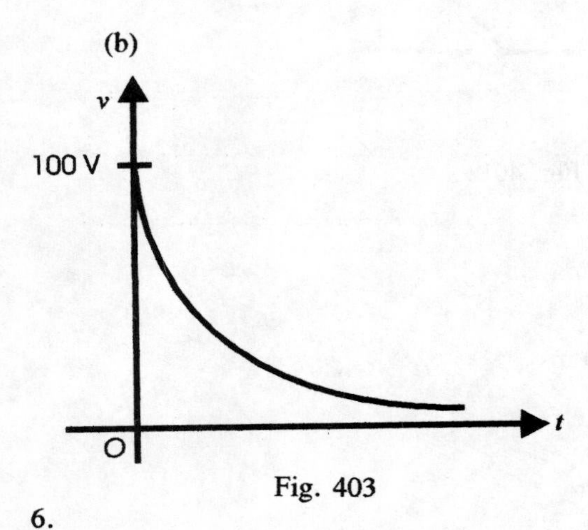

Fig. 403

(c)

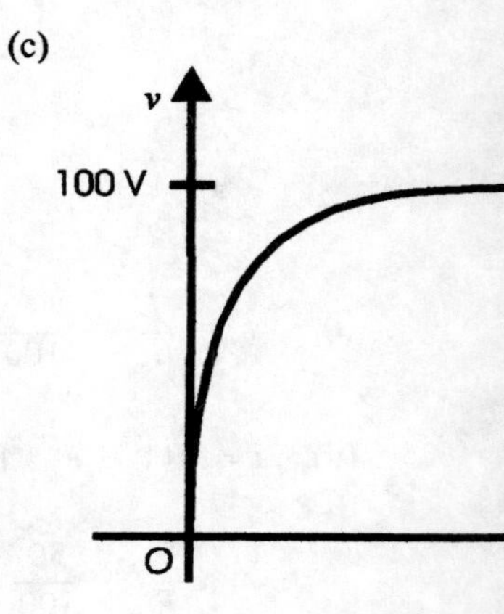

Fig. 404

6.

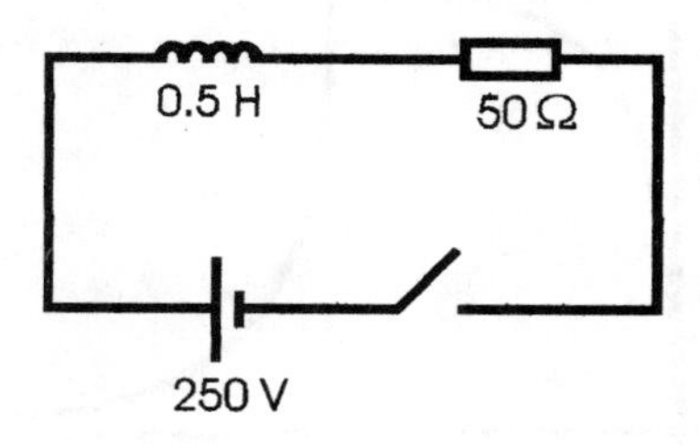

Fig. 405

(a) $\tau = \frac{L}{R} = \frac{0.5}{50} = 10$ ms

(b) $I = \frac{V}{R} = \frac{250}{50} = 5$ A

(c) Initial rate of current growth $= \frac{V}{L} = \frac{250}{0.5} = 500$ A/s

(d) $W = \frac{1}{2}LI^2 = \frac{1}{2} \times 0.5 \times (5)^2 = 6.25$ J

(e) $i = I\left(1 - e^{-\frac{t}{\tau}}\right) = 5\left(1 - e^{-\frac{0.01}{10 \times 10^{-3}}}\right) = 3.160$ A

7. The time constant, $\tau_1 = RC = 1 \times 10^6 \times 5 \times 10^{-6} = 5$ s

$$v_C = V\left(1 - e^{-\frac{t}{5}}\right)$$

t (s)	0	2	5	10	20	25	30	50
v_C (V)	0	16.2	31.6	43.2	49.1	49.7	49.9	50

Specimen calculation $\quad v_C = 50\left(1 - e^{-\frac{t}{5}}\right)$

when $t = 0$, $v_C = 0$

when $t = 2$s, $v_C = 50\left(1 - e^{-\frac{2}{5}}\right) = 16.5$ volts

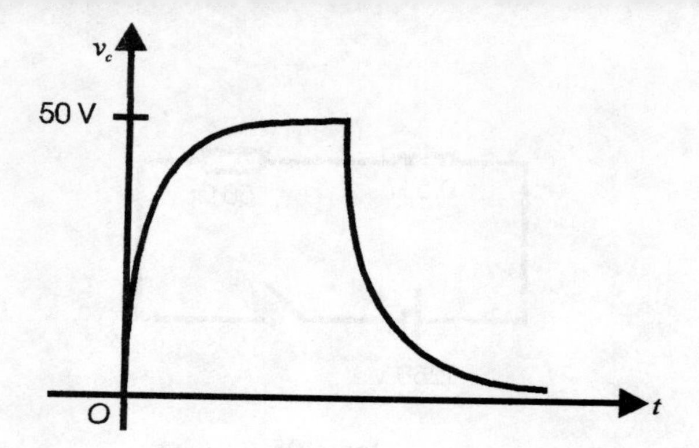

Fig. 406

Initial fall of the charging current

$$i = \frac{V}{R} = \frac{50}{1 \times 10^6} = 50\ \mu\text{A}.$$

Discharging

$$R_T = 1 \times 10^6 + 680 \times 10^3 = 1.68 \times 10^6 = 1.68\ \text{M}\Omega$$

$$\tau_2 = 1.68 \times 10^6 \times 5 \times 10^{-6} = 8.4\ \text{s}.$$

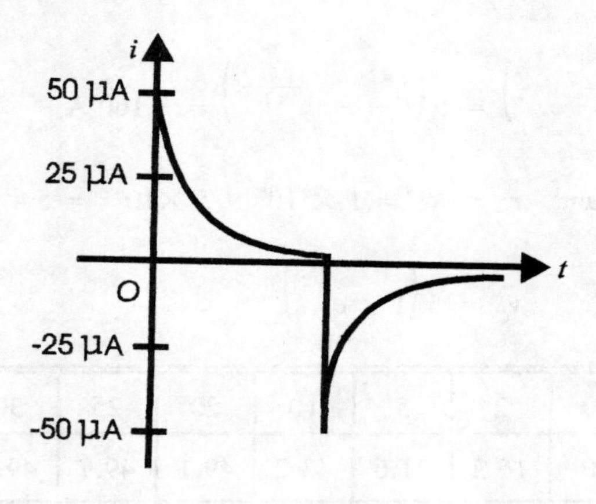

Fig. 407

SOLUTIONS 4

1.

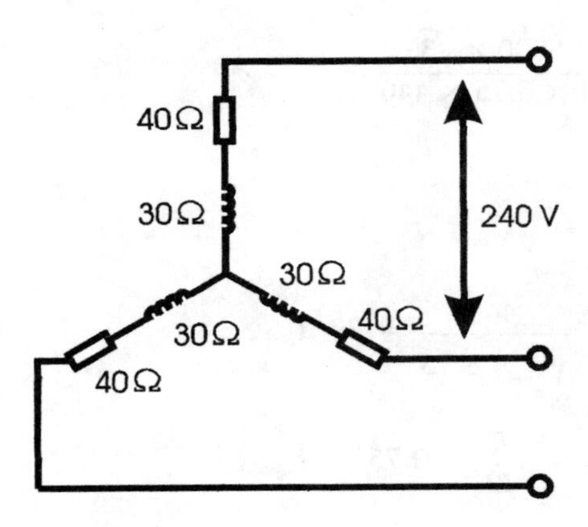

Fig. 408

$$|Z| = \sqrt{40^2 + 30^2} = 50\ \Omega$$

$$I_L = I_P = \frac{V_P}{|Z|} = \frac{139}{50} = 2.78\ \text{A}$$

$$V_L = \sqrt{3}\ V_P \Rightarrow V_P = \frac{240}{\sqrt{3}} = 139\ \text{volts}$$

(i) $I_P = 2.78$ A

(ii) $I_L = 2.78$ A

(iii) $V_P = 139$ V

(iv) $V_L = 240$ V

(v) $P = 3 I_P V_P \cos\phi = 3 \times 2.78 \times 139 \times 0.8 = 927.2$ W

$$\cos\phi = \frac{R}{Z} = \frac{40}{50} = 0.8$$

2. (a) $P = 3\, I_P\, V_P \cos\phi = 5000$

$$I_P \frac{V_L}{\sqrt{3}} = \frac{5000}{3 \times 0.75}$$

$$I_P = I_L = \frac{5000 \times \sqrt{3}}{3 \times 0.75 \times 440} = 8.75 \text{ A}$$

$$I_P = \frac{V_P}{|Z|}$$

$$|Z| = \frac{V_P}{I_P} = \frac{440}{\sqrt{3} \times 8.75} = 29\ \Omega$$

$$\cos\phi = \frac{R}{|Z|} = \frac{R}{29} = 0.75$$

$$\boxed{R = 21.8\ \Omega}$$

$$R^2 + X_L^2 = 29^2$$

$$X_L^2 = 29^2 - 21.8^2$$

$$X_L = \sqrt{29^2 - 21.8^2} = 19.1\ \Omega = 2\pi 50 L$$

$$L = \frac{19.1}{2\pi 50} = 61 \text{ mH}$$

$$\boxed{L = 61 \text{ mH}}$$

(b) $V_P = 440$

$$P = 3\, V_P\, I_P \cos\phi = 5000$$

$$I_P = \frac{5000}{3 \times 440 \times 0.75} = 5.05 \text{ A}$$

$$I_L = \sqrt{3}\, I_P = \sqrt{3} \times 5.05 = 8.75 \text{ A}$$

$$|Z| = \frac{V_P}{I_P} = \frac{440}{5.05} = 87.1\ \Omega$$

$$\cos\phi = \frac{R}{|Z|} = 0.75$$

$$R = 0.75 \times 87.1 = 65.4\ \Omega$$

$$R^2 + X_L^2 = 87.1^2$$

$$X_L = \sqrt{87.1^2 - 65.4^2} = 57.5\ \Omega = 2\pi 50 L$$

$$L = \frac{57.5}{100\pi} = 183\ \text{mH}.$$

3.

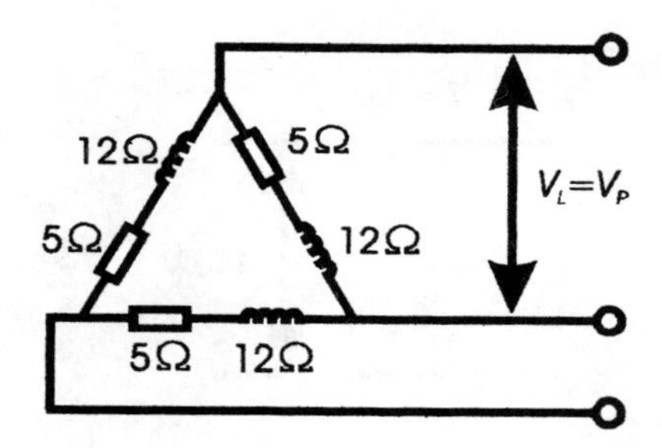

Fig. 409

$$P = \sqrt{3}\, I_L\, V_L \cos\phi = 1200$$

$$|Z| = \sqrt{5^2 + 12^2} = 13\ \Omega$$

$$\cos\phi = \frac{5}{13} = 0.3846153$$

$$I_L\, V_L = \frac{1200}{\sqrt{3} \times 0.3846153} = 1801$$

$$I_P = \frac{V_P}{|Z|} = \frac{V_L}{|Z|}$$

$$I_L\, V_L = 1801$$

(a) $I_L = \sqrt{3}\ I_P$

$I_L I_P |Z| = 1801$

$\sqrt{3}\ I_P I_P 13 = 1801$

$$I_P = \sqrt{\frac{1801}{13\sqrt{3}}} = 8.94 \text{ A}$$

$I_L = \sqrt{3}\ 8.94 = 15.5$ A

(b) $\cos\phi = 0.385$

(c) $V_L = V_P = I_P|Z| = 8.94 \times 13 = 116$ volts

4.

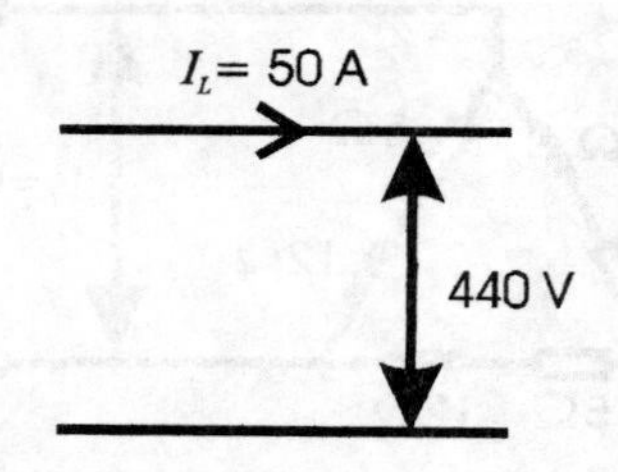

Fig. 410

$\sqrt{3}\ I_L V_L \cos\phi = 30000$

$$\cos\phi = \frac{30000}{\sqrt{3} \times 50 \times 440}$$

$$= 0.787$$

SOLUTIONS 5

1.

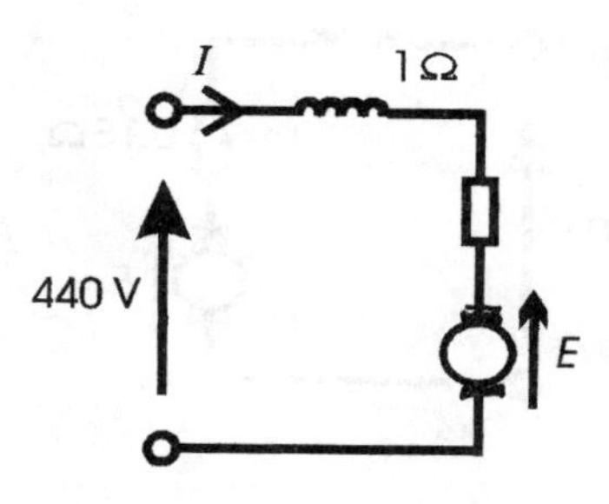

Fig. 411

$I_1 = 20$ A, $N_1 = 1260$ rev/min, $T_1 = 325$ Nm

$$\phi_1 = \phi$$

$I_2 = 40$ A, $\phi_2 = \phi + 50\,\% = 1.5\phi$

(a) $E_1 = 440 - (20 \times 1) = 420$ V

$E_2 = 440 - (40 \times 1) = 400$ V

$$E \propto N\phi \quad \therefore\ N \propto \frac{E}{\phi}$$

$$\text{Thus } \frac{N_2}{N_1} = \frac{E_2}{E_1} \times \frac{\phi_1}{\phi_2}$$

$$\text{or } N_2 = \frac{400}{420} \times \frac{\phi}{1.5\phi} \times 1260 = 800 \text{ rev/min}$$

(b) $$T \propto I\phi \quad \therefore\ \frac{T_2}{T_1} = \frac{I_2}{I_1} \times \frac{\phi_2}{\phi_1}$$

$$\text{or } T_2 = \frac{40}{20} \times \frac{1.5\phi}{\phi} \times 325 = 975 \text{ Nm}$$

2.

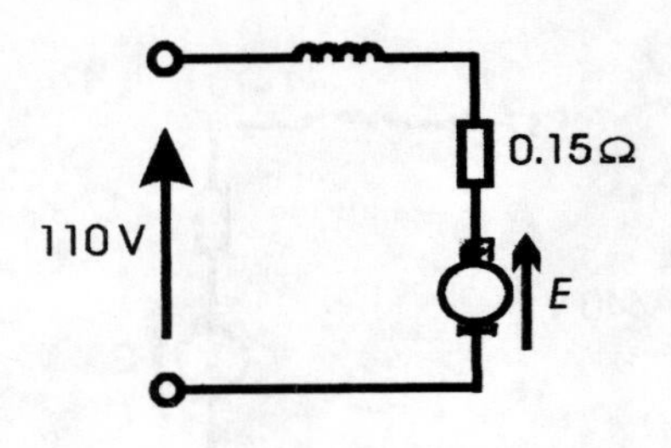

Fig. 412

$T_1 = T_F, \ N_1 = 1200$ rev/min., $I_1 = 24$ A

$\phi \propto I$

$T_2 = 0.25\,T_F$

(a) $T \propto I\phi \propto I^2$

$$\therefore \frac{T_2}{T_1} = \left(\frac{I_2}{I_1}\right)^2 \quad \text{i.e. } I_2 = I_1\sqrt{\frac{T_2}{T_1}}$$

$$\text{or } I_2 = 24 \times \sqrt{\frac{0.25\,T_F}{T_F}} = 12 \text{ A}$$

(b) $E_1 = 110 - (24 \times 0.15) = 106.4$ V

$E_2 = 110 - (12 \times 0.15) = 108.2$ V

$$E \propto N\phi \qquad \therefore N \propto \frac{E}{\phi}$$

$$\frac{N_2}{N_1} = \frac{E_2}{E_1} \times \frac{\phi_1}{\phi_2} = \frac{E_2}{E_1} \times \frac{I_1}{I_2}$$

$$\text{i.e. } N_2 = \frac{108.2}{106.4} \times \frac{24}{12} \times 1200 = 2440 \text{ rev/min.}$$

3. $$E = V - I(R_A + R_F)$$

(a) $E_1 = 235 - (13 \times 0.87) = 113.69$ V

$E_2 = 125 - (16.7 \times 0.87) = 110.47$ V

$E_3 = 125 - (20.6 \times 0.87) = 107.08$ V

$E_4 = 125 - (25 \times 0.87) = 103.25$ V

$$EI = \frac{2\pi NT}{60}$$

$$\therefore T = \frac{60 \times EI}{2\pi N}$$

i.e. $$T_1 = \frac{60 \times 113.69 \times 13}{2\pi \times 1275} = 11.07 \text{ Nm}$$

$$T_2 = \frac{60 \times 110.47 \times 16.7}{2\pi \times 1050} = 16.78 \text{ Nm}$$

$$T_3 = \frac{60 \times 107.08 \times 20.6}{2\pi \times 900} = 23.4 \text{ Nm}$$

$$T_4 = \frac{60 \times 103.25 \times 25}{2\pi \times 790} = 31.2 \text{ Nm}$$

(b) At 900 rev/min, gross torque, $T_3 = 23.4$ Nm

Shaft torque, $T = T_3 -$ loss torque

i.e. $T = 23.4 - 4.05 = 19.35$ Nm

Output power, $$P_0 = \frac{2\pi NT}{60}$$

$$= \frac{2\pi \times 900 \times 19.35}{60} \text{ W}$$

$= 1.824$ KW

4.

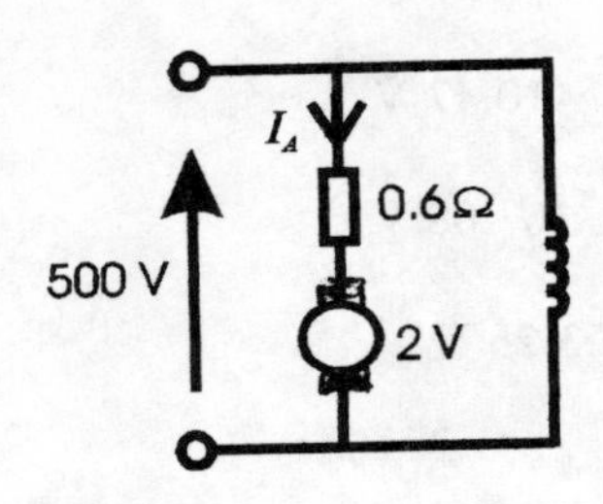

$N_1 = 800$ rev/min

$I_{A1} = 42$ A

$\phi_1 = \phi \qquad \phi_2 = 0.75\phi$

$T_1 = T$

Fig. 413

(a) $T_2 = T_1 \qquad T \propto I_A \phi \qquad \therefore I_A \propto \dfrac{T}{\phi}$

$$\frac{I_{A2}}{I_{A1}} = \frac{T_2}{T_1} \times \frac{\phi_1}{\phi_2} \quad \text{or} \quad I_{A2} = \frac{T_1}{T_1} \times \frac{\phi}{0.75\phi} \times 42 \text{ A} = 56 \text{ A}$$

$$E_1 = 500 - (42 \times 0.6) - 2 = 472.8 \text{ V}$$

$$E_2 = 500 - (56 \times 0.6) - 2 = 464.4 \text{ V}$$

$$E \propto N\phi \qquad \therefore N \propto \frac{E}{\phi} \quad \text{or} \quad \frac{N_2}{N_1} = \frac{E_2}{E_1} \times \frac{\phi_1}{\phi_2}$$

i.e. $N_2 = \dfrac{464.4}{472.8} \times \dfrac{\phi}{0.75\phi} \times 800 = 1048$ rev/min

(b) $T_2 = 0.8\,T_1$

$$I_{A2} = \frac{T_2}{T_1} \times \frac{\phi_1}{\phi_2} \times I_{A1} = \frac{0.8\,T_1}{T_1} \times \frac{\phi}{0.75\phi} \times 42 \text{ A} = 44.8 \text{ A}$$

$$E_2 = 500 - (44.8 \times 0.6) - 2 = 471.12 \text{ V}$$

$$N_2 = \frac{E_2}{E_1} \times \frac{\phi_1}{\phi_2} \times N_1 = \frac{471.12}{472.8} \times \frac{\phi}{0.75\phi} \times 800 = 1063 \text{ rev/min}$$

5.

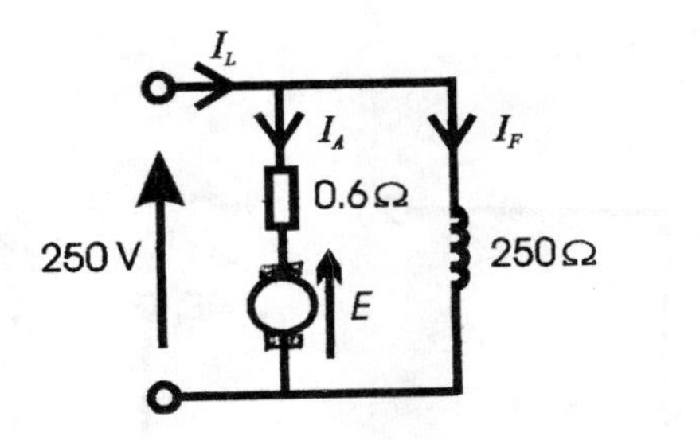

$I_{L1} = 2.5$ A

$N_1 = 1500$ rev/min

$I_{L2} = 10$ A

Fig. 414

$I_{F1} = \dfrac{V}{R_F} = \dfrac{250}{250} = 1$ A $\qquad \therefore\ I_{A1} = I_{L1} - I_{F1} = 2.5 - 1 = 1.5$ A

$E_1 = 250 - (1.5 \times 0.6) = 249.1$ V.

On full load, $I_{A2} = 10 - 1 = 9$ A

$\therefore$ counter-e.m.f., $E_2 = 250 - (9 \times 0.6) = 244.6$ V

$E \propto N\phi \qquad \therefore\ N \propto \dfrac{E}{\phi} \propto E$ since ϕ and I_F are constant

$\dfrac{N_2}{N_1} = \dfrac{E_2}{E} \qquad \therefore\ N_2 = \dfrac{E_2}{E_1} \times N_1$

i.e. $N_2 = \dfrac{244.6}{249.1} \times 1500 = 1473$ rev/min

6. (a) $T \propto I_A\phi \qquad T_1 = 95$ Nm $\qquad I_{A1} = 50$ A, $I_{A2} = 25$ A

For constant field current, ϕ is constant

$\therefore\ T \propto I_A \qquad \therefore\ T_2 = \dfrac{I_{A2}}{I_{A1}} \times T_1 = \dfrac{25}{50} \times 95 = 47.5$ Nm

(b) $T \propto I_A\phi \qquad \therefore\ T_2 = \dfrac{I_{A2}}{I_{A1}} \times \dfrac{\phi_2}{\phi_1} \times T_1$

Now $\phi_2 = 1.2\phi_1$

$\therefore\ T_2 = \dfrac{25}{50} \times \dfrac{1.2\phi_1}{\phi_1} \times 95 = 57$ Nm

SOLUTIONS 6

1. $I_0 = \sqrt{I_c^2 + I_m^2} = \sqrt{0.5^2 + 1^2} = 1.12 \text{ A}.$

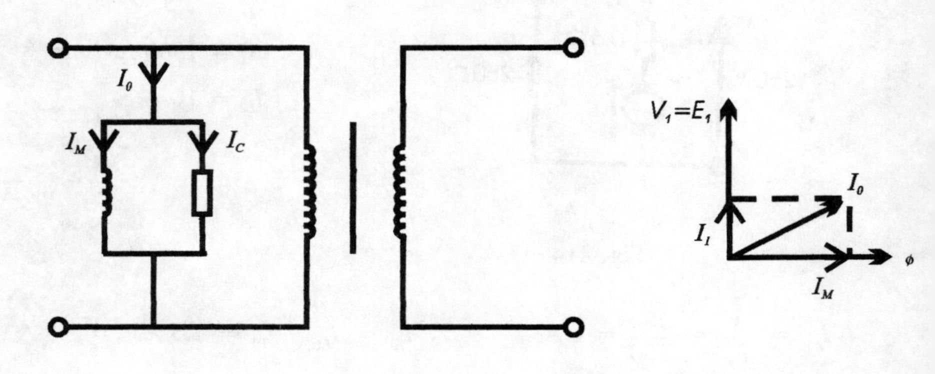

Fig. 415

2. (i) $I_0 V_1 \cos \theta_0 = 50$

$$2 \times 240 \times \cos \theta_0 = 50$$

$$\cos \theta_0 = \frac{50}{480} = 0.104$$

$$\theta_0 = 84°.$$

(ii) $I_c = I_0 \sin 6° = 0.209 \text{ A}$

$$I_m = I_0 \cos 6° = 1.99 \text{ A}.$$

3. $$\frac{V_2}{I_2} = R_L = \frac{\frac{N_2}{N_1} V_1}{\frac{N_1}{N_2} \times I_1} \quad \text{where} \quad \frac{V_2}{V_1} = \frac{N_2}{N_1} = \frac{I_1}{I_2}$$

$$R_{in} = \frac{V_1}{I_1} = \left(\frac{N_1}{N_2}\right)^2 R_L = 10^2 R_L$$

$$R_{in} = \frac{1}{100} \times 8400 = 84 \ \Omega.$$

4. $$\eta = \frac{\text{output power}}{\text{input power}} = \frac{\text{input power} - \text{losses}}{\text{input power}}$$

$$= \left(1 - \frac{\text{losses}}{\text{input power}}\right) \times 100$$

$$= \left(1 - \frac{1000}{10000}\right) \times 100 = 90\ \%.$$

5.

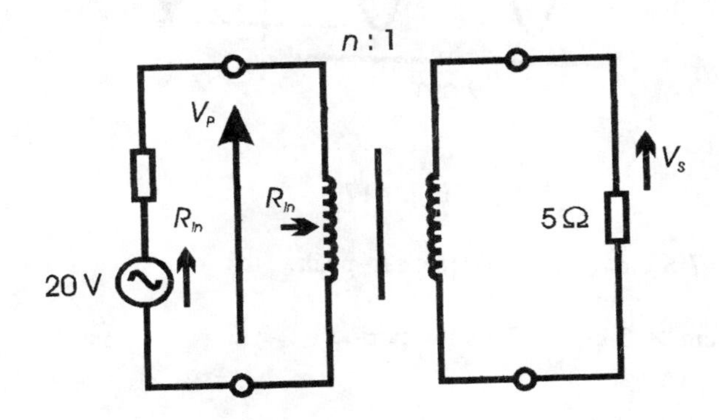

Fig. 416

(i) $605 = n^2 R_L$

$$n^2 = \frac{605}{5} = 121$$

$$n = 11$$

(ii) $I_P = \dfrac{20}{2 \times 605} = 0.0165\ \text{A}$

(iii) $I_S = I_P \times n = 0.0165 \times 11 = 0.1818\ \text{A}$

(iv) $V_P = 20 - I_P R_{in} = 20 - 0.0165 \times 605 = 10\ \text{V}$

(v) $V_S = \dfrac{V_P}{n} = \dfrac{10}{11} = 0.909\ \text{V}$

(vi) $P_{LOAD} = I_S^2 R_L = (0.1818)^2 \times 5 = 0.165\ \text{W}.$

SOLUTIONS 7

1.

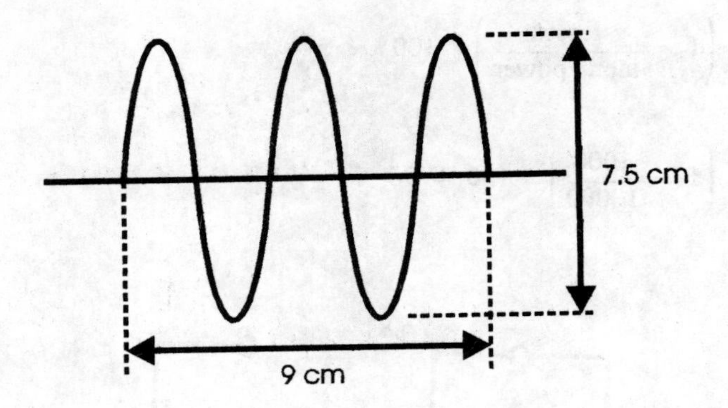

Fig. 417

$5 \text{ V/cm} \times 7.5 \text{ cm} = 37.5 \text{ V}$ peak-to-peak.

$T = 1\ \mu\text{s/cm} \times 3 \text{ cm} = 3\ \mu\text{s}$ the period

$$f = \frac{1}{3 \times 10^{-6}} = \frac{10^6}{3} = \frac{1}{3} \text{ MHz.}$$

2. See text.

3. (i) 2:1

(ii) 3:1

(iii) 1:3

(iv) 1:1

4. $\frac{50R}{50 + R} \cdot I = 40$ where R is in KΩ and I is in mA.

$$\frac{50R}{50 + R} 10 = 40$$

$$50R = (50 + R)4 = 200 + 4R$$

$$46R = 200$$

$$\boxed{R = 4.35 \text{ K}\Omega}$$

5. $L_x = PQC = 10 \times 100 \times 50 \times 10^{-6} = 50000 \times 10^{-6} = 0.05$ H

$$R_x = \frac{PQ}{R} = \frac{10 \times 100}{100} = 10\ \Omega$$

6. Minimum values

$$L_x = \frac{PL}{Q} = \frac{\left(P - \frac{1}{100}P\right)}{Q + \frac{1}{100}Q} . L = \frac{P}{Q}\frac{0.99}{1.01} . L = 0.98\frac{PL}{Q}$$

$$R_x = \frac{P}{Q}R = \frac{P - \frac{1}{100}P}{Q + \frac{1}{100}Q}\left(R - \frac{1}{100}R\right) = \frac{P0.99}{Q1.01}R0.99 = \frac{PR}{Q}0.97$$

Maximum

$$L_x = \frac{P}{Q}L = \frac{P1.01}{Q0.99}L = 1.02\frac{P}{Q}L$$

$$R_x = \frac{P}{Q}R = \frac{P1.01}{Q0.99}1.01\ R = 1.03\frac{PR}{Q}$$

7. (a) 20000 Ω/V × 10 = 200 KΩ

(b) 200 Ω/V × 300 = 60 KΩ.

ELECTRICAL AND ELECTRONIC PRINCIPLES III

Miscellaneous

PHASE TEST. 1 (1 - 8 inclusive)

1.

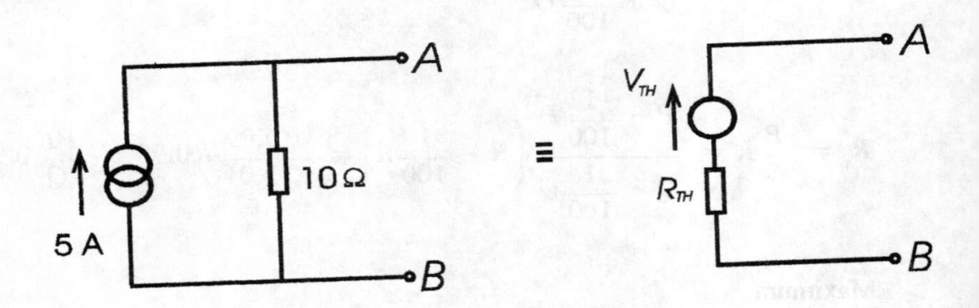

Fig. 418

Determine V_{TH} and R_{TH}, show clearly your solution.

2.

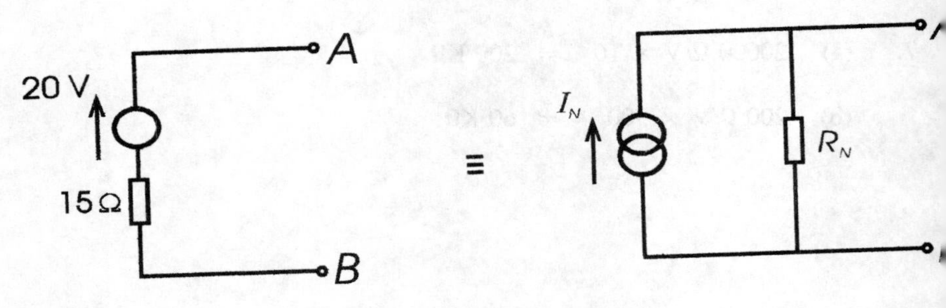

Fig. 419

Determine I_N and R_N, show clearly your solution.

3.

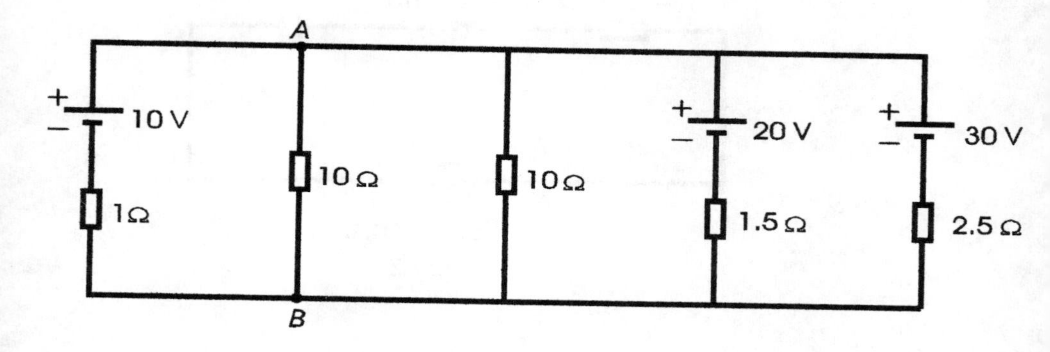

Fig. 420

Use the constant current generator principle for each cell and hence calculate the p.d.

4.

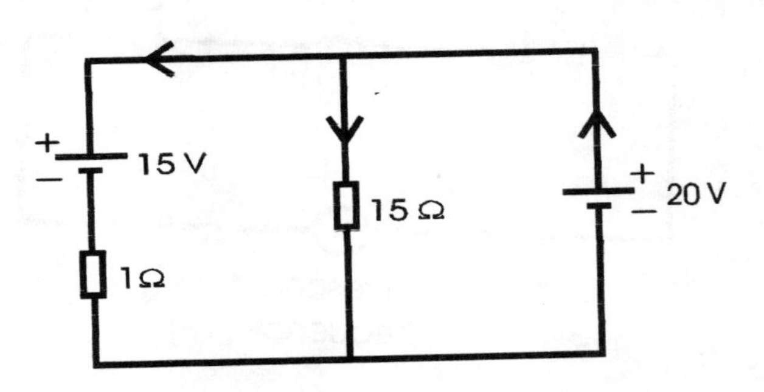

Fig. 421

Use the superposition principles in order to calculate the current through the 15 Ω load.

5.

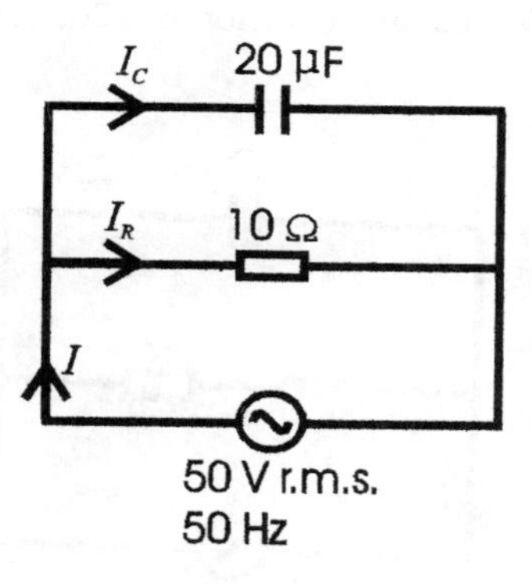

Fig. 422

Determine the currents:- (i) I_C (ii) I_R (iii) I and the total impedance of the circuit.

6.

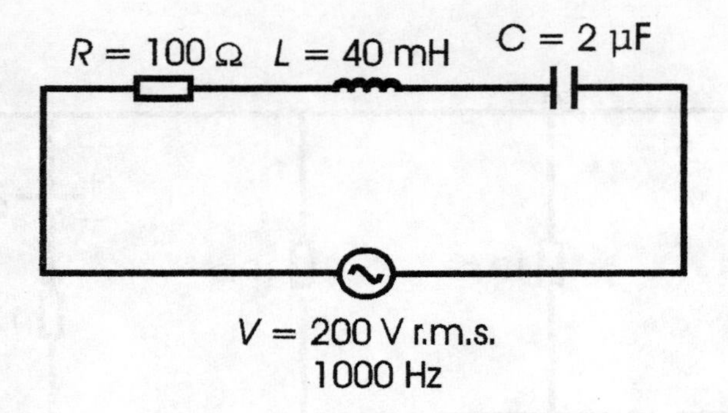

Fig. 423

Calculate the total current.

7.

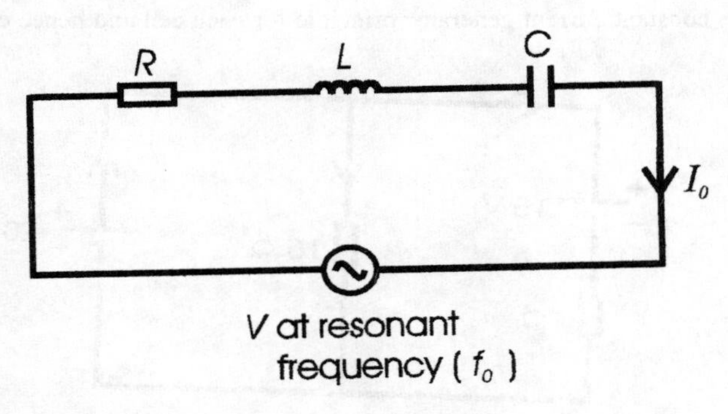

Fig. 424

(i) What is the phase angle between the current I_o and the voltage V(taking current as the reference).

(ii) If $V = 40$ mV and the Q-factor is 120, find the voltage across the capacitor.

8.

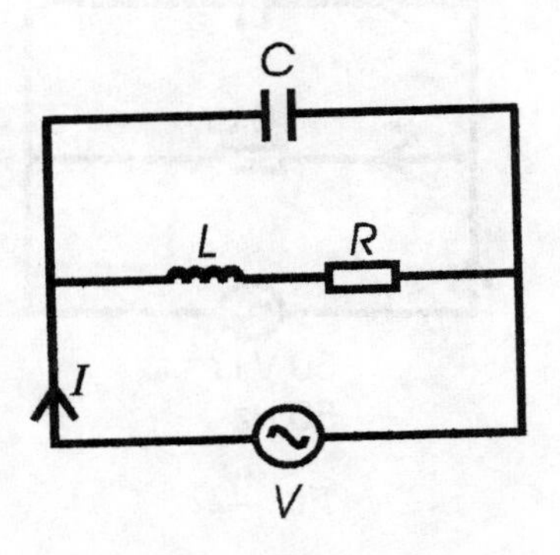

Fig. 425

(i) Determine the resonant frequency of the circuit.

$$\left(f_o = \frac{1}{2\pi}\sqrt{\frac{1}{LC} - \frac{R^2}{L^2}}\right)$$

(ii) What is $\frac{L}{RC}$?

(iii) If $Q = 180$, $I = 25\ \mu A$ at resonance, calculate the value of I_C.

(iv) If $V = 100$ volts, then find the supply current to the circuit.

END TEST. 1 (9 - 13 inclusive)

9. A coil of inductance 8 mH and resistance R is connected in parallel with a capacitor of capacitance C.

The circuit is at resonance, where $f_o = 3446$ Hz, the supply current and voltage are given as 200 mA and 80 V respectively. The Q-factor is given as $\sqrt{3}$.

Determine:- (i) the current through the capacitor
(ii) the current through the coil
(iii) the dynamic resistance
(iv) the resistance R
(v) the capacitance C.

10. (a) Determine the impedance of the circuits at 100 Hz

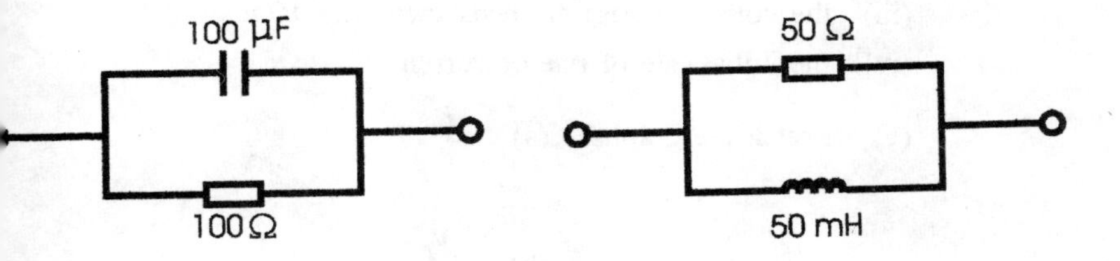

Fig. 426

(b)

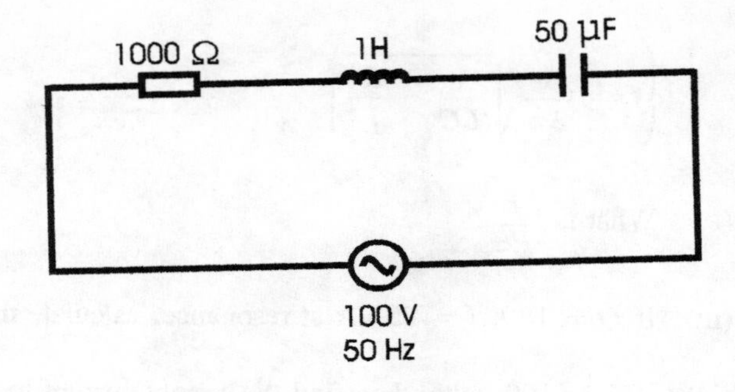

Fig. 427

Calculate:- (i) the power factor of the circuit
(ii) the impedance of the circuit
(iii) the current in the circuit.

11. Three coils having resistance 10 Ω and inductive reactance 15 Ω are connected
(i) in star
(ii) in delta.

If the line voltage of a 3-phase 415 V supply, calculate in both cases
(a) the line and phase currents
(b) the line and phase voltages at the load
(c) the total power consumed in each case.

12. A coil of inductance 5 H and resistance 125 Ω is suddenly switched across a 100 V d.c. supply.

Calculate:- (i) the time constant of the circuit
(ii) the current after 15 ms
(iii) the voltage across the resistance after 10 ms
(iv) the initial rate of rise of current
(v) sketch the graphs (a) $\frac{i}{t}$

(b) $\frac{V_L}{t}$

(c) $\frac{V_R}{t}$.

13. Give a simple sketch of the core of a transformer. Why must the core be laminated?

List the losses in a transformer.

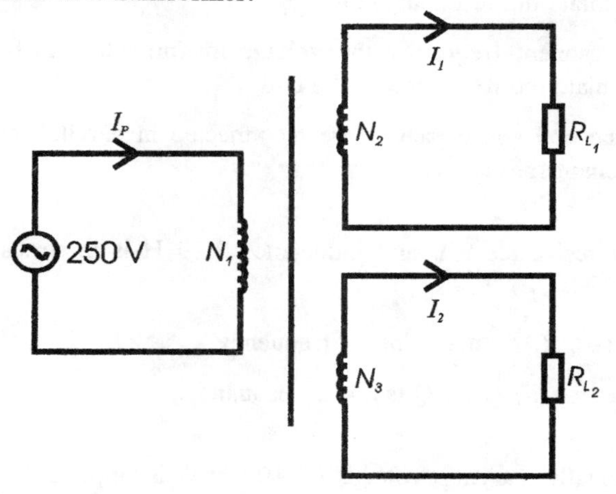

Fig. 428

Given that:- $N_1 = 10{,}000$ turns, $N_2 = 60{,}000$ turns

$N_3 = 2{,}000$ turns, $R_{L1} = 1{,}000\ \Omega$, $R_{L2} = 200\ \Omega$

Calculate (i) V_1 (ii) V_2

(iii) I_1 (iv) I_2

(v) total power drawn from the supply

(vi) input resistance of primary winding

(vii) primary current I_P

How is the efficiency of a transformer measured?

14. A circuit consists of a capacitor, an inductor, and a 20 ohms resistor connected in series to an 8 volts a.c. supply.

When the resonant frequency is 2275 Hz the Q-factor of the circuit is unity.

Determine:- (a) the values of the capacitor and the inductor,

(b) to what frequency the supply must be adjusted to ensure that the voltage across the capacitor is twice the voltage across the inductor.

Give answer correct to one decimal place.

15. A series a.c. circuit comprises a coil of inductance 12 mH and a capacitor of capacitance 0.025 μF.

(a) Calculate the resonant frequency.

(b) At resonant frequency the voltage magnification is found to be 5.6, calculate the resistance of the coil.

(c) If the coil and capacitor are reconnected in parallel, calculate the new resonant frequency.

16. A coil of resistance 8 Ω and inductance 0.19 H is in series with a 200 μF capacitor.

Calculate:- (i) the resonant frequency,

(ii) the *Q*-factor at resonance.

17. A sinusoidally varying voltage of 100 mV is applied to a series circuit containing $R = 5\ \Omega$, $L = 1.592$ mH and $C = 159.2$ nF.

Calculate:- (a) the supply current at a supply frequency of 9 KHz.

(b) the supply frequency for series resonance to occur,

(c) the supply current at resonance,

(d) the *Q*-factor at resonance,

(e) the voltage across the capacitor at resonance.

18. An inductor of 0.001 H, a resistor of 5 ohms and a capacitor of 0.004 μF are all connected in series. If a variable frequency signal generator of 1 V terminal voltage is connected across the series circuit.

(a) What is the frequency of the signal generator that will cause maximum current to flow,

(b) what is the voltage developed across the capacitor, with conditions as in (a) above,

(c) what is the magnification factor of the circuit?

(d) Sketch a graph of the impedance of the circuit against frequency.

19. The diagram below, shows a STAR connected generator and its DELTA connected load.

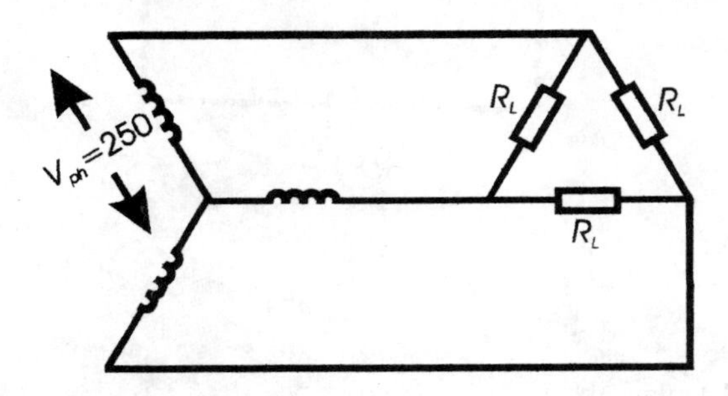

Fig. 429

The generator phase voltage is shown in the diagram.

Calculate:-
- (i) Generator line voltage
- (ii) load current in each branch if $R_L = 100\ \Omega$
- (iii) Line current
- (iv) Power in each phase
- (v) Total power supplied by the generator.

PHASE TEST. 2 (20 - 24 inclusive)

20.

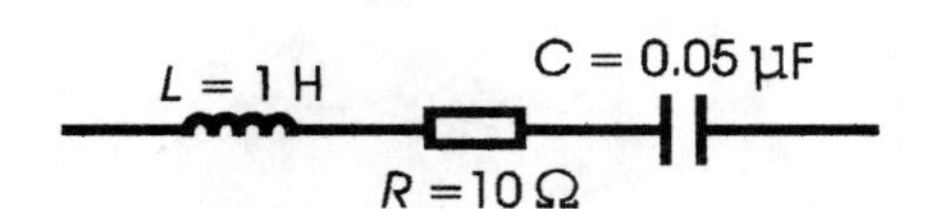

Fig. 430

Find the impedance of this circuit at the following frequencies:- 100 Hz, 1,000 Hz, 1,000,000 Hz. Hence sketch a graph of Z against f. What is the resonant frequency?

21.

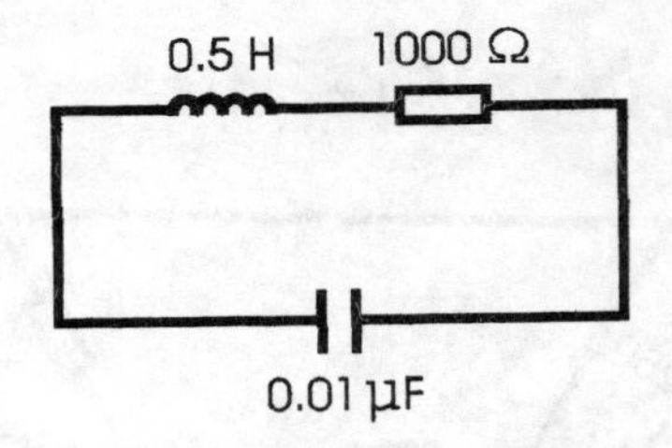

Fig. 431

(a) Calculate the resonant frequency.

(b) What is the *Q*-factor at this frequency?

(c) Sketch the impedance/frequency graph.

22.

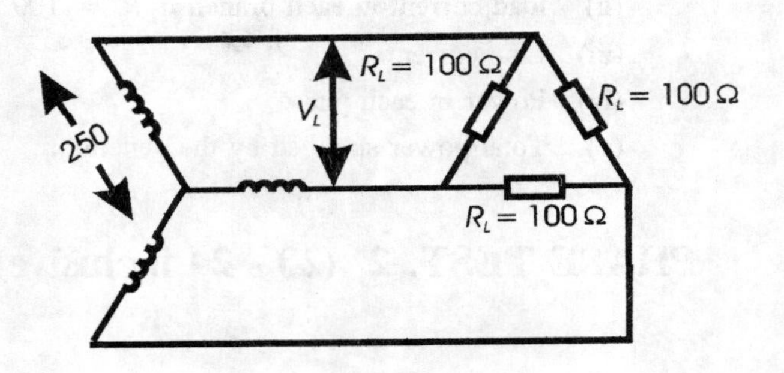

Fig. 432

(i) Calculate V_L

(ii) the load current

(iii) the line current

(iv) the total power.

23.

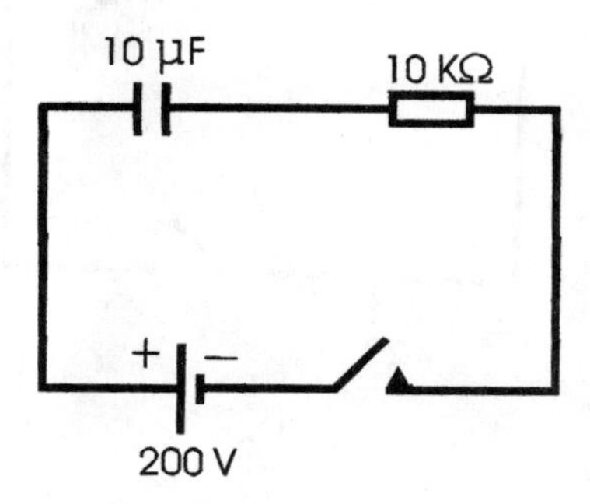

Fig. 433

(i) Calculate the time constant.

(ii) The switch is closed draw a graph of q against t.

(iii) Calculate the initial value of current.

24. Discuss the losses of the transformer. For an ideal transformer write down the following equations:-

(i) The input resistance looking into the primary if R_L is the load and $\frac{N_1}{N_2}$ is the turns ratio.

(ii) The input volt amperes and output volt amperes.

PHASE TEST. 3 (25 - 28 inclusive)

25. A fully discharged capacitor is connected in series with a 50 KΩ resistor to a 50 V d.c. supply. Given that the initial rate of rise of voltage across the capacitor is 25 V/s, calculate:-

(i) τ (ii) C (iii) v_C and v_R 0.5 s after closing the switch.

26.

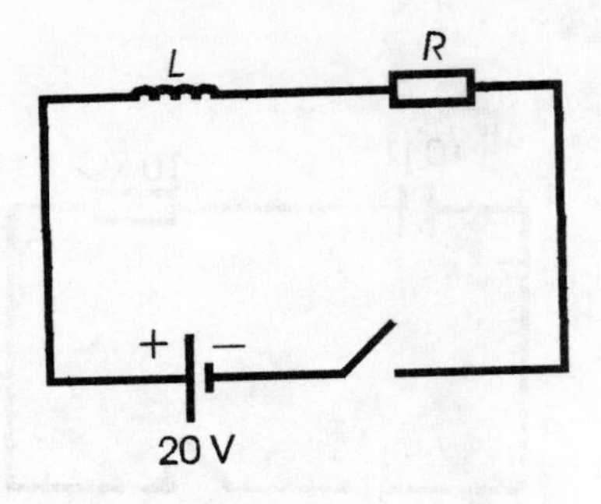

Fig. 434

(i) Calculate i, if $R = 50\ \Omega$, $L = 20$ mH, 0.1 ms after closing the switch.

(ii) What is the value of $L\dfrac{di}{dt}$ at $t = 0$ and at very large time after closing the switch.

27. Three identical coils each of resistance 9 Ω and inductance 38.2 mH are connected in star across a 415 V, 50 Hz, 3 phase supply.

Calculate:- (i) the phase voltage
(ii) the line current
(iii) the power factor
(iv) the total power dissipation.

28. Two sine waves are displayed on a *CRO* as follows:-

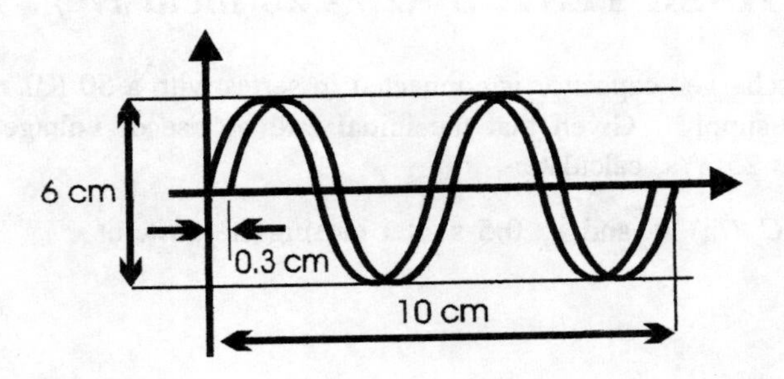

Fig. 435

The X amplifier setting is 10 μs/cm. The Y amplifier setting is 50 mV/cm.

Calculate:- (i) the r.m.s. values of the voltage

(ii) the frequency

(iii) the phase difference between the two voltages.

PHASE TEST. 4 (29 and 30)

9.

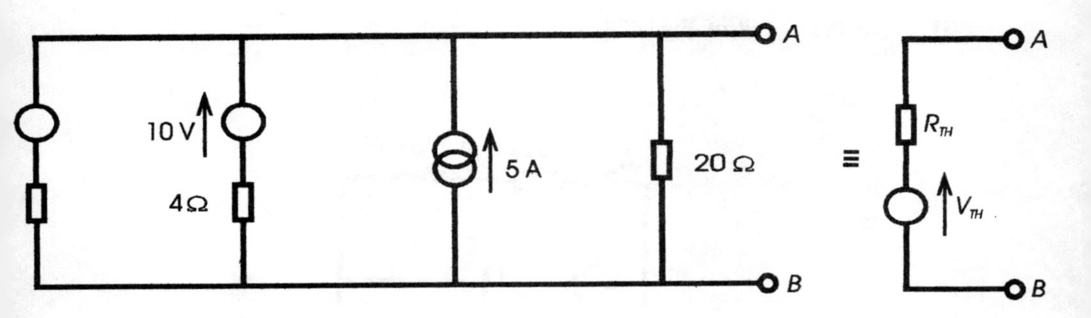

Fig. 436

(i) Determine R_{TH} and V_{TH}

(ii) Determine the load to be connected between A and B for maximum power in it.

(iii) Determine the maximum power in W in the load.

0.

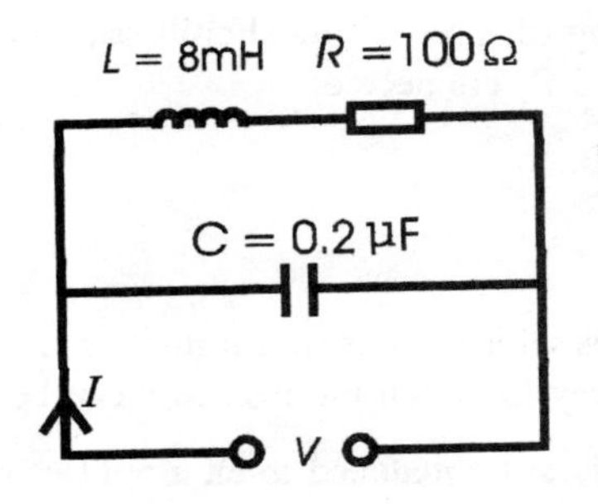

Fig. 437

(i) Determine f for minimum current drawn for the supply (I).

(ii) The Q factor at the circuit at this frequency

(iii) R_d, the dynamic resistance

(iv) V, the supply voltage.

PHASE TEST. 5 (31 - 35 inclusive)

31. For the circuit of Fig. 438

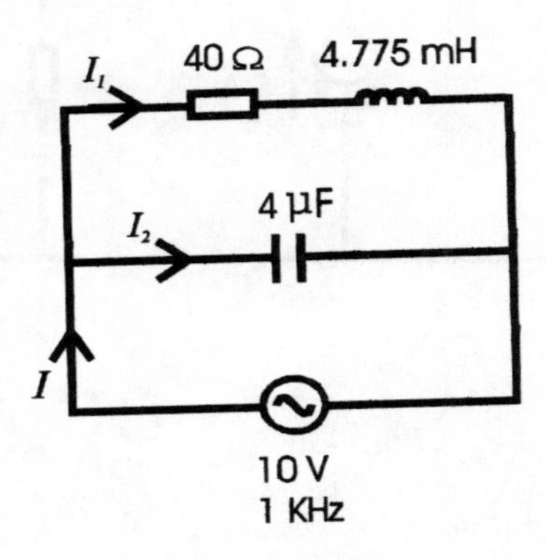

Fig. 438

(a) Calculate (i) the current I_1 and its phase angle relative to the supply voltage;

(ii) the current I_2.

(b) Determine the supply current I_3.

32. Calculate the line current which would be taken from a 415 V, 50 Hz 3 phase supply by a load consisting of three identical coils each of inductance 12.73 mH and resistance 3 Ω, connected:-

(a) in star

(b) in delta.

33. (a) Name the iron losses which occur in a transformer. In each case state, with reasons, one way in which the iron loss can be reduced.

(b) A 4 Ω loudspeaker is to be matched to an amplifier of output resistance 5 KΩ. Calculate the turns ratio of the transformer required to achieve this matching.

34. A 100 nF capacitor is connected in series with a 200 KΩ resistor across a 50 V d.c. supply. Calculate:-

(a) the time constant

(b) the initial value of current

(c) the initial rate of change of current

(d) at a time equal to the time constant:-

(i) the current

(ii) the p.d. across the capacitor

(iii) the p.d. across the resistor.

35. A coil having resistance of 10 Ω and inductance 15.92 mH is connected in series with a 6.37 μF capacitor.

This series circuit is connected across a 1 V variable frequency supply.

Calculate:- (a) the frequency to which the supply must be adjusted for resonance to occur,

(b) the current at resonance,

(c) the power supplied at resonance,

(d) the *Q*-factor at resonance.

END TEST. 2 (36 - 41 inclusive)

36. An inductor of inductance 0.50 H takes a current of 1.75 mA from a 6.0 V, 1.0 KHz supply.

Calculate the resistance of the inductor.

Calculate the capacitance of the capacitor required to be connected in parallel with the inductor to produce resonance at 1 KHz, and determine the total current taken from the supply.

Draw a phasor diagram to scale, showing the voltage, the two branch currents and the total current.

(Ans. 1373 Ω, 99×10^{-9} F).

37. A 415 V, 3-phase supply delivers a current of 12.6 A at a power factor of 0.86 lagging to a delta-connected motor.

Calculate:- (i) the total active power input to the motor,

(ii) the impedance, resistance and reactance per phase of the motor winding.

(Ans. (i) 4.496 KW (ii) 57.08 Ω, 49 Ω, 27.28 Ω.)

38. A capacitor of capacitance 2.5 μF is connected in series with a resistor of resistance 400 KΩ to a d.c. supply of 120 V. Calculate the time constant of the circuit.

    Construct graphically the curve of voltage across the capacitor from zero time, when the capacitor is fully discharged, using a time scale of 0 to 2.3 s.

    From your graph read:-

    (i) the voltage across the capacitor at a time of 1.3 s,

    (ii) the time needed for the voltage to reach 80 V.

    (Ans. 1s (i) 90 V (ii) 1 s)

39. Draw and label a phasor diagram for a transformer on no load, showing flux, primary terminal voltage, no-load current and phase angle. Resolve the no-load current into its magnetising and core-loss components.

    A transformer has 492 turns on the primary winding and 640 turns on the secondary winding, which has a centre tap.

    (a) When a p.d. of 240 V a.c. is applied to the primary winding, calculate the secondary voltage between

    (i) the end terminals,

    (ii) one end terminal and the centre tap.

    (b) Neglecting the no-load current, determine the primary current if a current of 5.0 A is taken from

    (i) the terminals (a) (i) above,

    (ii) the terminals (a) (ii) above.

    (Ans. (a) (i) 312 V (ii) 156 V (b) (i) 3.84 A (ii) 1.92 A)

40. A d.c. separately excited motor has a no-load speed of 1200 rev/min. when 110 V is applied to both armature and field circuits. Calculate the no-load speed when:-

    (i) the armature voltage is 120 V and the field voltage is 110 V,

    (ii) the armature voltage is 110 V and the field voltage is 100 V,

    (iii) the armature voltage is 100 V and the field voltage is 120 V

    Assume that the resistance of the field circuit remains constant and the flux is proportional to field current.

41.

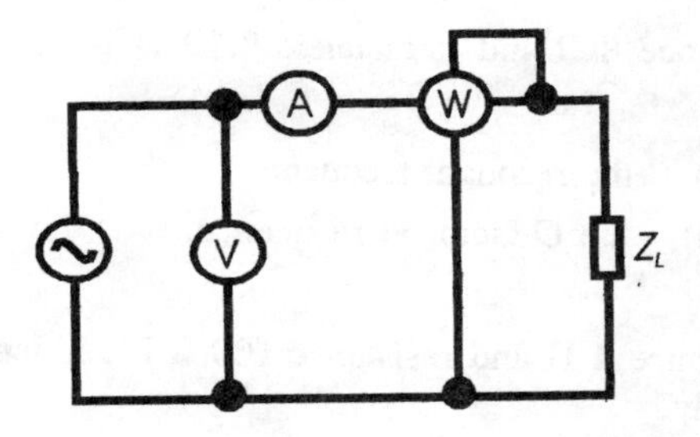

Fig. 439

In the circuit shown, the voltmeter reads 240 V, the ammeter reads 4.0 A and the wattmeter reads 750 W. Calculate the power factor of the load from the above readings.

If the voltmeter is scaled 0 - 300 V, the ammeter is scaled 0 - 5 A and the wattmeter is scaled 0 - 1500 W, what are the highest and lowest possible values of power factor, assuming that all the instruments are within their specified accuracy class of 1.0%?

(Ans. 0.78, p.f. varies)

END TEST. 3 (42 - 46 inclusive)

42. Calculate the current in the 5 Ω resistor of Fig. 440 using Thevenin's theorem.

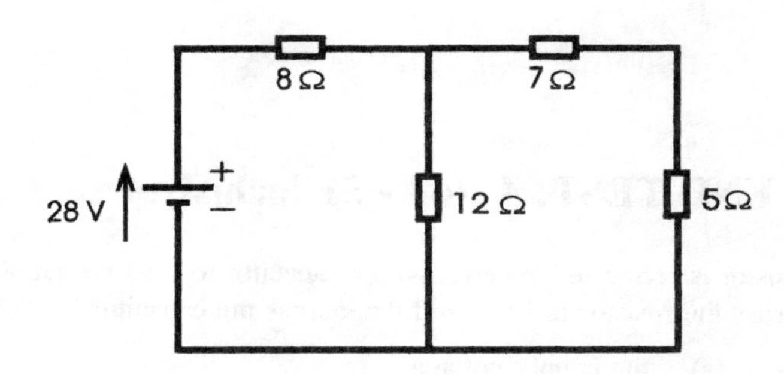

Fig. 440

43. A coil takes 1.6 KW and 3 KVAr when connected to a 240 V, 50 Hz supply.

Calculate the resistance and inductance of the coil.

44. A coil of resistance 8 Ω and inductance 0.19 H is in series with a 200 μF capacitor.

Calculate:- (i) the resonant frequency,

(ii) the Q-factor at resonance.

45. A coil of inductance 4 H and resistance 100 Ω is suddenly switched across a 50 V, d.c. supply.

Calculate:- (i) the time constant of the circuit,

(ii) the current after 10 milliseconds.

46. A voltmeter of resistance 500 Ω/V is used on the 100 V range to measure the voltage between X and earth in the circuit of Fig. 441.

Calculate the percentage error in the reading.

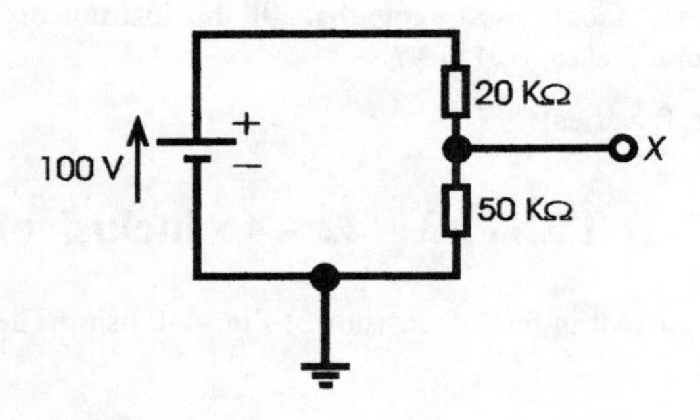

Fig. 441

END TEST. 4 (47 - 51 inclusive)

47. A 30 Ω resistor is connected in series with a capacitor to a 50 Hz supply. The voltage across the resistor is 15 V and that across the capacitor is 20 V.

Calculate:- (a) the supply voltage,
(b) the power developed,
(c) the power factor.

48. A 1500 pF capacitor is in parallel with a coil of inductance 1 mH and resistance 100 Ω. The circuit is connected to a supply of 100 V at the resonant frequency.

Calculate:- (i) the resonant frequency,
(ii) the dynamic impedance
(iii) the Q-factor.

49. Apply Norton's theorem to calculate the current I in the circuit shown in Fig. 442.

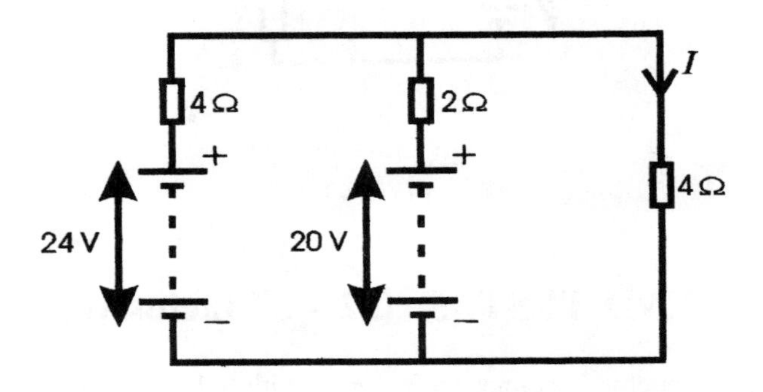

Fig. 442

50. A capacitor of 2.2 μF is charged via a 100 MΩ resistor from a 300 V d.c. supply.

Calculate:- (i) the time constant of the circuit
(ii) the voltage across the capacitor after 2 mins.

51. A voltmeter of f.s.d. 10 V and resistance 500 Ω/V is connected across the 2.5 KΩ resistor in the circuit of Fig. 443. It reads 5 V.

Calculate:- (i) the value of R,
(ii) the true voltage across the 2.5 KΩ resistor (i.e. without the voltmeter).

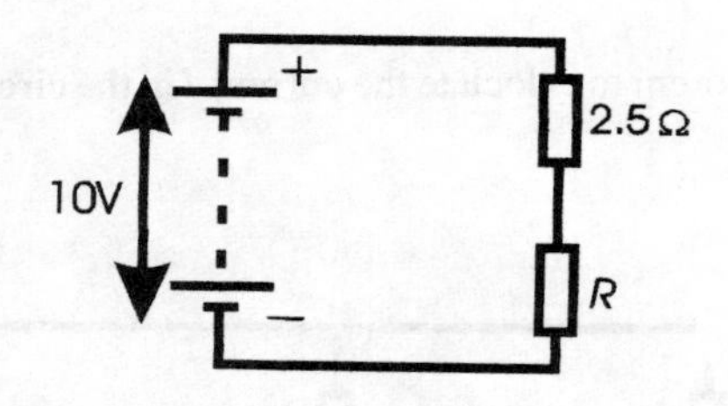

Fig. 443

END TEST. 5 (52 - 57 inclusive)

52. A 10 μH inductor is connected in series with a 1 nF capacitor across 1 mV variable frequency supply. The frequency is adjusted until a maximum current of 200 μA is obtained.

At this frequency, calculate:- (i) the coil resistance

(ii) the frequency,

(iii) the circuit Q factor

(iv) the p.d. across the capacitor.

(Ans. (i) 5 Ω (ii) 1.59 MHz (iii) 20 (iv) 20 mV)

53. (a) Explain why power factor correction is advantageous in low power factor industrial circuits.

(b) A 5 KVA, 400 V, 50 Hz single phase induction motor has a full load power factor of 0.82 lagging. Calculate the value of the parallel capacitor required to give an overall power factor of unity. Sketch the complete phasor diagram.

(Ans. 56.9 μF)

54. Three identical coils, each of resistance 9 Ω and inductance 38.2 mH are connected in star across a 415 V, 50 Hz, 3 phase supply.

Calculate:- (i) the phase voltage,

(ii) the line current

(iii) the power factor
(iv) the total power dissipation

(Ans. (i) 240 V (ii) 16 A (iii) 0.6 lag (iv) 6.90 KW)

55. A 40 KΩ resistor is connected in series with a 50 nF capacitor across a 20 V d.c. supply.

Calculate:- (i) the circuit time constant,
(ii) the initial value of current,
(iii) the initial rate of change of current,
(iv) the current through the capacitor after 2 ms.
(v) the p.d. across the resistor after 2 ms.

(Ans. (i) 2 ms (ii) 500 μA (iii) 0.25 A/s (iv) 184 μA (v) 7.36 V)

56. (a) Explain briefly why power transformer cores are laminated.

(b) A sinusoidal source of e.m.f. 50 V and internal resistance of 15 Ω is connected to the 500 turn primary of a transformer.

The 100 turn secondary is connected across a 1 Ω resistor.

Sketch the circuit diagram and calculate:-

(i) the resistance reflected into the primary winding
(ii) the primary current,
(iii) the turns ratio which would be required to give maximum power transfer.

(Ans. (i) 25 Ω (ii) 1.25 A (iii) 3.87.)

57. A separately excited 500 V d.c. motor has an armature circuit resistance of 0.4 Ω.

When the armature current is 10 A, the speed is 1500 rev/min, $E = 496$ V.

Calculate the speed when the armature current rises to 50 A and the pole flux decreases by 2 % and $E = 480$ V.

(Ans. 481 rev/min)

END TEST. 6 (58 - 61 inclusive)

58.

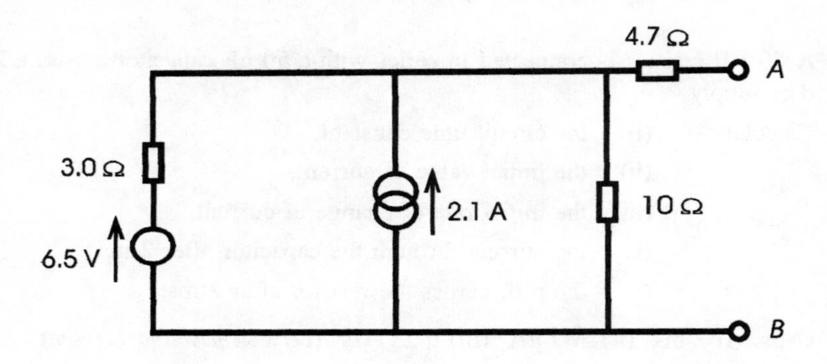

Fig. 444

(a) Draw the Thevenin generator equivalent to the circuit shown at terminals *AB*.

(b) Draw the Norton generator equivalent to the circuit shown at terminals *AB*.

(c) What is the resistance of a resistor connected between terminals *A* and *B* to which maximum power can be delivered?

59. A series a.c. circuit comprises a coil of inductance 12 mH and a capacitor of capacitance 0.025 μF.

(a) Calculate the resonant frequency.

(b) At resonant frequency the voltage magnification is found to be 5.6, calculate the resistance of the coil.

(c) If the coil and capacitor are reconnected in parallel, calculate the new resonant frequency.

60. (a) A fully discharged capacitor is connected in series with 1 megohm resistor to a 110 V d.c. supply. If the initial rate of rise of voltage across the capacitor is 27.5 V/s, determine:-

(i) the time constant,

(ii) the capacitance of the capacitor.

(b) A 110 V d.c. supply is switched on to a field circuit.

(i) If the initial rate of rise of current is 1.76 A/s, determine the inductance of the circuit.

If the time constant of the circuit is 3.125 s, determine:-

(ii) the resistance of the circuit,

(iii) the final current.

61.

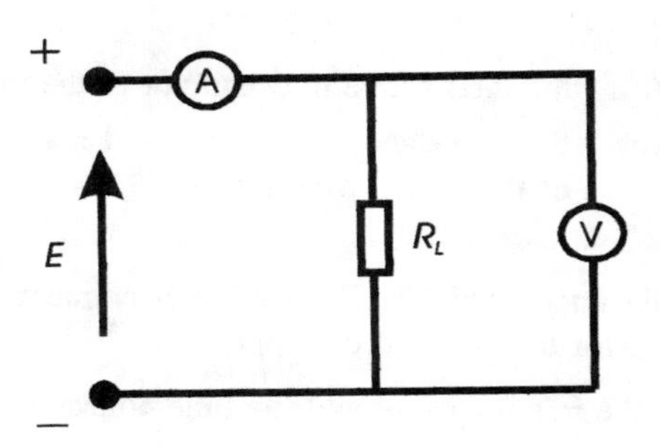

Fig. 445

In the circuit shown above, the 0 - 150 V moving coil voltmeter having a resistance of 1 kilohm/volt reads 110 V and the ammeter reads 110.7 mA.

When the moving coil voltmeter is replaced by a 0 - 150 V moving iron voltmeter, the readings are 100 V and 146.7 mA, respectively.

Determine:- (i) the resistance of the load resistor R_L and

(ii) the resistance of the moving iron voltmeter in terms of ohms/volt.

END TEST. 7 (62 - 68 inclusive)

62. (a) Explain why power factor correction capacitors are connected across the load in parallel rather than in series.

(b) A 400 V, 4 KVA, 50 Hz single phase induction motor has a full load power factor of 0.8 lagging. Determine the value of parallel capacitance which will give an overall power factor of unity.

63. A coil of inductance 100 μH and resistance 200 Ω is connected in series with a 100 pF capacitor across a variable frequency supply of constant p.d. 100 mV. The frequency is varied until the current is at maximum.

In this condition, determine:- (a) the frequency,
(b) the current,
(c) the p.d. across the capacitor,
(d) the p.d. across the coil
(e) the *Q*-factor.

64. A 415 50 Hz 3-phase supply delivers a line current of 23 A at a power factor of 0.92 lagging to a delta-connected motor.

Calculate:- (a) the total active power input to the motor,
(b) the impedance, resistance and reactance values per phase of the motor winding.

65. (a) An initially uncharged 10 μF capacitor is connected in series with a 50 KΩ resistor to a 20 V d.c. supply.

Calculate the initial current and the time constant.

(b) Either:- (i) Construct a graph of current against time, and use it to determine the current 0.8 s after connecting the supply,

OR:- (ii) Sketch a graph of current against time, and calculate the current 0.8 s after connecting the supply.

66. (a) A 50 KVA, 3300/240 V, single phase transformer is operated at normal voltage and full rated load.

Calculate:- (i) the secondary current,
(ii) the effective impedance of the load,
(iii) the primary current.

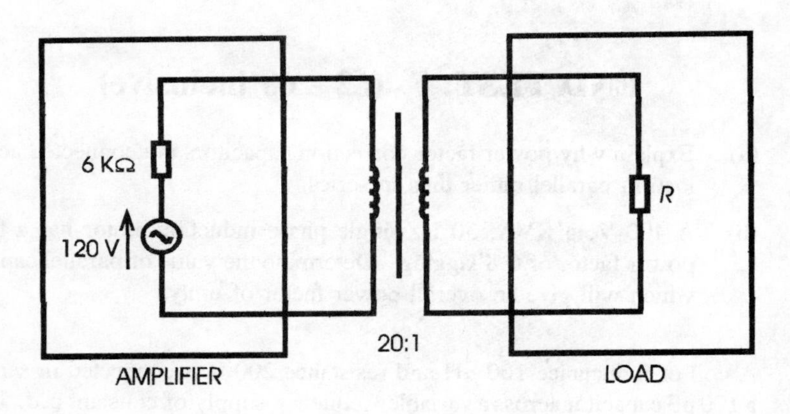

Fig. 446

(b) A matching transformer is being used to match the load resistance R to the amplifier whose equivalent output circuit is illustrated above.

Calculate:- (i) the load resistance R,

(ii) the power delivered to R.

67. (a) A d.c. generator has a generated e.m.f. of 250 volts at 1000 rev/min.

What will be the new generated e.m.f. at 1200 per min. if the field flux decreases by 5%?

(b) A 200 V d.c. shunt wound motor has an armature circuit resistance of 0.25 Ω and a field circuit resistance of 100 Ω.

Determine the efficiency when drawing a total current from the supply of 50 A. Assume the mechanical and iron losses total 500 W.

68.

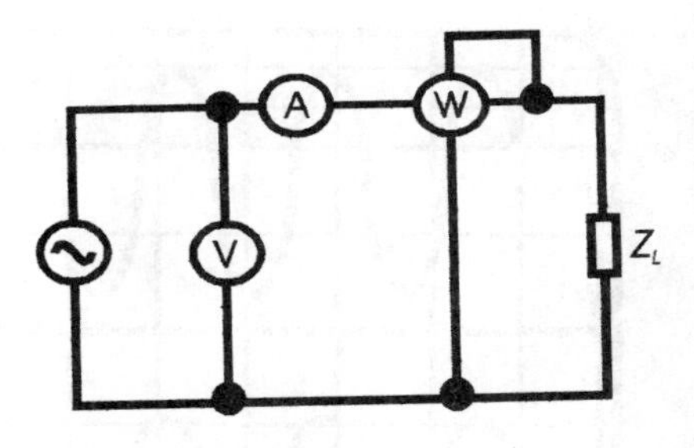

Fig. 447

(a) In the circuit shown above, the voltmeter reads 250 V, the ammeter reads 5.0 A and the wattmeter reads 1200 W.

Calculate the power factor of the load from the given readings.

(b)

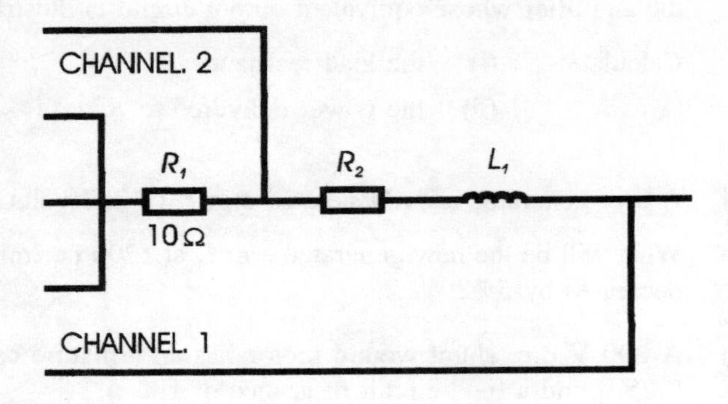

Fig. 448

The two channels of an oscilloscope are used to examine voltages in part of a network, as shown above. The resulting screen display is illustrated below:-

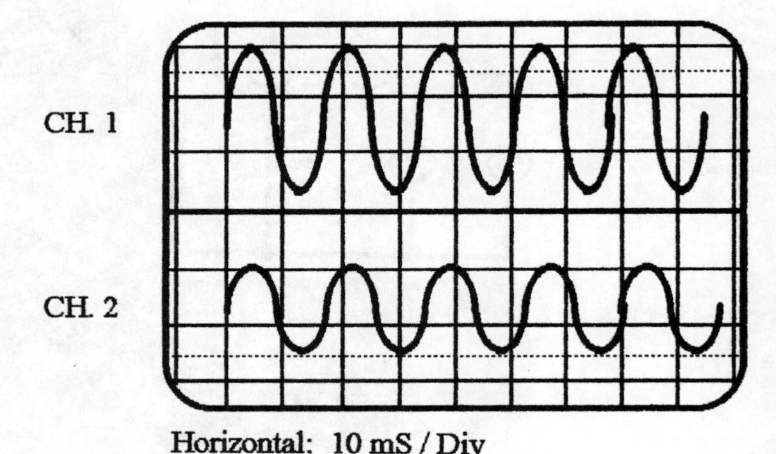

Horizontal: 10 mS / Div

Vertical Ch.1: 2 Volts / Div

Vertical Ch.2: 100 mVolts / Div

Fig. 449

Estimate:- (i) **the frequency of the voltage,**

(ii) **the peak to peak and r.m.s. values of the total voltage applied to Channel 1.**

(iii) **the peak to peak value of the current in the network.**

END TEST. 8 (69 - 72 inclusive)

69. (a) Determine the Thevenin equivalent circuit with respect to terminals *AB* in Fig. 450.

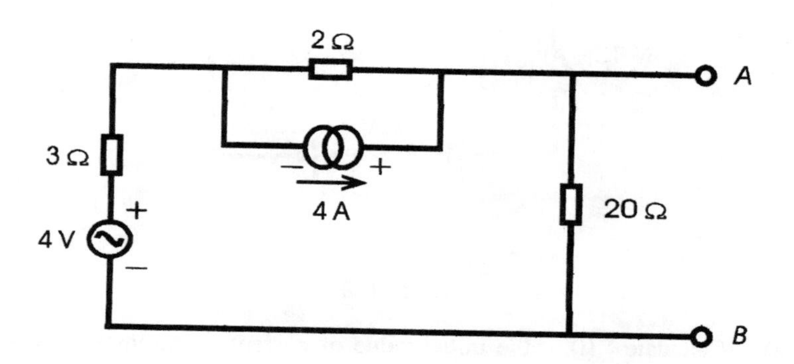

Fig. 450

(b) What value of resistive load will take maximum power from terminals *AB* and what is the value of this power?

70. For the circuit of Fig. 451.

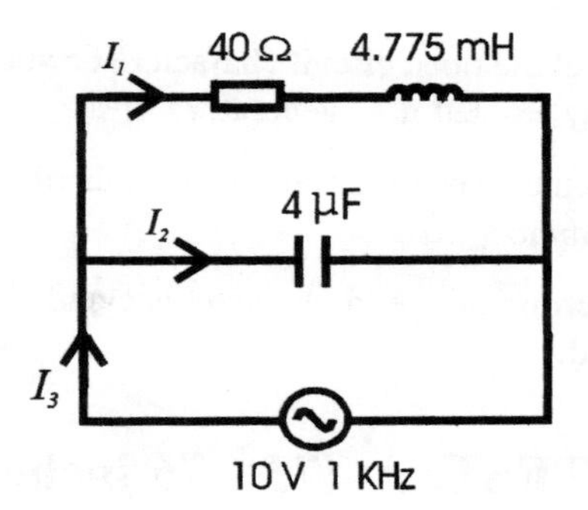

Fig. 451

(a) Calculate (i) the current I_1 and its phase angle relative to the supply voltage,

(ii) the current I_2.

(b) Determine the supply current I_3.

71. For the circuit of Fig. 452.

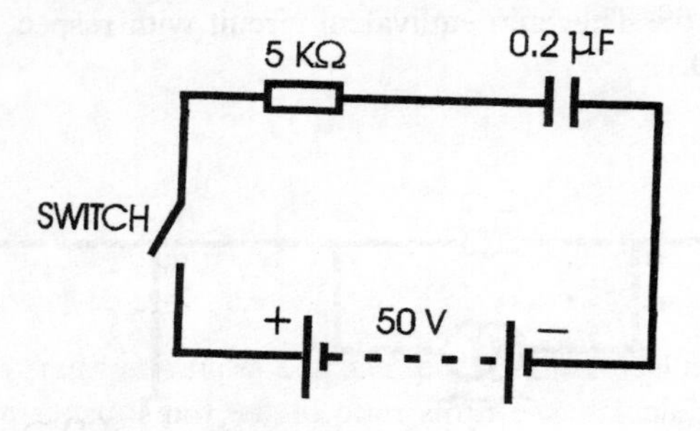

Fig. 452

(a) Calculate (i) the initial value of current at the instant the switch is closed,

(ii) the time constant of the circuit,

(iii) the capacitor voltage after a time equal to the time constant.

(b) If the current in the circuit given by $i = \frac{V}{R} . e^{-\frac{t}{CR}}$.

Calculate the capacitor voltage 0.5 ms after closure of switch.

72. (a) Sketch and label the open circuit characteristic and the load characteristic for a separately excited d.c. generator.

(b) Sketch the circuits for two different methods of speed control for a d.c. shunt wound motor.

Sketch the appropriate speed characteristic and state ONE advantage for EACH method.

END TEST. 9 (73 - 76 inclusive)

73. A coil inductance 100 μH and resistance 100 Ω is connected in series with a 100 pF capacitor across a 20 mV constant voltage variable frequency supply. The frequency is then varied until the current is at a maximum. In this condition, calculate:-

(a) the frequency,

(b) the current,

(c) the Q factor

(d) the bandwidth.

74. Calculate the line current which would be taken from 415 V, 50 Hz 3 phase supply by a load consisting of three identical coils, each of inductance 12.73 mH and resistance 3 Ω, connected:-

   (a) in star

   (b) in delta.

75. (a) Name the iron losses which occur in a transformer. In each case state, with reasons, one way in which the iron loss can be reduced.

   (b) A 4 Ω loudspeaker is to be matched to an amplifier of output resistance 5 KΩ. Calculate the turns ratio of the transformer required to achieve this matching.

76. A 100 nF capacitor is connected in series with a 200 KΩ resistor across a 50 V d.c. supply.

Calculate:-

   (a) the time constant

   (b) the initial value of current

   (c) the initial rate of change of current

   (d) at a time equal to the time constant

      (i) the current

      (ii) the p.d. across the capacitor

      (iii) the p.d. across the resistor.

ELECTRICAL AND ELECTRONIC PRINCIPLES III

PARALLEL CIRCUIT CONTAINING *C* AND *R*

The impedance, *Z*, of the circuit is given

$$Z = \frac{X_C R}{\sqrt{R^2 + X_C^2}}$$

PARALLEL CIRCUITS AND RESONANCE

RESONANT FREQUENCY AND DYNAMIC RESISTANCE OF PARALLEL TUNED CIRCUIT.

$$f_o = \frac{1}{2\pi}\sqrt{\frac{1}{LC} - \frac{R^2}{L^2}}$$

Q-FACTOR

Q stands for Quality

Q-factor is a measure of the quality of an electrical component.

Definition

$$Q = 2\pi \times \frac{\text{Maximum energy stored in one cycle}}{\text{Energy dissipated during one cycle}} \quad \ldots (1)$$

$$Q = \frac{1}{2\pi f C R_1}$$

Quality factor of the capacitor $= \dfrac{\text{reactance}}{\text{resistance}}$ where reactance $= \dfrac{1}{2\pi f C} = X_C$

$$Q = \frac{X_C}{R_1}$$

THE *Q*-FACTOR OF A COIL

$$Q = 2\pi \frac{\text{Maximum energy stored in one cycle}}{\text{Energy dissipated during one cycle}}$$

$$= 2\pi . \frac{\frac{1}{2} L I_m^2}{\frac{I^2 R}{f}} = \frac{2\pi \frac{1}{2} L \left(\sqrt{2}\, I\right)^2}{\frac{I^2 R}{f}} = \frac{2\pi f L}{R} = \frac{X_L}{R}$$

$$\boxed{Q = \frac{\text{reactance}}{\text{resistance}}}$$

EQUIVALENT CIRCUITS OF AN INDUCTOR

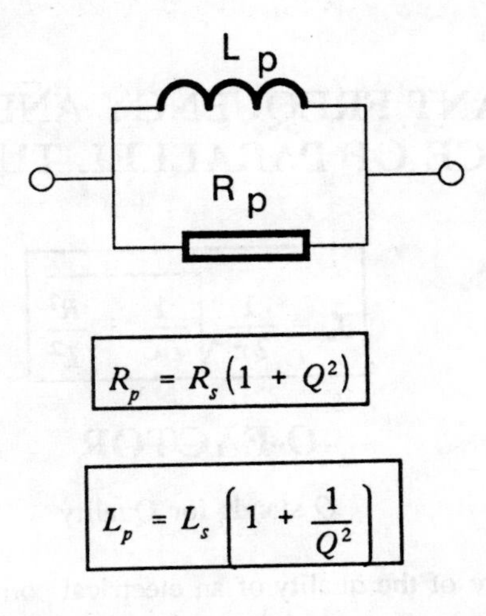

$$\boxed{R_p = R_s\left(1 + Q^2\right)}$$

$$\boxed{L_p = L_s\left(1 + \frac{1}{Q^2}\right)}$$

Thévenin-Helmholtz's and Norton's Theorems

OPEN CIRCUIT VOLTAGE

$$V_{O/C} = V_{TH} = E$$

SHORT CIRCUIT CURRENT

$$\text{Output resistance} = \frac{\text{Open Circuit Voltage}}{\text{Short Circuit Current}}$$

$$\boxed{R_{TH} = R_N = \frac{V_{TH}}{I_N}}$$

MAXIMUM POWER TRANSFER THEOREM

$\boxed{R = r}$ condition for maximum power

$$P_{max} = \frac{E^2 r}{(2r)^2} = \frac{E^2}{4r}.$$

TRANSFORMER MATCHING

$$\boxed{R_{in} = \left(\frac{N_1}{N_2}\right)^2 R_L}$$

CHARGING UP A CAPACITOR

$$C = \frac{i}{\frac{dv}{dt}} = \frac{\text{instantaneous current}}{\text{the rate of change of voltage across the capacitor}}$$

$$\boxed{i = Ie^{-\frac{t}{\tau}}}$$

$$\boxed{v_C = V\left(1 - e^{-\frac{t}{\tau}}\right)}.$$

$$\boxed{q = Q\left(1 - e^{-\frac{t}{\tau}}\right)}$$

TIME CONSTANT

$\boxed{i = Ie^{-\frac{t}{\tau}}}$ if $t = \tau$, $i = Ie^{-1} = 0.368\,I$

$\boxed{v_C = V\left(1 - e^{-\frac{t}{\tau}}\right)}$ if $t = \tau$, $v_C = V\left(1 - e^{-1}\right) = 0.632$ V

The gradient at A, that is, at $t = 0$

$$\left(\frac{di}{dt}\right)_{t=0} = -\frac{I}{\tau}.$$

If $t = 0$, $\boxed{\left(\frac{dv_C}{dt}\right)_{t=0} = \frac{V}{\tau}}.$

The initial rate of growth of charge.

If $t = 0$

$$\left(\frac{\mathrm{d}q}{\mathrm{d}t}\right)_{t=0} = \frac{Q}{\tau}.$$

At $t = 0$

$$\left(\frac{\mathrm{d}v_R}{\mathrm{d}t}\right)_{t=0} = -\frac{V}{\tau}.$$

The initial rate of change of voltage across R.

PURE INDUCTOR IN SERIES WITH A RESISTOR

Magnetizing the coil.

$$i = I\left(1 - e^{-\frac{t}{\tau}}\right)$$

the growth of current

$$v_L = Ve^{-\frac{t}{\tau}}$$

$$v_R = V\left(1 - e^{-\frac{t}{\tau}}\right)$$ the growth of voltage

THREE-PHASE SYSTEMS

STAR CONNECTION

$$V_L = V_P\sqrt{3}$$

$$I_P = I_L$$

For Δ connection

$$I_L = \sqrt{3}\, I_P$$

$$V_L = V_P$$

For either Y or Δ

$$P = 3V_P I_P \cos\phi$$

$$P = \sqrt{3}\, V_L I_L \cos\phi$$

DC MACHINES

EMF AND TORQUE

$$\boxed{E \propto N\phi}$$

where N is the speed in rev/min = 60 n.

$$\boxed{T \propto I_a \phi}$$

DC MOTORS

SHUNT MOTOR

$$\boxed{E = V - I_a R_a}$$

$$\boxed{I_f = \frac{V}{R_f}}$$

$$\boxed{I_L = I_a + I_f}$$

R_a = armature-circuit resistance

R_f = field-circuit resistance

SERIES MOTOR

$$I_L = I_a = I_f$$

$$E = V - I_L(R_a + R_f)$$

DC GENERATORS

SHUNT GENERATOR

$$I_f = \frac{V}{R_f}$$

$$I_a = I_f + I_L$$

$$E = V + I_a R_a$$

SERIES GENERATOR

Field coils wound with a few turns of wire of a large cross-sectional area.

Hardly ever used except for special purposes, e.g. boosters.

Boosters are series generators connected in series with d.c. feeders. They inject a voltage into the feeder to compensate for the volt drop. Both volt drop and injected voltage are proportional to current in the feeder, so the compensation is adequate.

LOSSES IN DIRECT-CURRENT MACHINES

EFFICIENCY

$$\text{Efficiency, } \eta = \frac{\text{output}}{\text{input}}$$

which may also be written:-

$$\eta = \frac{\text{input} - \text{losses}}{\text{input}} = 1 - \frac{\text{losses}}{\text{input}} = \frac{\text{output}}{\text{output} + \text{losses}}$$

TRANSFORMERS

Ideal Transformer

$$\frac{N_1}{N_2} = \frac{E_1}{E_2} = \frac{V_1}{V_2} = \frac{I_2}{I_1}$$

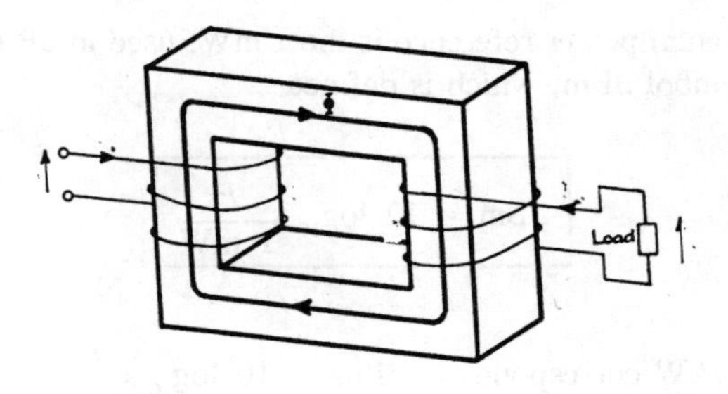

MEASURING INSTRUMENTS AND MEASUREMENTS

The Decibel.

$$\text{bel} = \log_{10} \frac{P_{out}}{P_{in}}$$

$$\text{decibel} = 10 \log_{10} \frac{P_{out}}{P_{in}}$$

$$\text{or dB} = 10 \log_{10} \frac{P_{out}}{P_{in}}$$

$$\frac{P_{out}}{P_{in}} = \text{power ratio}$$

The decibelmeter reference level (dBm)

The most useful datum power reference is the 1 mW, used in dB measurements and is given by the symbol dBm, which is defined

$$\boxed{\text{dBm} = 10 \log_{10} \frac{P}{1 \text{ mW}}}$$

A power level of P W corresponds to dBm $= 10 \log_{10} \dfrac{P}{10^{-3}} = 10 \log_{10} 1000P$,

a power level of 100 mW corresponds to dBm $= 10 \log_{10} \dfrac{100 \times 10^{-3}}{1 \times 10^{-3}} = 20$,

a power level of 0.2 mW corresponds to dBm $= 10 \log_{10} \dfrac{0.2}{1} = -6.99$.

The dBm is widely used in telecommunications systems in conjunction with a standard impedance value of 600 Ω. 0 dB corresponds to the voltage at which 1 mW is developed across 600 Ω resistance.

Maxwell's Inductance Bridge

$$R_x = \frac{PQ}{R}$$

$$L_x = CPQ$$

Both sets of equations are independent of the frequency.

The former relies on the ratio arm and the latter on the product arm.

This bridge is used to measure small Q-factors.

Hay's Inductance Bridge

This bridge measures high Q-factors.

$$L_x = PQC$$

$$R_x = \frac{PQ}{R}$$

ELECTRICAL AND ELECTRONIC PRINCIPLES III

INDEX